Synthese Library

Studies in Epistemology, Logic, Methodology, and Philosophy of Science

Volume 449

The aim of *Synthese Library* is to provide a forum for the best current work in the methodology and philosophy of science and in epistemology. A wide variety of different approaches have traditionally been represented in the Library, and every effort is made to maintain this variety, not for its own sake, but because we believe that there are many fruitful and illuminating approaches to the philosophy of science and related disciplines.

Special attention is paid to methodological studies which illustrate the interplay of empirical and philosophical viewpoints and to contributions to the formal (logical, set-theoretical, mathematical, information-theoretical, decision-theoretical, etc.) methodology of empirical sciences. Likewise, the applications of logical methods to epistemology as well as philosophically and methodologically relevant studies in logic are strongly encouraged. The emphasis on logic will be tempered by interest in the psychological, historical, and sociological aspects of science.

Besides monographs *Synthese Library* publishes thematically unified anthologies and edited volumes with a well-defined topical focus inside the aim and scope of the book series. The contributions in the volumes are expected to be focused and structurally organized in accordance with the central theme(s), and should be tied together by an extensive editorial introduction or set of introductions if the volume is divided into parts. An extensive bibliography and index are mandatory.

More information about this series at https://link.springer.com/bookseries/6607

Catalina González Quintero

Academic Skepticism in Hume and Kant

A Ciceronian Critique of Metaphysics

Springer

Catalina González Quintero
Department of Philosophy
Universidad de los Andes
Bogotá, Colombia

ISSN 0166-6991 ISSN 2542-8292 (electronic)
Synthese Library
ISBN 978-3-030-89749-9 ISBN 978-3-030-89750-5 (eBook)
https://doi.org/10.1007/978-3-030-89750-5

This Springer imprint is published by the registered company Springer Nature Switzerland AG
The registered company address is: Gewerbestrasse 11, 6330 Cham, Switzerland

To Bertha and Nono
Thesis and antithesis

I am indeed a rash person to attempt to join issue with a pupil of the Academy who is also a trained orator!
An Academic unversed in rhetoric I should not have been much afraid of, nor yet an orator, however eloquent, who was not reinforced by that system of philosophy; for I am not disconcerted by a mere stream of empty verbiage, not yet by the subtlety of thought if it is expressed in a jejune style.

Cicero, *De Natura Deorum*

Acknowledgments

This book is the result of several years of research and numerous courses taught at Universidad de los Andes (Bogotá, Colombia) on ancient and modern skepticism. I want to thank the University for its academic and financial support. Funding from the School of Social Sciences, in particular, enabled me to conduct research in Berlin (Germany) to gather primary and secondary sources, and to take a research leave at DePaul University (Chicago, USA), where I finished writing the manuscript in Spanish. I am grateful to María del Rosario Acosta and the Department of Philosophy at DePaul University for this invitation, without which my stay would not have been possible.

Many people participated in the different stages of writing this manuscript. I want to thank them all, even if I do not mention someone specifically here. I am deeply indebted to the late Rudolf Makkreel, for giving me the gift of a heterodox reading of Kant, and Steven Strange, for introducing me to Cicero's skepticism. My special thanks to Peter Cousins, who patiently prepared the translation of this book into English. I am grateful to Luis Carlos Suárez for his help in searching for sources and handling the references of Kant's *Gesammelte Schriften*; to Felipe González for his translation of Greek and Latin terminology; to Andrea Lozano, Ignacio Ávila, and Mariluz Restrepo for their reading and commenting on early drafts; to Manuel Cortés for his arduous work confirming bibliographical references and editing the Spanish manuscript; and to Alfonso Correa Motta for his comments on the first two chapters of the book. I cannot stress enough my gratitude to Allison Wolf for not only reading, commenting, and editing the entire book (a massive effort, indeed) but also giving me the confidence I needed to publish it in English. I am especially grateful to Lisímaco Parra for giving me the courage to pursue a career in philosophy and for our long (and sometimes heated!) conversations about Kant, Hume, and religion. My friends know how much this book owes them for their constant encouragement during these years. Among them, am particularly grateful to Antonio Sánchez, Marcela and Santiago Villegas, and Marcela García for their support at the most challenging moments of writing this manuscript. Finally, I want to thank my very dear family, my siblings Alberto and María Leonor, as well as their spouses and children, for their love and care. And, of course, Algarabía, for always living up to her name!

Contents

About the Author

Catalina González Quintero earned her Ph.D. in philosophy from Emory University in the USA and is currently Associate Professor of Philosophy at Universidad de los Andes in Bogotá, Colombia. Her research focuses on the history of modern philosophy, particularly in the areas of skepticism and rhetoric. She has published various articles and book chapters on the philosophy of Cicero, Kant, Hume, and Vico in US, European, and Latin American journals as well as in compilations published by Springer, Palgrave, De Gruyter, SUNY Press, Lawrence Erlbaum, Universidad Nacional de Colombia, and Universidad de los Andes.

Abbreviations

Aristotle

Metaph.	*Metaphysics*
Rh.	*Rhetoric*

Aulus Gellius

AN	*Noctes Atticae (Attic Nights)*

Cicero

Acad. I	*Academica* Book I
Acad. II	*Academica* Book II *(Lucullus)*
Ad. Att.	*Ad Atticus (Letters to Atticus)*
DND	*De Natura Deorum (On the Nature of the Gods)*
De Or.	*De Oratore (On the Orator)*
Fato	*De Fato (On Fate)*
Fin.	*De Finibus (On Moral Ends)*
Inv.	*De Inventione (On Invention)*
Off.	*De Officiis (On Duties)*
Or.	*Orator (Orator)*
TD	*Tusculanae Disputationes (Tusculan Disputations)*

Descartes

AT	*Ouvres (Adam-Tannery edition)*

Diogenes Laërtius

Diog. Laërt.	*Lives of Eminent Philosophers*

Epicurus

LM	*Letter to Menoeceus*

Eusebius of Caesarea

Praep. Evang. *Praeparatio Evangelica (Preparation for the Gospel)*

Photius

Bibl. *Bibliothèque (Bibliotheca)*

Hume

DNR *Dialogues concerning Natural Religion* (followed by N.K. Smith's edition numeration)

E. *Essays*

EHU *Enquiry Concerning the Human Understanding* (followed by Selby-Bigge's numeration)

EPM *Enquiry Concerning the Principles of Morals*

Letters *The Letters of David Hume*

LG *Letter to a Gentleman to his Friend in Edinburgh*

NHR *Natural History of Religion*

THN *Treatise of Human Nature* (followed by Selby-Bigge's edition numeration)

Kant[1]

KpV *Kritik der praktischen Vernunft (Critique of Practical Reason)*

KrV *Kritik der reinen Vernunft (Critique of Pure Reason)* (followed by A/B numeration)

KU *Kritik der Urteilskraft (Critique of Judgment)*

VL *Vorlesungen über Logik (Lectures on Logic)* (followed by name of the specific set of lectures).

P *Prolegomena zu einer jeden künftigen Metaphysik* (Prolegomena to Any Future Metaphysics)

C *Correspondence*

NF *Notes and Fragments*

Lucretius

Lucr. *De Rerum Natura (On the Nature of Things)*

Plato

Phd. *Phaedo*

Plutarch

Adv. Col. *Adversus Colotem (Reply to Colotes)*

[1] References to Kant's works follow the Academy Edition: *Kants gesammelte Schriften. Ausgabe der Königlich Preußische Akademie der Wissenschaften (Ak.)* For the most cited works, abbreviations will also be inserted.

Sextus Empiricus

M VII	*Adversus Mathematicos VII (Against the Logicians)*
M IX	*Adversus Mathematicos IX (Against the Physicists)*
Pyr.	*Pyrrhoniae Hypotyposes (Outlines of Scepticism)*

Chapter 1
Introduction

Abstract The purpose of this book is to examine the influence of the skepticism of the New Academy on David Hume's and Immanuel Kant's critiques of metaphysics. The traditional interpretation of the connection between these philosophers and skepticism is that both were influenced by Pyrrhonism and its modern adaptations. This influence became most evident in their treatment of traditional epistemological issues, such as the status of causal inferences and the belief in the external world. In this book, I present a different approach. Contrary to standard interpretations, I argue that Academic skepticism, not Pyrrhonism, most significantly influenced these authors, particularly their critiques of metaphysics. Both philosophers engaged with skepticism, not merely to question the principles of experience or to lay out a solid ground for science, but especially to refute dogmatic metaphysics and Christian theology. To this end, they employed skeptical methods and argumentative strategies in a novel, unprecedented way. However, they did not seek to undermine reason and ground religion on faith and revelation, as other modern Pyrrhonian skeptics and fideists had done before them. Instead, their critiques of metaphysics aimed at safeguarding rational autonomy in religious matters. Put differently, they both used skeptical means to examine the justification of metaphysical claims and to determine, in this way, which resulting beliefs could be held non-dogmatically and for practical purposes.

The purpose of this book is to examine the influence of the skepticism of the New Academy on David Hume's and Immanuel Kant's critiques of metaphysics. The traditional interpretation of the connection between these philosophers and skepticism is that both were influenced by Pyrrhonism and its modern adaptations. This influence became most evident in their treatment of traditional epistemological issues, such as the status of causal inferences and the belief in the external world. In this book, I present a different approach. Contrary to standard interpretations, I argue that Academic skepticism, not Pyrrhonism, most significantly influenced these authors, particularly their critiques of metaphysics.

In the following pages, I will explore the New Academy's impact on Hume's and Kant's views about the types of justifications that constitute knowledge and belief of metaphysical entities. These views underlie each authors' critique of traditional

C. González Quintero, *Academic Skepticism in Hume and Kant*, Synthese Library 449, https://doi.org/10.1007/978-3-030-89750-5_1

metaphysics, on which orthodox Christian doctrines were primarily based. I argue that Academic skepticism played a vital role in both authors' attacks on theological, religious, and superstitious beliefs. In fact, both philosophers engaged with skepticism, not merely to question the principles of experience or to lay out a solid ground for science, but especially to refute dogmatic metaphysics and Christian theology. To this end, they employed skeptical methods and argumentative strategies in a novel, unprecedented way. However, Hume and Kant did not seek to undermine reason and ground religion on faith and revelation, as other modern Pyrrhonian skeptics and fideists had done before them. Instead, their critiques of metaphysics aimed at safeguarding rational autonomy in religious matters. Put differently, they both used skeptical means to examine the justification of metaphysical claims and to determine, in this way, which resulting beliefs could be held non-dogmatically and for practical purposes. For, even though we tend to think that an Enlightened critique of traditional metaphysics should entail an agnostic or atheistic renunciation of religious convictions – and sometimes it did, as in the case of the philosophers associated with the radical Enlightenment (Israel 2001, pp. 329–444) – it seems that, for both Hume and Kant, this was not a wholly feasible or desirable end. Both thinkers aimed at a much more modest but still crucial outcome—that once traditional metaphysics were shown unjustifiable from a cognitive point of view, rational agents held their religious beliefs autonomously but not dogmatically. In other words, Hume and Kant's critiques sought to deter religious dogmatism, with the practical consequences of superstition, blind obedience to authority, and irrational enthusiasm. But they did not seek to extirpate the all too human – and, for them, *rational* – tendency to believe in the existence of God and also, at least for Kant, in human freedom and the immortality of the soul. These were, for them, metaphysical beliefs worth being held for practical purposes, notwithstanding their insufficient justification. But they should also be critically examined and never sustained as *dogmata*. Thus, for both philosophers, metaphysical skepticism should convey neither a complete suspension of belief nor a wholly irrational adoption of religious faith. Maintaining or denying religious beliefs should continue to be a rational choice, even if justified solely on practical grounds. And this was a very Academic way to approach metaphysical issues.

To begin to make these points more transparent, I will now briefly elaborate on the claim that it was Academic skepticism and not Pyrrhonism that provided the grounds for these authors' critiques of metaphysics and briefly describe how it applied to each case.

1.1 Academic Skepticism in the Enlightenment

In recent decades, a newfound interest has emerged in the reception of Academic skepticism during modernity. As contemporary historians of skepticism have begun to reconsider the conventional view that the revival of Pyrrhonism gave birth to modern skepticism,[1] it has become increasingly more apparent that Academic skepticism strongly influenced both modern skeptics and non-skeptics. J. R. Maia Neto has argued, for instance, that despite the new interest that the works of Sextus Empiricus arose in the seventeenth-century intellectual *milieu*, Academic skepticism continued to play an essential role after the Renaissance:

> Further research on the history of skepticism in the period has shown that the role of Cicero's *Academica* and of Academic skepticism, in general, continued to be considerable. First, Cicero's works were important items in the university curricula. Although Cicero's works of rhetoric and morals were those more read, the students could certainly learn about Academic skepticism through at least some of them. Second, many editions of *De Natura Deorum*, a work of Cicero in which Academic skepticism is quite central, were published during the seventeenth century. Third, features of Academic skepticism, exhibited above all in Cicero's *Academica*, are basic to the skepticism revived in the late Renaissance by Montaigne and Charron; were related to the new science in the first half of the seventeenth century by Gassendi, were used in the anti-cartesian reaction of the late seventeenth century by Rapin, Huet, Bayle, and above all Foucher; and were radicalized in the eighteenth century by Hume (Maia Neto 1997, p. 200).

As Maia Neto asserts, we have new historical evidence pointing to the modern reception of Academic skepticism through Cicero's works, which exerted an important influence on seventeenth- and eighteenth-century philosophers. P. Smith and S. Charles have recently documented this influence on a wide variety of modern authors, such as Sanches, Charron, Montaigne, and Bacon (Charles and Smith 2013, pp. xviii–ix). According to them, the Academic concept of "the probable," closely related to rhetorical persuasion and practical advisability, was employed by the modern method of experimental science as a device to examine inductive hypotheses (Charles and Smith 2013, p. xiii). In this way, Academic probabilism became an ally with modern science against the established Aristotelian outlook:

> In fact, criticism of Aristotelian science and interest in rhetoric go hand in hand, in so far as the concepts of certitude and truth are replaced by the concepts of probability and verisimilitude [...] Whereas the Pyrrhonist tries to bring about suspension of judgment, the Academic tries to establish that one side of a given question has more probability than the other [...] Accordingly, we see that a number of major philosophers seem to have used Academic scepticism more than is usually recognized (Charles and Smith 2013, p. ix).

Given the scientific meaning we give today to the concept of probability, most accounts of Academic skepticism's adaptations in modern philosophy focus on this notion's usefulness for evaluating empirical hypotheses. However, it is essential to bear in mind that the New Academy's criterion of the "persuasive" or "probable" (*to pithanon*) impression was also, and perhaps mainly, applied by Cicero and other Academics to metaphysical issues such as the existence of God, the immortality of the soul, the highest-end or *summum bonum*, and the problem of freedom and fate. Some of Cicero's most widely read dialogues during the seventeenth and eighteenth centuries, such as *De Natura Deorum* (*DND*), *Tusculan Disputations* (*TD*), *De Finibus* (*Fin.*), and *De Fato* (*Fato*), were devoted to assessing the "probability" or

[1] A view, of course, advanced by the pioneering works of Richard Popkin. See: Popkin 1979, 2003.

"persuasiveness" of the metaphysical claims held by the different Hellenistic schools. Cicero usually began these dialogues by asserting that, although reaching theoretical certainty on such matters was impossible, the skeptical analysis would show the views of one of the schools to be sufficiently "probable" and practically useful. Thus, the Academic criterion of probability was employed to skeptically evaluate metaphysical claims in ways that modern philosophers could easily recognize but are difficult for us to grasp, given the direct connection we make today between probability and empirical research. This connection has led contemporary historians of skepticism to overlook the role that Careneades's criterion could have played in modern discussions beyond the field of empirical science. Given this lack of attention, in this book, I attempt to show precisely how both Hume and Kant used the probabilistic approach of Academic skepticism to evaluate traditional metaphysics.

1.2 The Enlightened Critique of Metaphysics: Furthering Religious Autonomy

Hume and Kant adopted Academic skepticism to criticize traditional metaphysics and undermine Christian theology while safeguarding practical autonomy in religious matters. I suggest that their reasons for choosing Academic skepticism over Pyrrhonism were related to each school's practical orientation, that is, to the answer they each gave to the problem of how to act while skeptically examining an issue and suspending judgment about it. Scholars have called this problem the "*apraxia* objection,"—i.e., the dogmatists' common accusation that suspension of judgment would entail the skeptics' inability to act. To answer this objection, the ancient skeptical schools established different practical criteria to orient action *vis-à-vis* their purported *epochê*. Specifically, to continue acting, the Pyrrhonists decided to adopt subjective impressions, i.e., the so-called "appearances," and to non-dogmatically follow the laws and customs of their communities. The New Academy, on the contrary, recommended "approving" for practical purposes the claims that best withstood skeptical examination, that is, those that were sufficiently "persuasive" or "probable" (*to pithanon*).

In modernity, these criteria modeled different attitudes regarding religious matters. Pyrrhonism's acceptance of subjective appearances and abidance by the rules of society inspired a form of irrational "fideism" in early modern philosophers of this skeptical inclination (Popkin 1979, pp. 18–41; Popkin et al., 1997, pp. 1–17). In contrast, I argue in this book that the Academic's practical response was favored by philosophers of the Enlightenment, who advocated rational, practical autonomy also in religious matters. I suggest that the New Academy's recommendation to adopt, non-dogmatically and only for practical purposes, the most persuasive or probable religious claims after skeptical examination was much better suited to the Enlightenment's goal of recognizing the limits of human reason while grounding both scientific and practical knowledge on this very faculty. I believe that both Hume and Kant must have perceived Pyrrhonian fideism – with its irrational leap of faith and blind obedience to ecclesiastical authority – as an unnecessary and dogmatic renunciation of reason. For them, although setting limits to theological dogmatism was a necessary task to give a secular foundation to morality, it entailed neither embracing agnosticism nor founding religion on irrational faith. Instead, their projects involved an awareness of the modest but still legitimate place that religious belief had in the lives of rationally autonomous individuals. In other words, both Hume and Kant thought that a sort of rational creed, screened from superstition and dogmatism by skeptical examination – a reasonable and non-dogmatic religious belief – was a desirable outcome. And, as a result, they appropriated the moderate (or, to say it in Hume's terms, "mitigated") skepticism of the Academy, even though their conclusions, as we will see, were diametrically different.

1.3 Hume's and Kant's Academic Skepticism

While scholars have sufficiently acknowledged Hume's indebtedness to Cicero, particularly in his *Dialogues concerning Natural Religion* (*DNR*), the relationship between Hume and Academic skepticism continues to be obscured by interpretations that identify Hume's skeptical commitments with Pyrrhonism on the grounds of his epistemic naturalism. My suggestion, in this respect, is that to understand Hume's indebtedness to Academic skepticism, it is essential to distinguish the epistemic claims that pertain to his *normative theory of belief* from those that constitute his *descriptive theory of belief* or his so-called *naturalism*. The former directly relate to Carneades's epistemic criteria to determine the "persuasiveness" or "probability" of impressions, while the latter can be linked to Pyrrhonism but may also simply depict the natural attitude toward belief. Even though many commentators have referred to both the normative aspects of Hume's theory of belief and his preference for Academic skepticism, very few have explored to the necessary degree of detail the connection between these two features (Fosl 2020, pp. 311–337; Smith 2007, pp. 105–126). I seek to show in this book how Hume's normative theory of belief – involving his typology of beliefs into "knowledge, proofs, and probabilities," and his "rules for judging about causal inferences" – closely follows Carneades's epistemic criteria of the "persuasive," "examined," and "irreversible" impressions, and is employed in the *Dialogues concerning Natural Religion* to determine the legitimacy of the belief in God derived from the argument from the intelligent design of the universe.

Thus, I claim that Hume retrieved from Academic skepticism the very foundation of this normative theory of belief and constructed his critique of religion on its grounds. The *Dialogues* offer an example of how the theory is applied to reveal the strengths and weaknesses of natural theism's arguments and decide the belief that should be considered most probable and suited to guiding action. This belief, for Hume, is that a god, creator of the universe, who is "remotely similar" to the human mind, does exist. However, Hume thinks that this claim should not be further extended to include a theistic characterization of God's nature, that is, to provide a list of divine attributes, and, more importantly, that it should not ground any religious practice. Given Hume's insistence on the harmful historical consequences of revealed and popular religion, I argue that the belief he recommends as the most probable or persuasive one is, as Gaskin suggested, a form of non-dogmatic, anti-religious *deism* (Gaskin 1976, pp. 301–311, 1983, pp. 160–173).

Kant's relationship to ancient and modern skepticism has been much less debated than Hume's. It is usually discussed in connection with classical epistemological issues, such as Hume's problem of causality, Cartesian or Berkeleyan Idealism regarding the external world, and Bayle's conception of the antinomies (Forster 2008, pp. 4–5; Guyer 2008, pp. 23–70, 2014, pp. 10–15; Smith 2013, pp. 247–263, 2008, p. 465). None of these accounts has considered Kant's possible contact with Academic skepticism, either via other modern philosophers, like Hume or Bayle, or through Cicero himself. This lack of attention gives us an incomplete understanding of Kant's anti-metaphysical stance, developed in the Transcendental Dialectic of the *Critique of Pure Reason* (*KrV*), where he employs the methods of ancient and modern skepticism to assess traditional metaphysical claims. Although scholars have acknowledged Kant's repeated claim throughout the *Critique* that a "skeptical method" should be used to examine such metaphysical issues, they have overlooked that this method involves an important Academic feature: a *practical verdict* or *practical way of resolving* the theoretical opposition between thesis and antithesis. This practical resolution consists in theoretically suspending judgment on the matter examined, while non-dogmatically adopting some metaphysical beliefs for practical purposes. I argue that this feature of Kant's skeptical method can be read as an Academic response to the *apraxia* objection. After showing the skeptical opposition of the Antinomies' theses and antitheses, Kant's solution to the cosmological issues involves the recognition that, even though it is impossible to attain knowledge on these questions, the examined beliefs in God, freedom, and the immortal soul can be maintained at the practical level to promote the "highest good," that is, the idea of a future world in which an agreement between happiness and morality is possible. Given that these beliefs are practical consequences of morality and also important incentives for it, they are both rational and practically advisable. For Kant, then, unlike Hume, religious practices should be preserved for moral reasons, or, in other words, a *moral theism* should be autonomously but, again, non-dogmatically held, by the rational, Enlightened individual.

1.4 Outline of the Book

The structure of this book mirrors what I have described thus far. I devote two chapters of the book to each philosopher in discussion—Cicero, Hume, and Kant. The first chapter in each pair describes the author's skeptical method, while the second examines this method's application to metaphysical matters. Following this outline, in the first chapter, I present the fundamental aspects of the skepticism of the New Academy by offering a general analysis of Cicero's *Academica* (*Acad. I* and *II*) (45 BCE). I begin with an overview of the Middle and New Academies' skeptical arguments against Stoic epistemology and analyze their respective answers to the *apraxia* objection. I argue that Carneades developed the moderate form of skepticism that characterized the New Academy to give a robust response to this objection. After this, I describe the most prominent differences between the New Academy's and Pyrrhonism's answers to the *apraxia* objection, and conclude that Pyrrhonism advocates for a non-dogmatic practical heteronomy, while the New Academy supports an non-dogmatic practical autonomy.

The book's third chapter discusses Cicero's application of the New Academy's skeptical method to the Hellenistic schools' metaphysical theories on the nature of the Gods and the immortality of the soul, in *De Natura Deorum* and the first book of the *Tusculan Disputations*, respectively. I begin by describing the works' main arguments and then move to analyzing their different philosophical approaches. The first treatise exhibits a traditionally skeptical argumentation, with a claim being examined through arguments on both sides of the question, whereas the second dialogue is more constructive, as it argues for the plausibility of two opposed but complementary views on death. I explain this apparent incongruency by considering the role that Aristotelian rhetoric played on both the Academy and Cicero's thought and argue that it is probably because of this rhetorical influence that Carneades chose the criterion of the "persuasive" impression for practical orientation. To close the chapter, I compare the Academy's and Pyrrhonism's practical responses to religious matters and illustrate their differences with some passages from Thornton Wilder's novel *The Ides of March* (1948).

In the fourth and fifth chapters, I give an account of Hume's Academic skepticism. The influence of this school on Hume's thought, which Hume himself clearly expressed, has been long neglected by a mainstream interpretative tradition that, following Popkin's famous article "David Hume: His Pyrrhonism and his Critique of Pyrrhonism" (Popkin 1951, pp. 385–407), saw his naturalism as the best expression of modern Pyrrhonism. In contrast to this view, in the fourth chapter, I claim that Hume's Academic skepticism can be best appreciated if we recognize, alongside his naturalism, a *normative theory of belief* designed to evaluate the legitimacy of epistemological and metaphysical claims. I also refer there to Hume's diagnosis of the undesirable consequences of Pyrrhonism in the *Treatise* (*THN*) – namely *apraxia*, melancholy, and ostracism – and examine his praise of Academic skepticism in the first *Enquiry* (*EHU*). I conclude the chapter with a detailed account of the connection between Hume's normative theory of belief and Academic skepticism.

The fifth chapter examines how Hume applied this theory of belief to the theistic argument about the existence and nature of God in the *Dialogues concerning Natural Religion*. It begins with a general assessment of Hume's indebtedness to Cicero, by discussing his works, essays, and correspondence. After this, I draw out the most important similarities between Cicero's *De Natura Deorum* and Hume's *Dialogues*, and embark on a general analysis of Hume's skeptical arguments, showing how the normative theory of belief previously outlined is at work in this dialogue. I argue that Hume saw the belief in a God creator of the universe as a "probability" and not as "natural belief" or as a "proof," and observe how Hume used some of his rules for causal inferences to examine the analogy at the basis of the argument from the intelligent design of the universe. Finally, I approach the interpretative challenges of the *Dialogues*' last section, which are known in scholarly literature as the problems surrounding "Philo's reversal," and conclude that Hume's Academic skepticism inclined him to accept a form of non-dogmatic, anti-religious *deism*, as the most probable and practically helpful belief.

The sixth and seventh chapters are devoted to Kant's critique of metaphysics. In the sixth chapter, I discuss Kant's characterization of the "skeptical method" that he deemed useful for his critical project, and his appropriation of ancient and modern skeptical sources. I consider, first, Kant's relation to Pyrrhonism, both examining his remarks about this school in the *Lectures on Logic* and outline the possible modern sources for his acquaintance with this school. After this, I explore the modern discussions on Zeno of Elea's paradoxes concerning the infinite divisibility of time, space, and matter, which undoubtedly informed Kant's conception of the mathematical antinomies via Bayle. The second part of the chapter explores Kant's relation to Academic sources. I turn first to Kant's general references to Cicero, the Academy, and suspension of judgment in the *Lectures on Logic*, to conclude that, despite his initial preference for Pyrrhonism, in the last set of the *Lectures* Kant seems to adhere to a notion of suspension of judgment that is more akin to the New Academy's position. Finally, I summarize the scholarly discussion about whether it was Hume or the antinomies that produced Kant's "awakening from [his] dogmatic slumber," and claim that Hume's arguments, in the first *Enquiry*, on the metaphysical dispute about freedom and necessity – which Cicero's *De Fato* heavily influenced – were a crucial source of Kant's awakening and critical treatment of the antinomies.

In the seventh chapter, I examine Kant's application of his "skeptical method" to the cosmological problems of the Antinomies of Pure Reason. I argue that Kant's practical resolution of the antinomies may be interpreted as an Academic response to the *apraxia* objection in metaphysical and religious matters. To maintain this thesis, I reconstruct Kant's general view of a Dialectic of Pure Reason and his account of the cosmological issues, analyze Kant's theoretical solution to them, and discuss the insufficiency of this solution from a practical point of view. After this, I examine what I consider to be Kant's practical solution, both in the Canon of Pure Reason of the first *Critique* and the Postulates of Practical Reason of the second *Critique*, i.e., the *Critique of Practical Reason* (*KpV*). Mainly, I examine Kant's claim that it is legitimate to sustain the beliefs in freedom, God, and the soul's immortality for practical purposes. I finish the chapter by arguing that the epistemic attitude corresponding to Kant's practical belief – a belief that is justified only from a practical point of view – can be accurately interpreted as corresponding to the attitude the Academics called "approval."

In the book's conclusion, I stress what I take to be the most important legacy of Academic skepticism for the Enlightenment—the distinction between theoretical and practical justifications of metaphysical beliefs. In my view, the Enlightened goals of attaining rational, practical autonomy, furthering a secular morality, and recognizing the limits of human knowledge led both Hume and Kant to adopt Academic skepticism rather than Pyrrhonian fideism, since the former responded to the *apraxia* objection with a rational, autonomous, and non-dogmatic acceptance of some metaphysical and religious claims—a response that best suited their own Enlightened inclinations.

1.5 A Brief Word on Local Circumstances and Global Demands

Writing about European skepticism as a Colombian woman in Latin America is, indeed, a challenging endeavor. This is not only the case for the most basic reasons – namely limited library access to historical sources and secondary literature, scarcity of translations, and meager research funding – but also because it entails putting oneself in a paradoxical situation. In the whole region, the philosophical community is divided between global and local demands. At the global level, Latin American philosophers are pushed to focus on areas seen as "valuable" to the global community and our academic institutions are under constant pressure to measure themselves against high-quality standards. This means that we are compelled to discuss problems that belong to the mainstream philosophical tradition and enter a global market that far exceeds our capacities to compete. By contrast, at the local level, Latin American scholars are pressed to think about urgent political and social issues rather than engage in questions that have perplexed humanity for centuries. Under these circumstances, investigating topics such as the history of modern skepticism and its significance for the decline of metaphysics is either seen as a children's game, undeserving of real philosophical brains, or as an exorbitant luxury, only apt for an irresponsible and old-fashioned intellectual elite—in other words, it is considered either irrelevant or insubstantial.

I do not want to elaborate here on the justifications I have given colleagues and students through all these years for my work. I only want to say that the two options are, as the thesis and antithesis of a Kantian antinomy, based on a mere illusion of reason. The history of skepticism speaks to problems that are both globally relevant and locally acute. The epistemic issues of what constitutes knowledge as opposed to mere belief and the questions about what status should be accorded to religious creeds remain hotly contested despite the efforts to give them contemporary solutions. Clearly, these problems have practical consequences that affect as much the global community as our particular region. More broadly, social and cultural issues, like the revival of religious dogmatism and its recent alliance with political cynicism, require an urgent reassessment, or, at least, some interrogation, by means of skeptical presuppositions and methods. For, even though our societies' rapidly increasing and widespread political cynicism does not appear to need a new dose of skepticism, much could be learned, in fact, from the ancient and modern skeptical attitudes, which not only sought to protect us from sheer gullibility and deceiving authorities, but also invited us to beware of excessive suspiciousness and social distrust. This skeptical teaching is a lesson that equally applies to the Global North and the Global South and just as much now as it did in the past. For these reasons, I am proud to be part of a Latin American philosophical community that is increasingly aware of the contemporary relevance of the skeptical tradition and devotes itself to the study of its epistemological, metaphysical, and practical questions. I hope to have honored my fellow Latin American historians of skepticism throughout the pages of this book.

Finally, one methodological word. The reader will notice that I adopt the inclusive use of the pronouns 'she' and 'her' to refer to both genders, particularly when it comes to attributing philosophical attitudes, such as those of the Stoic and Academic sages. It has been my pleasure to imply that philosophical wisdom – if that exists at all – not only belongs to male canonical figures (particularly the stern, rational Stoic!) but also to female, non-canonical ones.

Bibliography

Charles, Sébastien, and Pínio J. Smith, eds. 2013. *Scepticism in the Eighteenth Century: Enlightenment, Lumières, Aufklärung*. Dordrecht: Springer.

Cicero, Marcus Tullius. 1950. *Tusculanae Disputationes* (*Tusculan Disputations*). Trans. J.E. King. Cambridge/London: Harvard University Press.

———. 2000. *De Natura Deorum* (*On the Nature of the Gods*). Trans. H. Rackham. Cambridge/London: Harvard University Press.

———. 2001a. *De Fato* (*On Fate*). Trans. H. Rackham. Cambridge/London: Harvard University

———. 2001b. *De Finibus* (*On Moral Ends*). Trans. Julia Annas. Cambridge: Cambridge University Press.

———. 2006. *Academica* (*On Academic Scepticism*). Trans. Charles Brittain. Indianapolis/Cambridge: Hackett Publishing Company.

Forster, Michael. 2008. *Kant and Skepticism*. Princeton: Princeton University Press.

Fosl, Peter. 1994. Doubt and Divinity: Cicero's Influence on Hume's Religious Skepticism. *Hume Studies* 20 (1): 103–120.

———. 2019. *Hume's Scepticism: Pyrrhonian and Academic*. Edinburgh: Edinburgh University Press.

———. 2020. *Hume's Scepticism: Pyrrhonian and Academic*. Edinburgh: Edinburgh University Press.

Gaskin, John C.A. 1976. Hume's Critique of Religion. *Journal of the History of Philosophy* 14 (3): 301–311.

———. 1983. Hume's Attenuated Deism. *Archiv für Geschichte der Philosophie* 65: 160–173.

Guyer, Paul. 2003. Kant on Common Sense and Scepticism. *Kantian Review* 7: 1–37.

———. 2008. *Knowledge, Reason, and Taste: Kant's Response to Hume*. Princeton: Princeton University Press.

———. 2014. *Kant*. London: Routledge.

Hume, David. 1896. *A Treatise of Human Nature*, ed. L.A. Selby-Bigge. Oxford: Clarendon Press.

———. 1902. *Enquiries Concerning the Human Understanding and Concerning the Principles of Morals*, ed. L.A. Selby-Bigge. Oxford: Clarendon Press.

———. 2007. *Dialogues concerning Natural Religion and Other Writings*, ed. Dorothy Coleman. Cambridge: Cambridge University Press.

Israel, Jonathan. 2001. *Radical Enlightenment. Philosophy and the Making of Modernity*. New York: Oxford University Press.

Kant, Immanuel. 1992. *Lectures on Logic* (*VL*). Trans. J. Michael Young. Cambridge: Cambridge University Press.

———. 1996. Critique of Practical Reason (KpV). In *Practical Philosophy*. Trans. Mary J. Gregor, 133–271. Cambridge/New York: Cambridge University Press.

———. 1998. *Critique of Pure Reason* (*KrV*). Trans. Paul Guyer and Allen Wood. Cambridge/New York: Cambridge University Press.

Maia Neto, Jose R. 1997. Academic Skepticism in Early Modern Philosophy. *Journal of the History of Ideas* 58 (2): 199–220.

Popkin, Richard H. 1951. David Hume: His Pyrrhonism and His Critique of Pyrrhonism. *The Philosophical Quarterly* 1 (5): 385–407.

———. 1979. *The History of Scepticism from Erasmus to Spinoza*. Berkeley: University of California Press.

———. 2003. *The History of Scepticism*. New York: Oxford University Press.

Popkin, Richard, de Olaso, Ezequiel, and Tonelli, Gorgio. 1997. *Scepticism in the Enlightenment, International Archives of the History of Ideas*. Dordrecht: Springer.

Smith, Plínio J. 2007. Hume y el escepticismo antiguo. *Signos Filosóficos* 19 (18): 105–126.

———. 2008. La Critique de la raison pure face aux scepticismes cartésien, baylien et humien. *Dialogue* 47: 463–500.

———. 2011. Cómo Hume se volvió escéptico. *Revista Internacional de Filosofía* 52: 71–84.

———. 2013. Kant's Criticism and the Legacy of Modern Scepticism. In *Scepticism in the Eighteenth Century: Enlightenment, Lumières, Aufklärung*, ed. Sébastien Charles and Pínio J. Smith, 247–263. Dordrecht: Springer.

Tonelli, Giorgio. 1997. Kant and the Ancient Sceptics. In *Scepticism in the Enlightenment,* eds. Richard H. Popkin, Ezequiel De Olaso y Gorgio Tonelli, 69–98. Dordrecht: Kluwer Academic Publishers.

Chapter 2
The Skepticism of the New Academy: From *Epochê* to Persuasion

Abstract This chapter offers a general overview of the two schools of ancient skepticism that had the most significant influence on the Enlightenment, namely, the Platonic Academy and Pyrrhonism. I begin by giving an account of how Arcesilaus of Pitane (c. 316–241 BCE) and Carneades of Cyrene (c. 214–128 BCE) – heads of the Middle and New Academies, respectively – attacked the Stoics' conception of knowledge and their criterion of truth, i.e., the *kataleptic* impression. After this, I draw a contrast between Arcesilaus's and Carneades's answers to the '*apraxia* objection'—namely the charge that *epochê* forced the skeptic to paralysis or inactivity. I move, then, to compare the Academics' and Sextus Empiricus's answers, the latter of which involved an undogmatic adherence to "appearances" and to the laws and customs of the skeptic's community. I conclude the chapter by suggesting that the skepticism of the New Academy aimed at the promotion of practical autonomy, while Pyrrhonism endorsed a sort of non-dogmatic but voluntary practical heteronomy.

Ancient skepticism is generally defined as the philosophical position that questioned human beings' ability to attain knowledge, opposed all types of dogmatism, and advocated suspending judgment (*epochê*) as a means to obtain peace of mind (*ataraxia*). It began with Pyrrho of Elis (c. 360–270 BCE) and was developed in Hellenistic antiquity by the Platonic Academy and the Pyrrhonian School, inspired by Socrates and Pyrrho, respectively. Both schools followed the dialectical method of examining arguments for and against a given thesis to attain their goals. Once the analysis showed that the arguments possessed equal logical validity or persuasive force (*isostheneia*), the skeptics concluded that assenting to the thesis was not rationally justified and, thus, suspended judgment.

However, as these schools developed over time, suspension of judgment or *epochê* was itself submitted to examination. Questions like: Is it possible to suspend judgment on all matters? And, if judgment is suspended on all matters, how is it possible to act? were constantly debated throughout Hellenistic antiquity. The famous *apraxia* objection – i.e., the charge that *epochê* forced the skeptic to paralysis or inactivity – was an essential piece of this discussion, often used by the dogmatists to refute and even mock the skeptics. As a result, some schools of

C. González Quintero, *Academic Skepticism in Hume and Kant*, Synthese Library 449, https://doi.org/10.1007/978-3-030-89750-5_2

skepticism decided to narrow the scope of *epochê*, concluding that, under certain conditions and circumstances, it was worth holding some beliefs in order to act, while others maintained a broad suspension but offered practical criteria or guidelines to avoid inactivity.

This chapter offers a general overview of the two schools of ancient skepticism that had the most significant influence on the Enlightenment, namely, the Platonic Academy and Pyrrhonism.[1] My goal here is not to resolve all the interpretative challenges faced by recent scholarship but only to outline their respective views on the matter. That being said, I will take a position on how and why these schools dealt with this problem, and, in this sense, I will partake in the current debate about the Ancient skeptical schools' scope of *epochê* by providing an interpretation that should shed light on these schools' aims and practical responses (Barnes 1997, pp. 58–91; Burnyeat 1997, pp. 25–57; Frede 1997, pp. 1–24). As we will see in the remainder of the book, this interpretation will be essential to understand Hume's and Kant's preference for Academic skepticism over Pyrrhonism in their Enlightened critiques of metaphysics.

I suggest that the most prominent contrast between these Ancient skeptical schools lies in the fact that Pyrrhonism adopted a broad-scope *epochê* – that is, it recommended suspending judgment on all matters examined[2] – while the Academy, specifically, the New Academy, endorsed a more circumscribed suspension, choosing to non-dogmatically approve the beliefs that better withstood skeptical examination to orient action. To explore this contrast, I begin by giving an account of how Arcesilaus of Pitane (c. 316–241 BCE) and Carneades of Cyrene (c. 214–128 BCE) – heads of the Middle and New Academies, respectively – attacked the Stoics' conception of knowledge and their criterion of truth, i.e., the *kataleptic* impression. After this, I draw a contrast between Arcesilaus's and Carneades's answers to the '*apraxia* objection.' On the one hand, Arcesilaus seems to have given two responses: either that the skeptic acts by following the *to eulogon*, namely what is "reasonable;" or that she is guided by mere impressions and impulses. On the other hand, Carneades seems to have argued that the skeptic should follow the beliefs that proved more persuasive or probable through skeptical examination; thus, his criterion for action was the renown *to pithanon.* I move, then, to compare the Academics' and Sextus Empiricus's answers,

[1] I restrict myself to Sextus Empiricus's version of Pyrrhonism. On the differences between Pyrrho's, Aenesidemus's, and Sextus Empiricus's Pyrrhonism, see: Alcalá 1996; Bolzani 2016, pp. 245–289; Bett 1997, 2000; and Conche 1994; among others. In general terms, it can be said, following Bett (1997, pp. 1–15), that Pyrrho's skepticism is a metaphysical skepticism or negative dogmatism; Aenesidemus's, a sort of relativism; and Sextus Empiricus's, an epistemological skepticism. The three, however, have enough features in common to constitute a unified tradition, that is, Pyrrhonism *simpliciter.*

[2] Although Pyrrhonian skepticism "yields to appearances" in order to take action, that is, it allows the skeptic to act by following some sort of mental representations (i.e., sense impressions and natural drives), I do not consider that this entails a reduction in the scope of *epochê*, insofar as these representations are not, properly speaking, "beliefs." See Sect. 2.2 of this chapter. For a contrary view, according to which Pyrrhonism is compatible with holding beliefs, see: Frede 1997, pp. 1–24.

the latter of which involved an undogmatic adherence to "appearances" and to the laws and customs of the skeptic's community. I conclude the chapter by suggesting that the skepticism of the New Academy aimed at the promotion of practical autonomy, while Pyrrhonism endorsed a sort of non-dogmatic but voluntary practical heteronomy. With this balance in mind, I argue in the following chapters that the modern adoption of an ancient school of skepticism crucially involved adhering to both the schools' methodological claims and practical orientation.

2.1 An Overview of Academic Skepticism

In the *Academica* (*Acad. I* and *II*),[3] Cicero outlined the skeptical system of the Middle and New Platonic Academies, from Arcesilaus of Pitane to Philo of Larissa (c. 159–84 BCE).[4] In general terms, this system comprised a wealth of skeptical arguments used to counter the dogmatic pretensions of Stoicism. It is said that the most prominent leaders of the Middle and New Academies, Arcesilaus of Pitane and Carneades of Cyrene, were involved in painstaking debates with their Stoic counterparts – Zeno of Citium (c. 333–264 BCE), Chrysippus of Soli (c. 281–208 BCE), and Antipater of Tarsus (c. Unknown–130 BCE) – about the nature of belief and knowledge.[5] In the course of these debates, the skeptical point of view of the school gradually developed.[6] Eventually, however, the dispute with Stoic dogmatism became a somewhat internal matter in the Academy. The last leader, Antiochus of Ascalon (c. 130–68 BCE), adopted Stoic claims concerning knowledge and argued that Plato held a positive doctrine and not a skeptical system, as the Middle and New Academies had assumed up to that point. Against Antiochus,

[3]The two extant books of the *Academica* are the first book of the second version (*Acad. I*) and the second book of the first version, entitled Lucullus (*Acad. II*).

[4]The First or Old Academy dates from its foundation by Plato in 387 BCE to the election of Arcesilaus as scholarch in 268 BCE. This Academy held a dogmatic position.

[5]It is argued that Arcesilaus and Zeno were, in their youth, co-disciples of Polemon at the Old Platonic Academy, but separated and became rivals when the former assumed a skeptical position and the latter founded Stoicism. According to Eusebius of Caesarea, "[. . .] when Arcesilaus saw that Zeno was a professional rival, and worth conquering, he shrank from nothing in trying to overthrow the arguments set forth by him" (*Praep. evang.*, XIV, 6). Cicero's testimony is more nuanced: "It was entirely with Zeno [. . .] that Arcesilaus set on foot his battle, not from obstinacy or desire for victory, as it seems to me at all events, but because of the obscurity of the facts that had led Socrates to a confession of ignorance [. . .]" (*Acad. I*, 44; see *Acad. II.*, 76). Regarding Carneades's opposition to Stoicism, Diogenes Laërtius attests that Carneades himself would say: "Had Chrysippus not existed, neither would I exist" (*Diog. Laërt.*, IV, 62) and Eusebius affirms that Antipater used to debate with Carneades, but only in writing (*Praep. Evang.*, XIV, 8). Cicero also refers to the argument Antipater put forward against Carneades's principle of inapprehensibility in *Acad. II*, 28–29.

[6]Perhaps the most important discussion on the influence of Stoicism on the development of Academic skepticism is P. Couissin's renowned article of 1929, translated into English by J. Barnes and M. Burnyeat (Couissin 1983, pp. 31–63).

Philo of Larissa appears to have defended, albeit with some qualifications, the skeptical position of the New Academy.[7] Given these disagreements, Antiochus named his school the "Old Academy," while Philo continued to call his the "New Academy." In Book II of the *Academica* (entitled *Lucullus*), Cicero gives a general overview of this dispute. The character called Lucullus sets out Antiochus's doctrine, and his opponent in the debate, Cicero himself, defends the return of the Academy to the point of view of Carneades, relayed to him mainly by Clitomachus (*Acad. II*, 78). This standpoint possesses a negative, or expressly skeptical, doctrine and a positive or constructive one.[8] The negative one is a revision of Arcesilaus's dialectical arguments against Stoic epistemology. It is well known that Carneades and his followers significantly refined Arcesilaus's arguments against Zeno to attack Chrysippus's equally improved form of Zenonian dogmatism. Although the variations of these arguments are, in themselves, an interesting topic of discussion, I believe that the positive part of Carneades's system, as described by Cicero, constitutes the New Academy's most important innovation, since in proposing it, Carneades distanced himself from Arcesilaus's radical skepticism and adopted a more moderate form of skepticism. This skepticism sustained that, although our mental faculties are unable to attain certainty, some beliefs that prove to be sufficiently persuasive or probable upon skeptical examination can be provisionally adopted to orient action.

Throughout the chapter, I focus on this aspect of the New Academy's skepticism. Still, some clarifications about the sources of Academic skepticism and the difficulties in interpreting them may be helpful before outlining the general arguments of the school. Drawing a clear picture of the principal tenets of the Academy is a challenging task to which recent commentators have devoted painstaking efforts. The main difficulties lie in the fact that the extant testimonies of Arcesilaus's and Carneades's views – which are to be found mainly in Cicero's *Academica*, Diogenes Laërtius's *Lives of Eminent Philosophers* (*Diog. Laërt.*) and Sextus Empiricus' *Outlines of Pyrrhonism* (*Pyr.*) and *Adversus Mathematicos* (*M VII* and *XIX*) – are either incomplete or give dissimilar (sometimes even contradictory) accounts of their skeptical systems, thus making it difficult to determine whether a specific position was in fact held by one of the scholarchs or was a reconstruction colored by the authors' interests.[9] In particular, interpreting the *Academica* faces many exegetical obstacles.

---

[7]It is said that Philo changed his mind and developed a dogmatic doctrine in his famous Roman Books. According to C. Brittain (2001, p. 17), there are three phases to Philo's thought, the second of which, called by Brittain "Philo and Metrodorus's phase," would be his Carneadean phase. An alternative discussion on Philo's position can be found in Tarrant (2007).

[8]With this characterization, I follow Zeller's classical indications (1923, Vol. 3, p. 518). It is noteworthy that V. Brochard (2002, p. 140) and C. Lévy (1992, p. 38) see this twofold description of Carneades's system as mistaken, since it emphasizes Carneadean probabilism to the detriment of his ethical and theological refutative efforts. I believe, however, that the division has a methodological and explanatory value. I will refer to this issue in more depth in Chap. 2.

[9]For example, some scholars argue that Cicero's position coincides more with that of Philo of Larissa than with that of Carneades (Thorsrud 2009, pp. 84–101; Görler 1997; and Tarrant 2007, p. 43).

To begin with, the treatise has survived only in fragmentary form,[10] although, fortunately, the existing texts are sufficiently extensive. Beyond this, the historical controversy described in the *Lucullus* between Antiochus of Ascalon and Philo of Larissa relies on sources that are unfortunately non-extant, making it impossible for contemporary readers to verify its details.[11] And, if the scarcity of sources was not an interpretative challenge enough, the treatise's dialogical structure poses yet other intricacies, such as disentangling which views belong to Carneades himself, which to his successors, which to Cicero's, and which are mere inventions put in the mouths of the dramatic characters to suit their historical personalities.[12]

One last interpretative concern is the difficulty in determining whether Arcesilaus and Carneades positively held the skeptical arguments described in Cicero's, Diogenes Laërtius's, and Sextus's testimonies, or they merely used them dialectically to attack the Stoics. As we will see in the following discussion, I agree with some commentators on the dialectical use of Arcesilaus's arguments (Frede 1997, p. 130), but I also believe that both scholarchs professed their skepticism honestly. An associated view is that Arcesilaus and Carneades were skeptics only exoterically while holding the positive doctrines of Platonism esoterically.[13] However, this view

---

[10] See footnote 3.

[11] Cicero says he used Clitomachus's treatises as his main source on suspension of judgment (*Acad. II.*, 98). Unfortunately, these texts have not survived either. Perhaps exaggerating Clitomachus's prolificacy, Diogenes Laërtius affirms that he wrote more than 400 treatises (*Diog. Laërt.* IV, 67).

[12] Cicero expresses doubts as to whether Carneades thought that the wise man could not hold opinions or granted some assent to them. According to his testimony, Clitomachus held the former view, while Metrodorus and Philo thought that the latter was the case: "This is the one disagreement still outstanding. The view that the wise person will not assent to anything has no part in this controversy: he would fail to apprehend anything and yet still have his opinions. In fact, this is said to have been the position approved by Carneades – although, since I trust Clitomachus rather than Philo or Metrodorus, I consider it a position he argued for rather than approved" (*Acad. II.* 78). This testimony has produced much perplexity among scholars. See, for example, Frede 1997, pp. 140–142; and Thorsrud 2009, pp. 80–81. The main difference between the Clitomachean and the Metrodorean interpretation of Carneades's skepticism seems to be, in a nutshell, that the former sees Carneades's probabilism as a skeptical stance, insofar as the sage would not assent to persuasive impressions but merely approve them, while the latter would understand it as a mild dogmatism since the sage would give her assent to persuasive impressions. For a detailed discussion of the distinction, see: Fosl 2020, pp. 39–76.

[13] According to the esoteric thesis, Arcesilaus and Carneades did indeed hold a doctrine but they only shared it with their closest pupils. Thus, in Sextus Empiricus: "And if one ought to credit also what is said about him [Arcesilaus], he appeared at the first glance, they say, to be a Pyrrhonist, but, in reality, he was a dogmatist; and because he used to test his companions by means of dubitation to see if they were fitted by nature for the reception of the Platonic dogmas, he was thought to be a dubitative philosopher, but he actually passed on to such of his companions as were naturally gifted the dogmas of Plato" (*Pyr.* I, 234). The thesis is famously attributed to Augustine of Hippo, who said: "[...] Arcesilaus, in my opinion, acted in a most prudent and useful way, since the evil was spreading widely, in concealing completely the doctrine of the Academy and in burying it as gold to be found at some time by posterity." (1951, p. 145). An analysis of this thesis can be found in Lévy (1978, pp. 335–348), Glucker (1978, pp. 296–306), and Dutton (2016, pp. 24–26). According to Dutton, Augustine thought that, "though Arcesilaus and his successors in the Academy employed the weapons of skepticism with consummate skill and unflagging vigor, they did so in the hope that they might one day lay them down and bring Plato's teachings back into the

seems to be either legend or some dogmatists' way of excusing the Academics from their otherwise immoral character.[14] Finally, there is the most prominent view among scholars – the one to which I am most sympathetic – according to which Arcesilaus and Carneades held their skeptical systems non-dogmatically, and these, on the whole, conform to Cicero's descriptions.[15] With these caveats in mind, I will begin to explore the *Academica* with the discussion about how the Academic skeptics refuted Stoic dogmatism.

2.1.1 Academic Arguments Against the Stoics

In general terms, Stoic epistemology held that the human mind can obtain knowledge from sense impressions and their corresponding mental representations (*phantasia*, in Greek; *visum*, in Latin).[16] However, since, as experience shows all too often, not all impressions are equally trustworthy, the Stoics believed that a criterion was necessary to distinguish true impressions from false ones, or impressions that led to true judgments from those that led to false ones.[17] Furthermore, the Stoics asserted that this criterion was a specific characteristic of some impressions, namely their clarity or evidence (in Greek, *enargeia*; in Latin, *perspicuitas* or *evidentia*) (*Acad. II*, 17), which attested to its correspondence with the object it represented. In other words, it was the mark that the impression was caused by this object and not by some other one: "[the impression] was, as Zeno defined it, [...]

---

open. Their skeptical practice was thus intended to slow and ultimately reverse the advance of materialism and empiricism within the larger community of philosophers to such a degree that those teachings might once again find fertile ground and take root." (Dutton, 2016, p. 25).

[14] According to Lévy (1978, p. 341), it is noteworthy that Cicero never referred to such an esoteric doctrine. Indeed, had Arcesilaus or Carneades held a doctrine of this kind, most probably their successors would have known it and Cicero would have recorded it. It is, therefore, more likely that the attribution of this doctrine began as a rumor among dogmatists, which spread in time, reaching out to Augustine.

[15] Here I follow the indications of Brochard (2002, pp. 145–151), Thorsrud (2002), Hirzel (1883), and Couissin (1983), among others. According to Brochard: "We can, therefore, say – not without reservations, it is true – that Carneades had renounced *epochê*, that he recognizes the legitimacy of certain beliefs, he is a probabilist" (2002, p. 148). Thorsrud also claims this: "If the dialectical interpretation of Arcesilaus and Carneades is correct, then Cicero seriously misunderstands, or at least misrepresents, his Academic predecessors" (2002, p. 4).

[16] In earlier translations, such as Rackham's, the words *phantasia* or *visum* were often rendered as "presentation" (Cicero, 2000). Following Brittain's translation (Cicero, 2006), however, I have chosen the term "impression," because it indicates more precisely the type of mental representation caused by the effect of an external object on the receptive capacity of human beings.

[17] The Stoics considered that impressions themselves, and not just the judgments that expressed them, could be true or false. Hence, on occasion, I will refer to "true or false impressions," but what should be understood is impressions leading to true or false judgments. Arcesilaus himself introduced the limitation of attributing truth solely to judgments – a limitation that today is mainstream (*M VII*, I, 154).

stamped and molded from its source in a way that it couldn't be from what wasn't its source" (*Acad. II.* 18).[18] The mind receives these impressions in a specific conceptual act of apprehension that they called *katalêpsis.*[19] The term refers to how the mind captures and takes a firm grip on impressions, which Zeno illustrated with the metaphor of a tightly-clenched fist:

> "He [Zeno] didn't put his trust in all impressions but only in those that revealed their objects in a special way. Since this kind of impression could be discerned just by itself, he called it 'apprehensible.' – Can you bear this?" "Of course," Atticus replied, "how else could you translate *katalêpton* [...]? – "But once it had been received and approved, he called it apprehension or 'grasp,' like something grasped by one's hand (In fact, that was his source for this term, since no-one had used this word before) [...]" (*Acad. I*, 41).

To the act of apprehending a *kataleptic* impression (as scholars call it today), the Stoics added a second mental act of assenting to it.[20] They considered this assent (in Greek, *sinkatathesis*; in Latin, *adsensione*) a voluntary act of subjectively affirming the impression,[21] that is, the activity through which the mind accepts or avows that the impression is caused by the object it represents and, thus, is true. The *kataleptic* impression – also called 'cognitive' or 'apprehensive' impression – was, then, the Stoic criterion for truth. Knowledge was obtained by assenting only to this kind of impressions, while opinion resulted from voluntarily giving one's assent to obscure, non-*kataleptic* ones (*Acad. I*, 41). Consequently, all arts and sciences – the whole domain of the Stoic sage's expertise – were based on *kataleptic* impressions. Otherwise put, scientific assertions and technical indications were considered knowledge only if grounded on apprehended impressions, and those who affirmed these impressions were seen as sages or experts (*Acad. II.* 22).

---

[18] Zeno also defines the *kataleptic* impression as that impression "stamped, impressed, and molded just as it is" (*Acad. II.* 77). Sextus Empiricus, in turn, offers the following description: "But when [the sense] is disturbed and somehow affected owing to the impact of things evident, then it indicates the objects. Therefore, the criterion must be sought in the affection of the soul caused by sensible evidence. And this affection must be both of itself and of the appearance which caused it, which affection is nothing else than the presentation" (*M VII. I*, 161).

[19] The Greek term *katalêpsis* is translated by Cicero as *comprenhensio* or *percipio*. Following standard usage, I call the mental activity in question "*kataleptic* impressions," "cognitive impressions," "apprehensive impressions," or more simply "apprehensions."

[20] Here I follow M. Chiesara's description of Chrysippus's theory of *katalêpsis*. According to Chiesara, there are important differences between Zeno's and Chrysippus's theories. For Zeno, assent was a mental act previous to apprehension or *katalêpsis*, while Chrysippus seems to have changed this order to respond to Arcesilaus's attack, according to which, if assent comes before apprehension, the wise Stoic assents to an empty representation. Chrysippus, then, claimed that assent follows *katalêpsis*: "[...] assent and apprehension thus become two different moments in a single process, which also allows Chrysippus to reject the accusation that Arcesilaus had directed at Zeno, of having assent precede apprehension and, therefore, to assent to that which is unknown" (Chiesara 2007, p. 61).

[21] " [...] He [Zeno] conjoined these – the impressions 'received' by the senses, so to speak – with the assent of our minds, which he took to be voluntary and have its source in us" (*Acad. I*, 40). Couissin has also discussed the voluntary nature of assent according to the Stoics (Couissin 1983, p. 35).

Furthermore, this definition of knowledge extended to practical matters. For the Stoics, leading a moral and, hence, happy life essentially involved distinguishing between *kataleptic* and *non-kataleptic* impressions about what is good or bad, and being moved to action only by the former. Thus, the Stoic sage's practical rationality was infallible and her decisions unfaltering:

> The study of the virtues also provides very strong confirmation that many things are apprehensible [...], so we claim that it, too, like the virtues, depends entirely on such apprehensions. Likewise, for wisdom, the art of life, which gives rise to the constancy of the wise. But suppose their constancy didn't depend on any apprehension or knowledge. Then I'd like to know [...] why the good man resolves to endure every torture or be wracked by intolerable pain rather than give up on an appropriate action or his word. Why would he impose such heavy constraints as these on himself if he did not rely on anything apprehended, known, or determined that would explain why this was fitting? (*Acad. II*, 23).

This is, in a nutshell, the Stoic doctrine of knowledge and practical wisdom. The Academics attacked it severely, questioning their assumption that there could be an infallible criterion to distinguish between true and false impressions. Their rejection of the *kataleptic* impression's criterion was based on the thesis that no inherent characteristic of an impression could indicate with absolute certainty its correspondence with the object that caused it. However, this criticism did not mean that the Academics denied the existence of sufficiently trustworthy impressions, nor that they maintained the impossibility of making true judgments. Instead, their point was that one could never be sure of whether any given impression was caused in fact by the object it represented, and hence, one could not assert with complete certainty whether a judgment was true or false. Put differently, they did not deny the possibility of truth, only the possibility of certainty. As a result, the sciences and arts – and any other human activity – could only be founded on more-or-less-clear impressions, which were the grounds of correspondingly more-or-less trustworthy beliefs.

The most critical Academic argument against Stoic dogmatism sought to demonstrate that for any impression resulting in a true judgment, there always was another impression, indistinguishable from the former, that led to a false judgment:

> There are four premises to the conclusion that nothing can be known or apprehended [...]. They are that [1] there are some false impressions; [2] those impressions aren't apprehensible; [3] when two impressions don't differ at all, it's not possible that one is apprehensible, while the other isn't; [4] there is no true impression derived from the senses that may not be paired with another impression that doesn't differ from it at all but isn't apprehensible. [...] The battle is entirely over the fourth premise (*Acad. II*, 83).

This argument is somewhat intricate and has given rise to many interpretations. Briefly, the Academics held that if, as the Stoics asserted that (i) there are false impressions, which, because of being so, are not apprehensible or *kataleptic*, then, (ii) when we cannot distinguish a false impression from a true one because of their likeness, it is not possible to apply the criterion. As a result, (iii) there is no way to distinguish between false and true impressions, and (iv) this amounts to saying that there are no *kataleptic* or cognitive impressions at all.[22] To support this conclusion, the Academics wielded various skeptical topics, including the indiscernibility of objects like two eggs or grains of sand, and the difficulty in distinguishing between dreams and reality (*Acad. II*, 84–90). Given these common cases, they concluded, impressions are always obscure or "inapprehensible."

In response to this attack, the Stoics claimed that any single entity is always, in principle, discernible from another (*Acad. II.* 85). If two impressions appear to be indistinguishable, it is either because our senses are not sufficiently trained to recognize their differences (like those of an expert in a particular art, a natural scientist, or the sage) or because some abnormal circumstances hinder this recognition (mental or sensory illness, atmospheric or physical conditions that make observation difficult, etc.) (*Acad. II.* 19). However, for the Stoics, the sage would always be capable of making the appropriate distinctions, since her senses and reason are perfectly healthy and finely sharpened by experience (*Acad. II.* 20).

The Academics remained unconvinced and insisted that to any impression, irrespective of its apparent clarity or evidence, another could always be opposed, which led to a false judgment. Again, their argument was based on the thesis that some degree of indiscernibility is inherent to impressions. Sextus Empiricus described the argument thus:

> So, then, once more, since there is no true presentation of such a kind that it cannot be false, but a false presentation is found to exist resembling every apparently true presentation [impression], the criterion will consist of a presentation which contains the true and false alike. But the presentation which contains them both is not apprehensive, and not being apprehensive, it will not be a criterion (*M VII. I*, 164).

Based on these arguments, the Academics advanced the so-called "principle of inapprehensibility (*akatalêpton*) of impressions." This principle held that all impressions were (or rather, seemed to be) inapprehensible, since no inherent characteristic could warrant their truthfulness. The positive formulation of the principle (that all impressions *are* inapprehensible) led Aenesidemus and Sextus to consider the Academics a sort of "negative dogmatists" who outright denied the very possibility of knowledge and abandoned inquiry even before beginning to undertake it. The observation appears in the opening passage of the *Outlines of Pyrrhonism* (*Pyr.*), in which Sextus classified the schools of philosophy according to their attitude regarding inquiry. He argued that the dogmatists and the Academics ceased the search

[22] A good description of how the Academics arrived at this quite general conclusion can be found in "Academic Arguments for the Indiscernibility Thesis," where C. Perin debates the scope of the Academic arguments on the indiscernibility of such objects as two twin brothers, two eggs, two pears, etc., and also of mental impressions obtained during wakefulness and sleep. Perin's view is that the Academics used these arguments to show how, in the case of an impression P of an object O, said impression P could have been caused by object M so that P would not be an apprehensive impression of O (for it would be false). The range of generality of the argument is given by the fact that it is founded on a counterfactual possibility (not only on a real possibility), which allows the refutation of the very definition of the Stoics' cognitive impression (Perin 2005, pp. 493–517).

altogether; the former, when they believed to have founded the truth, the latter even before they began the investigation, since they held that all impressions were inapprehensible. Only the Pyrrhonists (or skeptics), for Sextus, carried inquiry on:

> The natural result of any investigation is that the investigators either discover the object of the search, or deny that it is discoverable and confess it to be inapprehensible, or persist in their search. So, too, with regard to the objects investigated by philosophy, this is probably why some have claimed to have discovered the truth, others have asserted that it cannot be apprehended, while others again go on inquiring. Those who believe they have discovered it are the 'Dogmatists,' especially so-called – Aristotle, for example, and Epicurus and the Stoics and certain others; Cleitomachus and Carneades treat it as inapprehensible; the Sceptics keep on searching (*Pyr. I*, 1–4).

Before Sextus, we know that Aenesidemus also characterized the Academics as dogmatic skeptics. According to Photius's testimony, Aenesidemus wrote in his *Pyrrhonian Discourses*:

> [. . .] The philosophers of the Academy are dogmatists; they put forward certain principles as incontestable, and reject others without any hesitation; Pyrrho's followers practice doubt and are free of any statement of principle, and none of them, anyway, has said that everything could either be grasped or not, nor that things are such or such, or sometimes one thing and sometimes another (*Bib.* 169b–170a, my translation).[23]

Notwithstanding these testimonies, we have textual evidence pointing to the fact that the Middle and New Academies held the principle of inapprehensibility of impressions non-dogmatically and, consequently, did not cease inquiry before beginning it. The following passage in Cicero helps clarify the point:

> [. . .] To be sure, knowledge is always surrounded with difficulties, and the obscurity of the things themselves and weaknesses of our judgments is such that one can see why the earliest and most learned philosophers lost confidence in their ability to discover what they desired. Still, they don't give up, and we won't abandon our enthusiasm for investigation owing to exhaustion. Nor do our arguments have any purpose other than to draw out or formulate the truth or its closest possible approximation by means of arguing on either side (*Acad. II*, 7).

According to Cicero, the Academics did not give up inquiry beforehand, but their love for the truth led them to continue their research into both sides of a question until they could find the "closest possible approximation" to the truth. This testimony indicates that they did not hold the principle of inapprehensibility

[23]The difference also appears in Aulus Gellius: "But although the Pyrrhonians and the Academics express themselves very much alike about these matters, yet they are thought to differ from each other both in certain other respects and especially for this reason – because the Academics do, as it were, 'comprehend' the very fact that nothing can be comprehended, while the Pyrrhonians assert that not even that can by any means be regarded as true because nothing is regarded as true" (A*N*, XI, 5, 8). Another passage in Sextus Empiricus is important in this regard: "The adherents of the New Academy, although they affirm that all things are non-apprehensible, yet differ from the [Pyrrhonian] Sceptics even, as seems probable, in respect of this very statement that all things are non-apprehensible (for they affirm this positively, whereas the [Pyrrhonian] Sceptic regards it as possible that some things may be apprehended)." (*Pyr.* I, 226).

dogmatically—that is, they did not understand it as a true and certain judgment. Had they done this, the very principle would have been a judgment based on a *kataleptic* impression, and they would have contradicted themselves. But, since both Arcesilaus and Carneades were rather accomplished dialecticians, it is unlikely that they had made this crass mistake. Instead, it is more probable that they considered the principle itself as *non-kataleptic*, as Cicero claims. According to him, Arcesilaus saw the principle as a doubtful assertion, whose primary function was dialectical (*Acad. I*, 45), while Carneades took it as a merely persuasive or plausible claim.[24] He affirms:

> Given that a wise person can't have principles unless they are apprehended or known, [they say that] Carneades should at least allow that this principle itself is apprehended – i.e., that the wise person holds that nothing is apprehensible. As if the wise person has no other principles and could not live his life without principles! But just as he holds those as persuasive rather than apprehended principles, so with this one, that nothing is apprehensible (*Acad. II*. 109–110).[25]

Since the principle of inapprehensibility of impressions was not sustained as a dogmatic claim, embracing it did not commit the Academics to halt inquiry, like Sextus's famously asserts. As Bolzani argues, Sextus's characterization of Academic skepticism as a negative dogmatism probably had propagandistic aims:

> The Pyrrhonian accusation that the Academic are dogmatists, clearly expressed in the *Outlines of Pyrrhonism*, was probably dictated by the need to establish their skeptical

[24]H. Thorsrud affirms, and I agree with him, that characterizing the Academics as negative dogmatists is incorrect: "In particular, it is unclear whether it is fair to describe Carneades as negatively dogmatic, and permanently closed to the possibility of discovering the truth. Although Carneades does confidently argue and affirm that things are not apprehensible, in Cicero's account he does not claim to have apprehended this" (2009, p. 10). Also, G. Striker: "The point to which Sextus and Gellius refer [. . .] is, as one can see from Cicero, not correct: Arcesilaus and Carneades explicitly declared that they did not even know that they knew nothing, hence one might be inclined to think that the Pyrrhonists' attempt at demarcating themselves was no more a matter of school politics than of differences in content" (1996, p. 136). Some commentators believe, however, that there is a sense in which it really can be affirmed that the principle of inapprehesibility is an Academic *dogma*. Thus, for example, Tarrant argues that the term *dogma* (in Latin, *decretum*) refers more to a practical principle or rule of behavior than to a theoretical or scientific axiom. Therefore, the inapprehensibility of impressions could be sustained by the Academics as *dogma*, without the implication that it had to be considered as an apprehended principle (2017, p. 30). Barnes also examines in detail the meaning of the term *dogma* among Ancient skeptics, particularly in Sextus (1997, pp. 58–91). A dissimilar position can be found in Alcalá, who claims that Arcesilaus did hold the principle of inapprehensibility of impressions dogmatically (1994, p. 80).

[25]Cicero speaks here about Antiochus's and Antipater's critique of Carneades's principle of inapprehensibility of impressions. For Antiochus, being a practical principle, it was necessarily apprehended: "This gave rise to the demand Hortensius made, that you should at least admit that the wise person apprehends the claim that nothing is apprehensible. Antipater used to make the same demand: it is still consistent, he maintained, for someone affirming that nothing is apprehensible to say that this one claim is apprehensible though nothing else is. But Carneades resisted him more forcefully, saying that, far from being consistent, it was actually grossly inconsistent: someone claiming that nothing is apprehensible makes no exceptions; it follows necessarily that, since it hasn't been excepted, the claim itself can't be apprehensible, either" (*Acad. II*. 28).

> affiliation as the authentic one, separating the skeptic from any association with the dogmatism that, by the end of the Academy, culminated in the introduction of Stoicism. (Bolzani 2016, pp. 249).

Bolzani's assessment is, in my view, highly probable. Both Aenesidemus and Sextus became disappointed with the Academy's late introduction of Stoic doctrines, which may have led them to want to distinguish their skepticism from the Academy's by interpreting the principle of apprehensibility as an assertion of negative dogmatism.

The principle of inapprehensibility also served the purpose of refuting the Stoic's ideal of the sage. Cicero says that Arcesilaus employed the following syllogism against the Stoics: "(1) If the wise person ever assents to anything, he will sometimes hold an opinion; (2) but he will never hold an opinion; (3) so he won't ever assent to anything" (*Acad. II.* 67).[26] In other words, Arcesilaus claimed against the Stoics that, if they followed their own principle – namely that the Stoic sage only made judgments based on *kataleptic* impressions and, for this reason, never held opinions and never made mistakes – faced with the principle of inapprehensibility of impressions, they would have to suspend judgment in all matters. Indeed, if, as the Academics showed, there were no *kataleptic* impressions, to be consistent with their principles, they would only have one option left—to become skeptics.

Allow me to elaborate a little more on this argument. For the Stoics, opinions were the result of assenting to inapprehensible, obscure impressions. For this reason, while someone may have an opinion that, by sheer luck, turns out to be true, more often than not opinions are false descriptions of states of affairs. Thus, to ensure that her judgments were true and to always act correctly, the Stoic sage would only assent to *kataleptic* impressions. But if, as the Academics argued, there were no *kataleptic* impressions, then, to be consistent with her principles, she would have to become a skeptic and suspend judgment in all matters.

Given the broad and profound consequences of the Academic's principle of inapprehensibility, it comes as no surprise that the Stoics felt the need to attack the Academics using arguments that did not pertain to the epistemic realm but to the practical one. And, in fact, the most crucial argument against the skeptics was the so-called *apraxia* objection, namely that without assenting to impressions, the skeptics could not act. We should now turn to this argument. As I will attempt to show, this attack forced both Arcesilaus and Carneades to envisage a set of criteria for action consistent with their skepticism, while reducing the scope of the Academic *epochê*.

[26] Sextus sums up the Arcesilean argument thus: "[I]f the wise man shall assent the wise man will opine; for when nothing is apprehensible, if he assents to anything, he will be assenting to what is non-apprehensible, and assent to the non-apprehensible is opinion. So that if the wise man is in the class of assenters, the wise man will be in the class of those who opine. But the wise man, to be sure, is not in the class of those who opine (for, according to them, opinion is a mark of folly and a cause of sins); [. . .] And if this is so, he will necessarily refuse assent in all cases. But to refuse assent is nothing else than to suspend judgment; therefore, the wise man will in all cases suspend judgment" (M VII, 156–157).

2.1.2 Arcesilaus's Response to the Apraxia Objection

The *apraxia* problem or objection is the argument according to which suspending judgment necessarily brings about the skeptics' inability to act. The earliest formulation of the objection appears in Aristotle, who states in the *Metaphysics*:

> [...] And if he makes no judgment but thinks and does not think, indifferently, what difference will that be between him and the plants? – Thus, then, it is in the highest degree evident that neither any one of those who maintain this view nor anyone else is really in this position. For why does a man walk to Megara and not stay at home, thinking he ought to walk? Why does he not walk early some morning into a well or over a precipice if one happens to be in his way? Why do we observe him guarding against this, evidently not thinking that falling in is alike good and not good? Evidently, he judges one thing to be better and another worse. And if this is so, he must judge one thing to be man and another to be not-man, one thing to be sweet and another to be not-sweet (*Metaph.* 1008b 10–15).

If because of suspending judgment on all matters, the skeptic concludes that no particular assertion is more valid than its opposite, how can she act? How can she decide, for example, to stay safe on the path, if the impressions she obtains from the cliff are wholly indistinguishable from those of the valley and, following the skeptics' method, she has completely suspended judgment about whether or not she is facing a precipice?

In *Against Colotes*, Plutarch presents the objection in similar terms:

> But how comes it that the man who suspends judgment does not go dashing off to a mountain instead of to the bath, or why does he not walk to the wall instead of the door when he wishes to go out to the marketplace? (*Adv. Col.* 1122E).

As we can see from these initial examples, the objection was a common argument in antiquity and appeared in many anecdotes from ancient doxology. Diogenes Laërtius, for instance, claims that Pyrrho's suspension of judgment was such that, because he did not credit his senses, he would not deviate from his path even if he found in his way precipices, coaches, or dogs, thus his friends had to save his life constantly (*Diog. Laërt.* 9, 62). In the same vein, Eusebius of Caesarea relates that Lacydes – Arcesilaus's direct successor – was something of a miser and always left his provisions under lock and key when he left the house. Not understanding how he always arrived home to find the pantry empty, in a fit of desperation, he decided to join the Academy and embrace the principle of inaprehensibility of impressions. But one day, he found the cause for the disappearing supplies—his slaves would steal his food while he was out. In a fit, this time, of anger, he harshly accused his slaves of stealing his goods. The slaves responded that if he genuinely held Academic views, he could not punish them because, to do so, the accusation would have to be based on a *kataleptic* impression. Lacydes replied, "Of these things, my boys [...] we talk in our discussions one way, but we live in another" (*Praep. Evang*. XIV, 7; see *Diog. Laërt*. IV, 59). If the reader will indulge one last example, Eusebius tells us also how, on a particular occasion, Carneades encountered his pupil Mentor making love to his concubine and, instead of suspending judgment on the matter, expelled him from the Academy (*Praep. Evang*., XIV, 8). As we see from these anecdotes, the *apraxia* objection was already common currency against the skeptics during the Middle and New Academies (Frede 1997, p. 2).

For this reason, both Arcesilaus and Carneades offered what seems to be a solution to the objection. According to Sextus, Arcesilaus answered that, without assenting to any impression, the skeptic could still act by following the criteria of "the reasonable" (*to eulogon*). Sextus explains:

> Arcesilaus asserts that he who suspends judgment about everything will regulate his actions in general by the rule of "the reasonable," and by proceeding in accordance with this criterion, he will act rightly; for happiness is attained by means of wisdom, and wisdom consists in right actions, and the right action is that which, when performed, possesses a reasonable justification. He, therefore, who attends to "the reasonable" will act rightly and be happy (*M VII*. I, 158).

Interestingly enough, the Greek term used for this criterion, to *eulogon*, usually translated as "the reasonable," has a Stoic origin. Diogenes Laërtius affirms that the Stoics considered it a criterion used by ordinary people to act and defines it as "the proposition with the greatest claim to truth, as in, for example, 'I will be alive tomorrow'" (*Diog. Laërt.* 7, 76). Therefore, if Arcesilaus claimed that the skeptic oriented her life by the *to eulogon*, it seems that he offered a criterion for action that the Stoics considered unbecoming for the sage. This fact has led some commentators (Couissin 1983, p. 37; Striker 1996, pp. 92–105) to believe that the *to eulogon* was not really Arcesilaus's response to the *apraxia* objection, but rather one more of his dialectical weapons against Stoicism.[27] Arcesilaus' argument would go something like this: Given the principle of inapprehensibility, the Stoic sage would have to withhold assent to all impressions and, thus, would become a skeptic. Once a skeptic, however, to not remain inactive, she could follow the criterion of the "reasonable," which, for the Stoics, was a practical criterion used by ordinary people. As a result, suspending judgment in all matters and using the *to eulogon* to act, she would cease at once to be both "Stoic" and "sage." We can see in this attack Arcesilaus's great skill as a dialectician – he conjoined two syllogistic arguments with Stoic premises to demolish the very figure of the Stoic sage.

However, there is debate among scholars about whether Arcesilaus actually held the *to eulogon* as a criterion to act or simply used it as a dialectical weapon against the Stoics (Striker 1996, pp. 92–105; Thorsrud 2009, pp. 53–56; Alcalá 1994, p. 78). I think we should note that both interpretations have their advantages and disadvantages. In the dialectical one – which I subscribe – Arcesilaus did not believe that the skeptic should adopt the criterion but only put it forward as a final blow to the figure of the Stoic sage. The advantage of this interpretation is that it coheres with Arcesilaus's recommendation of a broad-scope *epochê*. Indeed, had he genuinely suggested that the skeptic maintained "the reasonable" or *to eulogon* to act, he would

---

[27] Here I follow Couissin's interpretation: "Reading between the lines of his argument, we sense a biting irony that will reduce the sage to a mere mortal. [. . .] Arcesilaus wanted to humble the Stoic sage, to show that in the end, he was not superior to the ordinary man or the Epicurean, who strives in all modesty to find a happy life through prudence" (1983, p. 37). A similar position, following Couissin up to a point, is to be found in Striker (1996, pp. 92–105).

have limited its scope, since he would have allowed for the adoption of some beliefs; however, no available source documents such a reduction. The disadvantage of this interpretation is that it does not seem to conform to Arcesilaus's position in the Academy. Had he maintained the criterion only dialectically, the *apraxia* problem would have remained unanswered. Still, it does seem that, as head of the Academy, Arcesilaus would have wanted to give a response to it.

With this in mind, it is worthwhile to ask whether Arcesilaus provided another criterion for action. The answer seems to be positive if we consider Plutarch's testimony in *Against Colotes.* Plutarch tells us that Arcesilaus responded to the *apraxia* objection by describing the skeptic's action as a result of mere sensation and impulse:

> The soul has three movements: sensation, impulse, and assent. Now the movement of sensation cannot be eliminated, even if we would; instead, upon encountering an object, we necessarily receive an imprint and are affected. Impulse, aroused by sensation, moves us in the shape of an action directed towards a suitable goal: a kind of casting weight has been put in the scale of our governing part, and a directed judgment is set afoot. So, those who suspend judgment about everything do not eliminate this second movement either but follow their impulse, which leads them instinctively to the good presented by sense (*Adv. Col.* 1122C–D).

Plutarch claims that, according to Arcesilaus, suspending judgment was not an obstacle for acting because action could be motivated by impression and impulse, without assent. I believe it is possible that this was Arcesilaus's genuine answer to the *apraxia* challenge and that he used the *to eulogon* criterion merely as a dialectical argument. After all, acting on judgments that are more true than false, even if it does not necessarily involve accepting *kataleptic* impressions, seems to require assenting to impressions and holding beliefs or opinions. But if Arcesilaus sustained instead that the skeptic could be moved to action by mere impressions and impulses, without assent, then the skeptic could continue to suspend judgment on all matters and would not need to hold any opinions in order to act. In other words, the skeptic would get "carried away" by her impressions and impulses, acting before even forming specific judgments on the corresponding states of affairs.[28]

---

[28]For the Stoics, impulse is instead the result of assent to an impression. Correa Motta describes it thus: "According to the Stoics, in the case of human beings, the impulse towards an object is only born when the mind assents to the corresponding impression. In other words, we only desire that which we represent to ourselves as desirable, that is, that which we judge to be good. Impulse [. . .] is a necessary (and, according to some perspectives, sufficient) condition for action. When assent is given to a cognitive impression, the impulse will be correct – it will correspond, as they would say, to our nature. That is the type of impulse that the sage will always have. When these conditions are not given, it will be incorrect. Anyway, and in all cases, without impulse there can be no action, and without assent (justified or otherwise), there will be no impulse." (2016, p. 105, my translation.) Arcesilaus's response, according to Plutarch's testimony, would then have to presuppose that assent is *not necessary* for impulse to occur.

Some commentators think that both responses to the *apraxia* objection are compatible,[29] and others, that Arcesilaus held both dialectically. For example, G. Striker claims that the argument registered by Plutarch is also founded on Stoic premises and, therefore, should be understood dialectically.[30] I think, however, that taking Plutarch's testimony as Arcesilaus's genuine response to the *apraxia* objection is exegetically helpful for various reasons.[31] First, even though the idea that the skeptic is led to action only by sensation and impulse undoubtedly entails a heterodox understanding of this figure, since it seems to divest the skeptic of practical rationality and moral responsibility (Striker 1996, p. 104), this hypothesis could explain why the Stoics often reproached Arcesilaus that his skepticism stripped them of reason and, with it, of their humanity (*Acad. II.* 37–39). Second, this interpretation is consistent with Arcesilaus's broad suspension of judgment or, in other words, it does not compromise Arcesilaus's scope of *epochê*. And, finally, it offers a very effective refutation of Stoicism, because it attacks one of the core presuppositions of the Stoics' theory of action – namely that assent is necessary for action – which, as we know, was Arcesilaus favorite skeptical endeavor. However, since we do not have sufficient evidence to give a conclusive answer to the problem, we should let

---

[29] Hankinson (1995, pp. 87–91) affirms that the passage in Sextus does not suggest that the criterion was held by Arcesilaus only dialectically and that the term *to eulogon* is used more widely than the Stoic definition suggests. However, he also accepts that it is difficult to make Sextus's and Plutarch's descriptions of Arcesilaus's solution compatible. On the other hand, Thorsrud (2009) considers that the two versions are compatible, since "the reasonable" is simply a criterion to justify action, once it has been carried out: "Arcesilaus maintains that just as purposeful action is possible without assent or belief, so too is reasonable action. All that Arcesilaus asks us to do is to provide an explanation of why it appeared reasonable after the fact. Given the social importance of explaining our actions, it is likely that the "reasonable justification" is aimed at convincing others that one's action was appropriate" (pp. 57–58). This argument is also made by Chiesara (2007, pp. 54–57) and Frede (1997, p. 136). On this point, however, I agree with Vogt, who claims that the criterion for "the reasonable" cannot be applied *after* an action has been taken since, then, it would not be a criterion: "An *ex post facto* justification cannot play the role of a criterion" (2010, p. 169).

[30] "Here again, I can see no good reason to consider Arcesilaus's rejoinder to the Stoic argument as a positive doctrine of his own. All he does is to maintain, possibly relying on Peripatetic doctrine, that assent is not necessary for action. [. . .] He seems merely to insist that an alternative is possible, perhaps that we could explain all action merely in terms of *phantasia* and *hormê*, but he offers no account of the difference between voluntary and, e.g., instinctive action." (Striker 1996, p. 104).

[31] An objection to this position is found in Frede, who argues that if Arcesilaus had held that impression and impulse were sufficient grounds for action, he would have been as dogmatic as the Stoics: "Such a diagnosis of the situation, however, involves overlooking that the skeptic, in this case, would be doing precisely what he usually criticizes the dogmatist for doing: he would be trying to deny what quite obviously is the case, viz., that actions presuppose beliefs, by relying on a theoretical, dogmatic argument which purports to show that action is possible even without assent to appearances, even without judgments. The claim that action does not presuppose belief, especially if based on an argument like the one outlined above is no less dogmatic than the dogmatic claim that action does presuppose belief." (1997, p. 6). As suggestive as it is, the argument, however, is not sound. Arcesilaus did not have to dogmatically hold this theory of action, just as he did not hold the principle of inapprehensibility of impressions dogmatically. He could have argued that the claim "assent is not needed for action, and the skeptic can act on mere impression and impulse" was not based on a *kataleptic* impression, and, therefore, he did not assent to it, and could put it into question when needed.

the question about Arcesilaus's criterion for action remain unsettled and turn now to Carneades's response to the *apraxia* objection.

2.1.3 From the Skepticism of Arcesilaus to the Fallibilism of Carneades

It may have been the *apraxia* objection along with Arcesilaus's alleged solutions that motivated Carneades to propose a moderate form of skepticism, which maintained the principle of inapprehensibility of impressions while considering that a complete suspension of judgment was neither necessary nor desirable (Hankinson 1995, p. 96). For, although Carneades continued and strengthened the Middle Academy's tradition of refuting the Stoics, his principal innovation seems to have been the introduction of the criterion of "the persuasive" or "the probable" (*to pithanon*) to reduce the scope of *epochê* and enable the skeptic to act.

Given Cicero's translation of the Greek *to pithanon* to the Latin *probabile*, it has been widely accepted that the skepticism of the New Academy was a kind of probabilism. However, as Cicero understood it, the term "probable" has very different connotations than we usually associate with it. Cicero used the terms '*probabile*' and '*veri simile'* more or less interchangeably (Glucker 1995, pp. 115–143),[32] so that "probability" was for him synonymous with "appearance of truth." It was not yet related to mathematical frequency, as became customary after Pascal and Leibniz adopted the term in the advancement of modern calculus (Niiniluoto 2000, pp. 145–170; Hacking 2006). For this reason, scholars disagree on whether or not Carneades should be considered a "probabilist." To avoid this problem, Burnyeat and Hankinson suggest that *probabile* be translated by "persuasive" or "plausible," and *veri simile* by its cognate "verisimilar," terms which more adequately convey the meaning "appearance of truth" that the ancient concept involved.[33] This word choice also preserves the connection with the rhetorical roots of the Greek term, which, as we will discuss, is at the basis of the concept (Burnyeat 1997, p. 33).

Beyond this terminological issue, the same interpretative challenge that we saw in the case of Arcesilaus's testimonies applies to Carneades's; namely, the difficulty in determining whether the scholarch genuinely held the criterion of the persuasive

---

[32] However, Glucker holds that there may exist a difference in nuance between the terms *probabile* and *veri simile*, leading Cicero to translate *pithanon* as *probabile* and *eikos* as *veri simile*.

[33] "Cicero's rendering of *pithanon* as "probabile" encouraged the English translation "probable," which has, in turn, nourished the fantasy that Carneades was dealing with a kind of "probabilism." To say that an impression is probably true is to issue a draft on the way things are likely to be: it implicitly introduces a commitment to the world independent of the senses, a position indeed difficult for a sceptic coherently to maintain." (Hankinson 1995, p. 111). Likewise, Burnyeat asserts: "Getting the translation right is the first step towards undoing the myth of Carneades as a proponent of probabilism" (1997, p. 33, n. 16).

impression or merely used it dialectically. As I mentioned before, most scholars maintain that this criterion is Carneades's actual innovation and the positive nucleus of his skeptical system. I subscribe to it as well.[34] I believe it is plausible that being pressed, on the one hand, by the logical refinement of Chrysippus's arguments, and, on the other, by Arcesilaus's broad suspension of judgment and insufficient answer to the *apraxia* objection, Carneades decided to change his skeptical strategy completely. Arcesilaus's argument against the Stoics could have been further elaborated by Carneades in the following way: Arcesilaus argued that (i) since the sage does not give assent to obscure impressions, she does not hold opinions, but (ii) given the principle of inapprehensibility, all impressions are of this sort, therefore, (iii) she must suspend judgment on all matters and become a skeptic. In turn, Carneades may have changed the conclusion to claim that the wise person, still a skeptic, did not have to suspend judgment entirely but could orient her life by non-*kataleptic* impressions, holding beliefs or opinions for practical purposes. However, with this change, Carneades would have narrowed Arcesilaus's scope of *epochê* and embraced a milder or less radical skepticism, which we will now describe in more detail.

If this is an accurate portrayal of Carneades's strategy, then his originality rested on advancing the figure of a sage who, based on persuasive impressions, holds judgments undogmatically and only for practical purposes. This sage, we could say, is a much more modest figure than the Stoic one because her most prominent characteristic is not infallibility or imperturbability but prudence. In fact, this sage seems to possess a practical rationality more akin to Aristotelean *phronesis*. According to Striker:

> If the wise man is he who always does as he should do, certainly no one would claim to be a sage [. . .], but it does not follow that the sage is superhuman. So, the 'Academic sage,' whom Cicero in some places contrasts with the Stoic [. . .] will perhaps be more cautious and prudent than the rest of us, but he, like all ordinary humans, will have to be content with what is plausible (Striker 1996, p. 108).

As Striker affirms, this sage would be an ordinary human being, well-trained in recognizing what is more plausible than not, and willing to use this calculus to orient herself in practical matters. As I claimed before, probably Carneades subscribed to the criterion of "the persuasive" (*to pithanon*) and did not simply use it for dialectical purposes. Moreover, it is plausible that he developed this criterion precisely in response to Arcesilaus (Striker 1996, p. 107). Let us recall that, according to the interpretative tradition, either Arcesilaus advanced "the reasonable" (*to eulogon*) as a skeptical criterion for action or argued that the skeptic acted only by sensation and impulse. Carneades may have found the *to eulogon* criterion helpful, even if Arcesilaus only held it dialectically, and may have decided to moderate the spirit of Academic skepticism by transforming the *to eulogon* into a somewhat similar

[34] See note 15.

criterion—namely, "the persuasive" (*to pithanon*) impression.[35] To support this hypothesis, we should first examine Cicero's account of the criterion:

> So many perceptual impressions deserve our approval, too, provided only that one remembers that none of them is such that there couldn't be a false impression not differing from it at all. Thus, the wise person will use whatever strikes him as persuasive (*probabile*) if nothing contrary to its persuasiveness presents itself; and the whole structure of his life will be governed in this way. After all, the wise person [. . .] also follows persuasive [impressions] in many cases – i.e., impressions that aren't apprehended or assented to, but are truth-like (*sed similia veri*). Indeed, if he didn't approve them, his whole life would be undermined (*Acad. II.* 99).

According to Cicero, Carneades argued that the sage orients her life by impressions that, while not *kataleptic*, are sufficiently persuasive or probable (*Acad. II.* 99). An impression of this kind has sufficient vivacity to be worthy of trust or credit (*Acad. II.* 101). The sage, of course, disregards non-persuasive impressions – ie., obscure ones – because they lead to judgments that are more likely false than true.[36] Now, we have to look at the two possible interpretations of this criterion, namely that Carneades either used it dialectically or held it positively. On the one hand, if the criterion of the persuasive impression was merely dialectical, Carneades would have used it, analogously to Arcesilaus's *to eulogon*, to suggest that the Stoic sage was not a sage. Since the criterion of the persuasive impression (*to pithanon*) was used by ordinary people, Carneades could have advanced it to imply that the Stoic sage's judgment was not more sophisticated than the average person's. On the other hand, if Carneades genuinely held the criterion, he must have thought either that following persuasive impressions was sufficient for a wise person or that this sole criterion was not enough, and some additional practical guidelines were necessary for the sage to evaluate judgments and act on their basis. And, indeed, Carneades considered a set of sub-criteria that the wise person should apply to determine the degree of persuasiveness of impressions. For this reason, I believe that Carneades sincerely advanced the *to pithanon* and did not just use it dialectically.

---

[35] I accept that this is a rather controversial claim. Brochard argues that the Arcesilean criterion of "the reasonable" (*to eulogon*) and the Carneadean one of "the persuasive" (*to pithanon*) are not comparable (2002, pp. 124–126). In the first place, because, at one point, Arcesilaus himself rejected "the persuasive" (*to pithanon*) as a criterion and, secondly, because the criterion of "the reasonable" (*to eulogon*) does not refer to impressions but to judgments, and, hence, it is based on reason and not on sense experience. Lévy (1992, p. 37) reprises this difference between the two criteria to argue for their incompatibility. Notwithstanding, for Glucker (1995, p. 127) the difference between the terms, although very clearly established by the early Stoics, became blurred over the centuries so that in Cicero's time they were used interchangeably. I believe that both criteria are very similar in their practical application and that this should have been noticed by Carneades.

[36] Sextus affirms: "And of the apparently true kind of presentation, one sort is the obscure [. . .] while the other sort, in addition to being apparently true, possesses this appearance of truth to an intense degree. And of these, again, the presentation which is obscure and vague will not be a criterion; for because of it's not indicating clearly either itself or that which caused it, it is not of such a nature as to persuade us or to induce us to assent (*synkatáthesin*). But that which appears true, and appears so vividly, is the criterion of truth according to the School of Carneades" (*M VII*, 171–173)

This set of sub-criteria, employed by the skeptic sage to assess the degree of persuasiveness of the impression, was built on the persuasive impression. Sextus gives their most extensive discussion:

> And respecting the probable impressions [the Academics] make distinctions: some they regard as just simply probable, others as probable and tested, others as probable, tested, and irreversible. For example, when a rope is lying coiled up in a dark room, to one who enters hurriedly it presents the simply "probable" appearance of being a serpent; but to the man who has looked carefully around and has investigated the conditions – such as its immobility and its color, and each of its other peculiarities – it appears as a rope, in accordance with an impression that is probable and tested. And the impression that is also "irreversible" or incontrovertible is of this kind. When Alcestis had died, Heracles, it is said, brought her up again from Hades and showed her to Admetus, who received an impression of Alcestis that was probable and tested; since, however, he knew that she was dead, his mind recoiled from its assent and reverted to unbelief. So then, the philosophers of the New Academy prefer the probable and tested impression to the simply probable and, to both of these, the impression that is probable, and tested, and irreversible (*Pyr.* I, 227–229).[37]

According to Sextus, the Academics claimed that an impression was more or less persuasive depending on whether or not it fulfilled the additional conditions of being carefully "tested" and "irreversible."[38] A "tested" impression is, as its name suggests, a persuasive impression whose trustworthiness stands up to careful examination. In *Against the Logicians* (*M VII*), Sextus offers a further example of this criterion—the sort of scrutiny applied to the lives of citizens who seek election to public office. He says that, just as the jury thoroughly inspect the candidates' lives before entrusting them with office, so too the wise skeptic can only consider as sufficiently trustworthy the impressions that withstand detailed examination (*M VII.* I, 182).

The following criterion, by which it is determined whether an impression is not only persuasive and carefully tested but also "irreversible," consists of observing whether it maintains a degree of coherence with related information or previously acquired beliefs on the matter. Thus, Sextus tells us that, since Admetus knew beforehand that Alcestis had died, even though he had a "persuasive" and "tested" impression of Alcestis, he did not believe she was alive. In this case, there was an opposing impression in Admetus's memory, that of Alcestis's death, which refrained him from accepting the present one. Another instance of this criterion's usage occurs when, to identify a human being as such, we first consider the set of characteristics contained in our pre-existing idea of humans. Only if these agree with the impressions we obtain in the present experience do we judge that what we perceive is

---

[37] The corresponding Greek terms are *pithanon, diexôdeumenê* y *aperíspastos.* R. Bury (Sextus Empiricus, 1933) translates them as "probable," "tested" and "irreversible impression" respectively. J. Annas y J. Barnes (Sextus Empiricus, 2000), as "plausible," "tested" and "stable." I use Bury's terms throughout my discussion to maintain the idea that the "irreversible" impression, as Cicero said, is not "hindered" by any other.

[38] "For all these factors together form the criterion – namely, the probable presentation, and that which is at once both probable and irreversible and besides these that which is at once probable and irreversible and tested" (*M VII*, 184).

indeed a human being. But if one of the impressions are opposed (if, for instance, this apparent human being has a horn on the head), our judgment is contradicted or diverted – pulled, as it were, in the opposite direction (*M VII.* 177–78). In this case, the impression is "persuasive" and "tested" but not "irreversible." Likewise, before arriving at a particular diagnosis, a physician examines first the patient's symptoms and their correlation. Fever, for example, should be accompanied by a sore throat to conclude that the patient has viral flu; but if one of the symptoms is absent or does not agree with the general description of the illness, the diagnosis's probability lessens, and the physician should avoid making a conclusive judgment on the matter (*M VII*, 179).

Before continuing with Carneades's argument, we should notice the specific meaning of the Greek terms. Couissin argues that *diexôdeumenê* and *aperispastos* come from Stoicism and have very precise methodological definitions. Chrysippus used the irreversibility (*aperispastos*) criterion to warn his students about the danger of debating for and against a thesis in the way sophists and skeptics did. According to him, this type of reasoning was detrimental to the education of young Stoics, because it could cause them to "divert" or "pull" in the opposite direction their assent to a *kataleptic* impression and, in this way, unnecessarily question a true claim.[39] In turn, the criterion of a "tested" (*diexôdeumenê*) impression is related to rhetorical argumentation. According to Couissin: "the *diexôdeumenon* would be that for which one could give a justification by detailed exposition" (1983, p. 48). The criterion refers to the argumentative exhaustivity expected from rhetorical discourses, where all possible reasons supporting a thesis are to be given, and the opposite ones are to be refuted. For this reason, when Sextus speaks about this criterion, he explicitly refers to forensic procedures:

> And it is because of this that, just as in ordinary life, when we are investigating a small matter we question a single witness, but in a greater matter several, and when the matter investigated is still more important we cross-question each of the witnesses on the testimony of the others – so likewise, says Carneades, in trivial matters we employ as criterion only the probable presentation, but in greater matters the irreversible, and in matters which contribute to happiness the tested presentation (*M VII.* I, 184).

These criteria are, of course, neither mutually exclusive nor necessarily accumulative.[40] In a given case, a set of relevant impressions may be tested to discover that one of them appears to be contrary to the others. But, upon further examination, it may be found that what initially seemed to be an opposing impression, in reality conforms with the rest of them. For this reason, the criteria are just guidelines to

[39] "Thus, Chrysippus says that by a persuasive expounding of contrary theses, the hearers are diverted from a thesis. Carneades suggests as a second criterion – the presentation which is persuasive and not diverted, that is, one that is not opposed by a contrary persuasion" (Couissin 1983, p. 49).

[40] They are presented by Sextus in inverse order in the two testimonies – in the *Outlines*, the persuasive and tested impression appears first, and the persuasive and irreversible impression afterward (*Pyr, I.* 227), while in *Against the Logicians*, the persuasive and irreversible impression is examined first, and the persuasive and tested impression in the second place (*M VII*, 184).

orient the skeptics in their examination of impressions and only provide the basis for assessing their degree of persuasiveness or probability. They should not, and this is an important point, be taken as epistemological conditions of knowledge—namely universally valid or objective criteria. If this were the case, by advancing them, Carneades would have abandoned the skeptical stance in favor of a mild dogmatism, as the Metrodorean interpretation suggests and seems to have been the case for some of Carneades's followers.[41] And, indeed, someone could object that it is not clear how these criteria provide an answer to the *apraxia* objection in a way that is *still skeptical*. One could think that sustaining beliefs that fulfill these criteria is as dogmatic as holding beliefs that satisfy the criterion of the *kataleptic* impression. How, then, does this method provide an answer to the *apraxia* objection while maintaining skepticism?

To better understand this, let us observe Cicero's example of how the Academic skeptic uses these criteria to act. He says that a sage wants to embark on a journey and, before doing it, examines the impressions associated with traveling:

> [...] when he steps into a boat, does the sage apprehend in his mind that he is definitely going to arrive? How could he? Still, should he set out from here to Puteoli, thirty stades away, in a tested vessel, with a good helmsman, and in calm weather like this, he would have the persuasive impression that he will arrive there safely. So, he will deliberate about what to do based on this type of impression (*Acad. II*, 100).

This example provides the basis to determine why, using Carneades's method, a "wise skeptic" would continue to be both "skeptic" and "wise." Before acting, our Academic sage analyzes the impressions about the traveling conditions. She has persuasive impressions that the boat is sturdy, the captain skilled, the weather calm, and the destination not too far away. Once she has carefully examined them, she concludes that none of the associated impressions pulls her judgment in another direction. Indeed, they appear to be consistent both among themselves and with her earlier opinions about traveling, acquired by personal experience and hearsay. As a result, she arrives at the highly persuasive impression that it is safe to set off. Unlike an ordinary human being, who would get carried away by untrustworthy impressions without testing them in advance, this person is a sage because she acts based on impressions that have been carefully assessed, is confident, and steadfast. But this sage is still a skeptic because she does not assent to the impressions and judgments that withstand skeptical examination. Her epistemic attitude or the mental act involved in her assessment of the impressions is a different one. Cicero refers to it with the term "approval:"

> [...] Since the person who keeps himself from assenting to anything nevertheless wants to move and act [...] there are still impressions of the kind that excite us to action; and likewise, there are still responses we can use when questioned on either side, by just following out impressions on the matter, provided we do so without assent [*sine adsensu*]. Yet not all impressions of this kind are approved [*adprobari*], but only those that aren't impeded by anything (*Acad. II*, 104).

[41] See note 12.

The difference between "assent" (*adsensus*) and "approval" (*adprobatio*) is essential to understand the epistemic attitude of the Academic sage and, thus, Carneades's skepticism as opposed to a mild dogmatism:[42]

1. Assent is the attitude that *takes for true* a *kataleptic* impression, that is, in which clear or evident impressions are judged to correspond with the object with no shadow of a doubt. In other words, assent involves epistemic certainty. On the contrary, approval is given to highly persuasive impressions, whose correspondence with the object is very probable but may still be subjected to doubt.
2. Assent has a *theoretical* character, while approval has only a *practical* one. The wise Stoic assents to *kataleptic* impressions and, being entirely certain that her judgments correspond to the particular objects they represent, determines herself to act. This is to say that theoretical certainty is always the basis of the Stoics' practical engagement. Instead, the Academic or skeptic sage approves persuasive impressions to act but does not judge that things are as the impression suggests. She maintains theoretical suspension of judgment while accepting the impressions examined for practical purposes.
3. Assent corresponds to claims about *how things are*, while approval only to claims about *how things appear to be*. An example of the former type of claim would be "it is safe to travel," while an instance of the latter would be, "it appears safe to travel." In other words, assent entails a more robust subjective assurance than approval, which does not involve giving up one's doubts on the matter.

With these definitions in mind, we can see that the attitude of the wise skeptic is extremely cautious. First, her actions are guided solely by impressions and judgments that stand up to exhaustive examination. Second, even after this examination, she does not give her assent to them—she does not hold them as beliefs. As Bett claims, assent implies "a stand on the truth of the impression" while approval "involves no commitment, one way or the other, as to whether the impression approved is really true or false" (1990, p. 10).[43] That this prudent sage will not take a stand on the truth of the impression means that she accepts that, in a future examination, the corresponding impressions and judgments may be falsified or found less persuasive. Thus, approval involves only provisionally adopting the judgments that seem to be persuasive enough to orient action.[44] Put differently,

---

[42] I take this characterization mostly from R. Bett's article (1990, pp. 3–20).

[43] Frede also describes the distinction in similar terms. He affirms that assent corresponds to "making a claim" and approval to "having a view": "To just *have a view* is to find oneself being left with an impression, to find oneself having an impression after having considered the matter, maybe even for longtime, carefully, diligently, the way one considers matters depending on the importance one attaches to them. [...] To make a claim, on the other hand, is to subject oneself to certain canons. It does, e.g., require that one should think that one's impression is true and that one has the appropriate kind of reason for thinking it to be true" (1997, p. 133).

[44] To my knowledge, except for Maia Neto's account of the influence of Academic skepticism on Descartes via Charron (Maia Neto 2014, pp. 109–113) the significance of this ancient skeptical school for the Cartesian provisional morals (*AT*, VI, 22) has not been sufficiently studied.

this sage has not completely overcome doubt. Approval is an epistemic attitude in which theoretical doubt is maintained while one holds the judgment practically, that is, to guide one's actions. The wise skeptic proposed here knows that the actual state of affairs may be different but approves the associated impressions and judgments to avoid inactivity. As a result, even if the theoretical questions remain unanswered, she responds to the practical ones. To such questions as, does the boat appear sufficiently safe to climb aboard? Or, does the climate seem favorable enough to set sail? she responds positively or negatively and acts accordingly. Therefore, approval should not be conflated with a 'soft' or 'weak' assent, which is still a dogmatic epistemic attitude, even if less robust.

I cannot sufficiently stress this point. The skepticism of the New Academy explicitly separates the theoretical and practical realms in the evaluation of impressions or claims. The resulting attitude of suspending a particular judgment in theory while adopting it for practical purposes is a very complex one, which accords in many respects with Hume and Kant's critique of traditional metaphysics and practical answers to these questions. We will return to this parallel in the following chapters. For the moment, I want to highlight that a theoretical dogmatism like the Stoics' could never accept this divide. For this school, theoretical certainty necessarily resulted in practical decision-making; and suspension of belief, in practical paralysis. One and the same epistemic act, assent, warranted both theoretical and practical knowledge. It would be preposterous for a Stoic, to think that one could suspend judgment theoretically while adopting a belief practically.[45]

With all of this in place, I believe we can now return to the main objective of this section, which is to draw the differences between Arcesilaus's and Carneades's answers to the *apraxia* objection. Carneades's "approval" cannot be identified with the psychological operations of having impressions and impulses which, after Plutarch, Arcesilaus considered sufficient for the skeptic to act. To *approve* an impression presupposes a process of careful evaluation that has nothing to do with the mechanical impulse of Arcesilaus's alleged answer. Carneades's approval is instead a rational attitude in which impressions are critically tested and autonomously adopted, albeit with reservations. As Vogt claims:

---

[45] However, as Correa Motta affirms, these two Carneadean notions, the persuasive impression and approval, are also derived from Stoicism: "The two elements which he brings into play [the *pithané* impression and its approval] constitute loans or refinements from the conceptual apparatus of his antagonists. On the one hand, the Stoics had the notion of *pithanós* and they used it, too, to classify impressions [. . .]. In their case, however, the notion corresponded to assent (the *pithané* impression was the very one that led to this). On the other hand, though it may not be completely clear that Carneades respected the threefold framework of the theory of Stoic action (impression/assent/impulse), changing assent for approval, he did maintain a degree of rational involvement on the part of the agent with the content of his representations." (Correa Motta 2016, pp. 123–124, my translation).

> [. . .] it does not seem to me that the Academics argue for, as it were, a "rationally blind" way of being active; their actions are guided by conscious engagement with impressions. [. . .] Action that follows the reasonable or persuasive is action for which the agent assumes responsibility and which she ascribes to her own, rational decision-making. (Vogt 2010, p. 171).[46]

Following Vogt, if this skeptic sage made a practical or moral mistake, she would not reply that, having acted from mere impulse, she bore no responsibility for her behavior. Instead, she would argue that, when deciding to act, she knew that she could make a mistake and would assume full responsibility for her errors. Having rejected the Stoic criterion for certainty, this kind of sage accepts that no rational analysis can free oneself from the possibility of error, but does not deem it advisable to act merely on natural impulses, as the Arcesilean skeptic, in Plutarch's interpretation, seems to accept.

Now, if we were to accept that the criterion of "the reasonable" (*to eulogon*) was Arcesilaus's genuine response to the *apraxia* objection, then we should ask ourselves whether or not there is a substantive difference between this criterion and Carneades's *to pithanon.* In my view, although there are semantic differences between the terms, they fade away in practice. The main difference between them is that the criterion of "the persuasive" applies directly to impressions, while that of "the reasonable" applies to judgments.[47] However, in the final analysis, following a persuasive impression or a reasonable judgment requires practical deliberation. Thus, perhaps the real difference between both answers to the *apraxia* objection is that Carneades's practical criterion is epistemologically more developed, given that it involves some sub-criteria for the assessment of impressions. However, if my hypothesis is correct and Arcesilaus only employed the criterion of "the reasonable" dialectically, then it is plausible that Carneades had given it a new form, taking Arcesilaus's argument as his point of departure to construct his own practical criterion. In so doing, he would have, so to speak, killed two birds with one stone. On the one hand, he would have responded to the Stoics, determining a *non-kataleptic* but sufficiently trustworthy impression as a criterion for action. And, on the other, he would have given a satisfactory and definitive reply to the *apraxia* objection.

2.2 Academics vs. Pyrrhonists on the Criterion for Action

Up to this point, I have argued that the Carneadean wise skeptic's attitude is very different from that of the Stoic sage since the skeptic does not claim to have theoretical or practical knowledge but acts on the grounds of persuasive impressions, to which she grants "approval" but not "assent." Interestingly, Carneades's response

---

[46] Vogt refers here to Carneades's followers when she speaks of "Academics."

[47] See note 35.

to the *apraxia* objection seems to bear some resemblance to Sextus's views on the matter. The latter argued that Pyrrhonists are led to action by "appearances" and follow the laws and customs of their community not to remain inactive (*Pyr.* I, 19–22). But before we begin the comparison between both answers to the *apraxia* challenge, we should pause for a moment and consider the historical link between the two schools.

The bond between Academic and Pyrrhonian skepticism is, unfortunately, obscure. It is generally accepted that the origin of Pyrrhonism is to be found in Academic skepticism, when Aenesidemus of Knossos, in the first century BCE, left the Academy and reclaimed Pyrrhonism as a reaction to the dogmatic probabilism of Carneades's successors, Philo of Larissa and Antiochus of Ascalon. Because of this origin, some of Aenesidemus's and Sextus's skeptical topics bear resemblance to the Middle and New Academies' (Burnyeat 1997, pp. 32–35; Frede 1997, pp. 146–147; Thorsrud 2009, p. 10; Striker 1996, p. 144 y 2010, p. 197).[48] It is also argued that Pyrrho had an influence on Arcesilaus and, because of this, the skepticism of the Middle Academy owed more to Pyrrho than to Socrates. Eusebius, for example, affirms:

> Mnaseas at least, and Philomelus, and Timon, the Sceptics, call him [Arcesilaus] a Sceptic, as they were themselves because he also overthrew truth and falsehood and probability. Therefore, although on account of his Pyrrhonistic doctrines he might have been called a Pyrrhonist, yet from respect for his lover, he submitted to be still called an Academic. He was, therefore, a Pyrrhonist, except in name: but an Academic he was not, except in being so-called (*Praep. Evang.* XIV, 6).[49]

According to Eusebius, Arcesilaus adopted Pyrrho's teachings via Timon and conveyed them to Carneades and his successors, among them Aenesidemus, who claimed a return to Pyrrho and delivered them to Sextus Empiricus. Some commentators, however, argue that this connection is not plausible. Chiesara, for instance, regards the sources linking Arcesilaus with Pyrrho as inconclusive and argues that, because Cicero never made the connection explicitly, it is probably not historically accurate (2007, p. 45). In addition, Bolzani stresses Arcesilaus link to the Socratic tradition, arguing that the Middle Academy's head was the first to "propose a skeptical notion of suspension of judgment" (2016, p. 255). Still, I believe that the

---

[48] Couissin says: "The [Pyrrhonian] skeptics were delighted to find in the Academic writings a whole battery of arguments against dogmatism. These arguments piled up in their archives, and sometimes no one could remember where they came from. This accounts for the anti-Stoic tone in the work of Sextus" (Couissin 1983, p. 57). For Frede, "[. . .] Pyrrhonism is not so much a revival of Pyrrho's philosophy, but a revival of classical Academic skepticism under the name of Pyrrhonism, to distinguish it from the dogmatism which Aenesidemus and Sextus Empiricus associated with the later skeptical Academy." (1997, p. 146).

[49] The description of Arcesilaus as a fanciful animal with the head of Plato, the tail of Pyrrho, and the trunk of Diodorus, attributed to Ariston and Timon, is frequently found in ancient doxology (*Diog. Laërt.* IV, 33; *Pyr.* I, 234; *Praep. Evang.* XIV, 5). Interestingly, Eusebius and Sextus found the same description useful to characterize Arcesilaus as a Pyrrhonian skeptic and as a Platonic dogmatist, respectively.

similarities between Pyrrhonism and the skepticism of Arcesilaus are significant, and at least one source – Sextus – seems to support their historical link. He says:

> Arcesilaus, however, who was, as we said, the president and founder of the Middle Academy, certainly seems to me to have shared the doctrines of Pyrrho, so that his way of thought is almost identical with ours. For we do not find him making any assertion about the reality or unreality of anything, nor does he prefer any one thing to another in point of probability or improbability, but suspends judgment about all. He also says that the end is suspension – which is accompanied, as we have said, by "quietude" (*Pyr.* I, 232).

The context of this quote is Sextus's distinction between Pyrrhonism and the probabilism of Carneades. To establish the contrast more clearly, he claimed that Pyrrhonism and the skepticism of Arcesilaus had in common the suspension of judgment. Sextus characterized the Pyrrhonist as one who, acknowledging the equal validity of contrary impressions and judgments, suspends assent and attains a state of imperturbability (*Pyr. I*, 8). As we have seen, the same can be attributed to Arcesilaus's skepticism. In the final sentence of the paragraph, however, Sextus introduced a qualification – it seems that Arcesilaus did not consider *ataraxia* to be the goal of *epochê*, as the Pyrrhonists ("we") did. Indeed, nowhere did Cicero claim that Arcesilaus recommended suspending judgment as a means for achieving peace of mind (Bolzani 2016, p. 255). Instead, following Socrates's example, Arcesilaus seems to have thought that the skeptics task was showing the dogmatists the unwarranted ease with which they accepted as true their own opinions or those of their mentors (*Acad. I.*, 16). Hence, his method was a kind of Socratic *elenchus*, leading to the suspension of unreflectively held opinions and the renewal of inquiry (Striker 1996, p. 148, n. 11; Sedley 1983, p. 16; Thorsrud 2009, p. 45).

Nonetheless, both Arcesilaus and the Pyrrhonists attached great weight to the suspension of judgment. Moreover, if we accept Plutarch's characterization of Arcesilaus's answer to the *apraxia* objection, we can see important similarities with Sextus's. Sextus argued that Pyrrhonism had a criterion for action, namely "the appearances:"

> For, as we said above, we do not overthrow the affective sense-impressions which induce our assent involuntarily; and these impressions are "the appearances." And when we question whether the underlying object is such as it appears, we grant the fact that it appears, and our doubt does not concern the appearance itself but the account given of that appearance – and that is a different thing from questioning the appearance itself. For example, honey appears to us to be sweet (and this we grant, for we perceive sweetness through the senses), but whether it is also sweet in its essence is for us a matter of doubt, since this is not an appearance but a judgment regarding the appearance. [. . .] The criterion, then, of the Sceptic school is, we say, the appearance, giving this name to what is virtually the sense-presentation. For since this lies in feeling and involuntary affection, it is not open to question (*Pyr.* I, 19–22).

Sextus claimed that the Pyrrhonists only debated about whether things were as they seemed to be, not whether they appeared to be as they did. The "appearances" were not called into question because they were simple affections or involuntary movements of the soul and were not taken as grounds for any theoretical assertion. However, they served as the basis for the skeptic's everyday behavior (Burnyeat

1997, p. 42). The similarity between this practical criterion and the impression and impulse which, according to Plutarch, Arcesilaus may have adopted as a practical criterion for action is, in my opinion, ostensible. As Striker argues:

> This reply to the inactivity argument was no doubt inspired by the Pyrrhonist's Academic predecessors. But it sounds a lot more like the first reply of Arcesilaus (quoted by Plutarch) than like the distinction between two kinds of assent introduced by Carneades because it eliminates assent altogether (Striker 2010, p. 204).

I absolutely concur. Indeed, as we have seen, in Plutarch's interpretation, Arcesilaus granted that the skeptics could be moved to act without assenting to any judgment, only by the involuntary mechanisms of impression and impulse. In the same way, it seems that the Pyrrhonists suspended judgment and oriented their actions by involuntary impressions and their derived impulses, not by assenting to or approving the relevant impressions, and holding beliefs or opinions.

Still, Sextus description of the sort of "appearances" that guide these skeptics' action is more complex than Arcesilaus's "impression" and "impulse":

> Adhering, then, to appearances we live in accordance with the normal rules of life, undogmatically, seeing that we cannot remain wholly inactive. And it would seem that this regulation of life is fourfold, and that one part of it lies in the guidance of Nature, another in the constraint of the passions, another in the tradition of laws and customs, another in the instruction of the arts. Nature's guidance is that by which we are naturally capable of sensation and thought; constraint of the passions is that whereby hunger drives us to food and thirst to drink; the tradition of customs and laws, that whereby we regard piety in the conduct of life as good, but impiety as evil; instruction of the arts, that whereby we are not inactive in such arts as we adopt (*Pyr.* I, 23–24).

Sextus argued that the Pyrrhonist acts on involuntary sense-impressions and mental representations—she suspends judgment but does not seek to rid herself of all mental contents (Burnyeat 1997, p. 40). She is led to satisfying her needs by vital instincts and drives, so that, if she is cold, she seeks shelter; if she is hungry, food. In addition to this, to act morally, she follows the impressions corresponding to what her community considers good and evil, appropriate or inappropriate, honorable and vile. And, finally, she learns an art; that is, she retains in her memory the necessary impressions and actions to develop a specific skill, most probably to assure her survival and avoid tedium.

The latter two observances, namely the "tradition of customs and laws" and the "instruction of the arts," seem to involve judgments and not only impressions and impulses. Sextus claimed that, by the former, the Pyrrhonists "regarded piety in the conduct of life as good, but impiety as evil" and, by the latter, they learned an art, which presumably involved adopting technical precepts. However, this "regarding piety as good and impiety as evil," as we will see in the next chapter, does not necessarily imply *believing* that one behavior is good and another bad, but only having the right sort of emotional disposition towards such religious practices. And, learning an art does not perforce entail considering technical instructions as true or false, right or wrong, but merely training oneself to master them.

In sum, it is noteworthy how, seeking *ataraxia* or imperturbability, the Pyrrhonist appears to have renounced a rational grounding of action, ceasing to determine her life's ends and adequate means autonomously. For this reason, the behavior of a Pyrrhonist has a good deal in common with that of a non-human animal, to the point that some commentators have argued that the school may have gone as far as to attempt to abandon human rationality (Striker 2010, p. 206; and Burnyeat 1997, p. 57). Of course, other interpretations underline the epistemic character of Pyrrhonian inquiry, namely its goal of searching for the truth (Perin 2010, pp. 7–32; Bolzani 2016, pp. 245–289). Be as it may, it is difficult to make compatible the two aims, *ataraxia* and the search for truth. Thus, I am more inclined to accept that the refutation of dogmatism is the only epistemic goal of Pyrrhonism.

Given these coincidences, I suggest that Pyrrhonism may have adopted some of the features of Arcesilaus's response to the *apraxia* objection as described by Plutarch. Following the appearances seems to me, at any rate, an adaptation of Arcesilaus's acting by impression and impulse. Yet, one last interpretation of Pyrrhonism's criterion for action needs to be addressed – that the Pyrrhonists' yielding to "appearances" is a rational attitude akin to the Carneadean "approval" of persuasive impressions, rather than just an involuntary response to impressions and natural impulses. It could be argued, for example, that the Pyrrhonist "approves" the sweetness of honey, without assenting to it, because it seems to her sufficiently persuasive, or that she "approves" the judgment passed down by tradition, according to which "it is evil to murder a fellow citizen," for this same reason. This is an interesting possibility, which would offer a less irrational interpretation of the Pyrrhonist's grounds for action. However, I do not believe it to be accurate. Sextus himself clarifies the point when he speaks about the difference between Pyrrhonism and the New Academy:

> The adherents of the New Academy [. . .] differ from us quite plainly in their judgment of things good and evil. For the Academics do not describe a thing as good or evil in the way we do; for they do so with the conviction that it is more probable that what they call good is really good rather than the opposite, and so too in the case of evil, whereas when we describe a thing as good or evil we do not add it as our opinion that what we assert is probable, but simply conform to life undogmatically that we may not be precluded from activity (*Pyr.* I, 226–27).

It is clear, then, that the Pyrrhonists opposed Carneades's approval of persuasive impressions because they thought that it led to dogmatism. During the skeptical examination, they did not accept the possibility that one impression could be more or less persuasive than the contrary, but, if anything, they focused on maintaining *isostheneia* by following the formula "not more this than that," among others (*Pyr.* I, 188–191). Sextus is emphatic when he says that they did not assert that a claim was probable but "undogmatically" abided by it, so they could act. Even though Sextus is mistaken in claiming that Carneades's approval of persuasive impressions is "dogmatic,"[50] since his intention in this passage is to draw the difference between both schools' answers to the *apraxia* objection, we should

[50] It may result from his contact with the Metrodorean interpretation of Carneades and the practice of some of his followers.

conclude that the Pyrrhonian following of appearances is not compatible with Carneades's approval.

Lest we were still not entirely convinced about this point, we may examine another argument in the *Outlines*, where Sextus refers explicitly to the New Academy's "approval:"

> And although both the Academics and the Sceptics say that they believe some things, yet here too, the difference between the two philosophies is quite plain. For the word "believe" (*peithesthai*) has different meanings: it means not to resist but simply to follow without any strong impulse or inclination, as the boy is said to believe his tutor; but sometimes it means to assent to a thing of deliberate choice and with a kind of sympathy due to strong desire, as when the incontinent man believes him who approves of an extravagant mode of life. [. . .] Therefore, Carneades and Clitomachus declare that a strong inclination accompanies their credence and the credibility of the object, while we say that our belief is a matter of simple yielding without any consent (*Pyr. I*, 229–230).

According to this paragraph, the Pyrrhonists did not "believe some things" in the sense of being persuaded of or approving appearances,[51] like Carneades and some of his followers did. As Striker affirms, although the Greek term *peithesthai* corresponds to the Latin *adprobare*, used by Cicero to refer to the attitude of Carneades's wise skeptic, in Sextus it means instead a way of "yielding" to appearances: "entirely passive and unquestioning, not based on any reasons at all" (2010, p. 205). This attitude, as we have seen, does not correspond to Carneades's approval of persuasive impressions. Furthermore, Sextus's description of the sort of "strong inclination" that leads the Academics to approve these impressions fails to do justice to Carneades's position. Indeed, the New Academy's approval can be described neither as a "strong inclination" to believe particular judgments nor as the tendency to believe something out of sheer convenience, like in the example provided by Sextus. His portrayal seems to be, at the least, inexact and, at the most, biased. Carneades's approval is rather a careful and critical acceptance of persuasive impressions and their derived judgments. Once again, this could be the product of Sextus's interpreting Carneades through the lenses of Metrodorus and other late scholarchs of the Academy, who had abandoned the skeptics' critical distance and assumed a form of dogmatism.

To finish this section with a clearer understanding of the schools' different answers to the *apraxia* objection, it may be helpful to imagine for a moment how a Pyrrhonist and a skeptic from the New Academy would investigate the impressions attached to the judgment "honey is sweet." The Pyrrhonist would begin by applying the adequate tropes to find pro and con arguments of equal weight. Let us say that she would choose to use the first two tropes and, in doing so, she would realize that for some animals and some humans, honey may seem to be bitter (*Pyr. I*, 36–49). As a result, she would suspend judgment about sweetness being an objective or real property of honey and would say: "It appears to me that honey is sweet, but I cannot speak to whether it really is. I cannot say either that honey's appearance of sweetness

[51] B. Mates's translation renders the expression as "they are persuaded" (Sextus Empiricus, 1996, p.122). It could also be translated as "they approve some things" as argued by Striker (2010, p. 205).

is more or less persuasive or plausible than its appearance of bitterness because the two assertions are equally valid according to the tropes applied." Afterward, she would proceed to act in conformity with these appearances—that is, following the impulse that the impression produces in her, if she likes what seems to her to be sweet, she would buy honey, spread it on some bread, and eat it. More importantly, though, in the process, she would have attained *ataraxia* by suspending any judgment on the thorny issue of whether honey is actually bitter or sweet.

The New Academy's alternative is entirely different. The wise skeptic would taste honey and have the persuasive impression that it is sweet. Then, she would examine the impression, first, by observing other impressions relative to it and questioning whether any of them contradicts or diminishes its probability. She would inquire, for example, if the honey she is examining remains sweet when served in a different dish, at a higher or lower temperature, etc. Carefully observing all these impressions, she would provisionally conclude that the impression of sweetness is both "persuasive" and "tested." Next, she would examine the most common and reputable opinions about honey to which she is aware. Finding that these opinions agree in considering honey sweet, she would conclude that the impression in question is "persuasive," "tested," and "irreversible," and, therefore, trustworthy enough. Finally, with all this information to hand, although she would not "assent" to the impression, she would feel sufficiently confident to "approve" it. Therefore, she would proceed to act according to the impression and, just like the Pyrrhonist, if she likes what seems to be sweet, she would go to the supermarket, buy honey, spread it on bread, and enjoy it. However, she would do none of this on mere impulses but on rational choice, approving the judgment that withstands scrutiny to guide her behavior. As for the theoretical search, she would still maintain her doubts about honey's real sweetness or bitterness, and would be willing to continue the examination and change her mind on the subject if other results were discovered in the future. Finally, she would probably not find *ataraxia* on these grounds. Still, since she was not looking for it in the first place, she would not want to get rid of the mixed emotions and value judgments that led her to begin such theoretical examination.

Given this characterization, someone could argue (as Aenesidemus and Sextus did) that she is no longer a skeptic but a mild dogmatist. But they would be wrong because this Academic skeptic would always maintain her theoretical doubts on the matter. As the Carneades of the Metrodorean interpretation, a mild dogmatist would have to accept probability as a criterion of truth and, thus, claim that the judgment "honey is sweet" amounts to knowledge, even if it does not involve certainty. But this was not Carneades's view, at least if we consider that approval is a qualitatively different epistemic attitude than assent.

On the other hand, the Pyrrhonist's obeying, yielding to, or merely following appearances is also a different attitude from both assent and approval. These skeptics follow the appearances and laws and customs of their society, not only with a critical distance but even with a complete rational disregard, given that they seek to suspend judgment about whether they are true or false, moral or immoral, appropriate or inappropriate, etc. Indeed, they do not "approve" them but only let themselves act guided by the natural impulses and affections that these appearances arouse in them. As we have already seen in the case of Arcesilaus's views, this difference between approving and yielding to impressions has important moral implications. I agree with Vogt in claiming that that the Carneadean criterion of the persuasive impression makes possible the attribution of responsibility to the agent, while the Pyrrhonian and Arcesilean criteria preclude it. This does not mean, however, that the Pyrrhonist and the Arcesilian skeptic necessarily act immorally, but only that they cannot be held responsible for their actions, at least in the way we usually attribute responsibility to rational agents. She argues:

> Adherence to reasonable or persuasive impressions leaves room for something like decision-making and accountability. The agent performs one action rather than another because it seems more reasonable or more persuasive to her to do so. But when it comes to Sextus, matters are different. [. . .] It is not his objective to show that Pyrrhonian action has the traits which are traditionally ascribed to the action of rational beings who are accountable for their actions. Rather, the aim is to show that the skeptic, as it were, goes through the motions of ordinary life. Sextus's skeptic is active, but he is not acting in the robust sense of the dogmatist's theory of human action (Vogt 2010, p. 171).

Given that the Pyrrhonists do not engage in moral deliberation, no rational justification for their actions can be demanded. They are not, in the final analysis, autonomous moral agents. They may be able to identify the actions that their community has determined to be "good" and "evil," and may also be able to inhibit their natural inclinations to transgress social norms and values, but their actions are not the result of rational, moral deliberation. And, indeed, they may still be moral agents because their upbringing gave them the right kind of habits and dispositions. This attitude is well portrayed in Michel de Montaigne's dictum concerning religion, "we are Christian for the same reasons that we are Perigordians or Germans" (2001, p. 700). Likewise, the Pyrrhonists are moral, insofar as being moral means not transgressing the rules of their community. In this point, I agree with Striker, who claims the Pyrrhonists' moral leanings, acquired during childhood's socialization processes, are so deeply rooted into their adult personality that they guide their actions mechanically, just as if they were physiological needs and drives (Striker 2010, p. 203).

However, it should be stressed that the Pyrrhonists' abidance by traditional values still involves a degree of critical distance. For this reason, this skeptic's attitude is not entirely identifiable with the behavior of a heteronomous agent, who uncritically and dogmatically assents to the dictates of society. In a way, we can consider the Pyrrhonist's attitude as a voluntary or active, but non-dogmatic, heteronomy. That is, this skeptic has voluntarily decided to renounce acting autonomously, given her impression that being heteronomous could lead her to attain the elusive *ataraxia*. However, her voluntary heteronomy is neither an autonomous choice since she has not decided on rational grounds and after careful examination that, with regards to *ataraxia*, it is better to be heteronomous than autonomous. As Sextus reminds us, the Pyrrhonists do not take a stance on whether suspending judgment is the real cause of *ataraxia*. All they claim is that peace of mind seems to them to follow *epochê*:

> We assert still that that Sceptic's end is quietude in respect of matters of opinion and moderate feeling in respect of things unavoidable. For the Sceptic, having set out to philosophize with the object of passing judgment on the sense-impressions and ascertaining which of them are true and which false, so as to attain quietude thereby, found himself involved in contradictions of equal weight, and being unable to decide between them suspended judgment; and as he was thus in suspense there followed, as it happened, the state of quietude in respect of matters of opinion (*Pyr. I*, 25–26).

Abiding by the laws and customs of her community seems to this skeptic helpful to maintain suspension of judgment, and, therefore, she is voluntarily and non-dogmatically heteronomous, but not autonomously heteronomous.

To finish this discussion on the differences between the New Academy and Pyrrhonism, I shall assess the previously-mentioned point of the two schools' differences in their attitude towards inquiry. Let us remember that, for Sextus, the Academics ceased inquiry altogether by assenting to the principle of the inapprehensibility of impressions, while the Pyrrhonists continued to investigate (*Pyr. I*, 1–4). However, we have also seen that neither Arcesilaus nor Carneades assented to this principle but found it also susceptible to doubt. Carneades could have approved it to act, that is, as a means to continue conducting research. As we have seen, the wise Academic skeptic is always prepared to accept the error of her previous judgments if doing so results from a yet more detailed examination of the impressions into question. For this reason, the principle does not entail giving up the inquiry but maintaining doubt precisely to renew the search and, as Cicero claimed, approach the truth (*Acad. II*, 7).

To better understand this problem, it is worth asking ourselves whether Sextus's notion of inquiry is the same as ours or the New Academy's. I answer in the negative on both fronts. In my view, Pyrrhonian inquiry aims at the refutation of dogmatism and the permanent renewal of *epochê* as a possible means to *ataraxia*. As Burnyeat describes it:

> To the extent that he [the Pyrrhonist] has achieved *ataraxia*, he is no longer concerned to enquire which [opinion] is right. When a thing appears in a certain light to him, that no more inclines him to believe it is as it appears than would the fact of its so appearing to someone else. It is merely one more impression or appearance to be noted. Thus, withdrawal from truth and real existence becomes, in a certain sense, a detachment from oneself (Burnyeat 1997, p. 41).

If this is an accurate description, Pyrrhonian inquiry is only an epistemic activity insofar as it engages in refuting dogmatism, that is, showing the dogmatists that they did not attain truth through the means they had thought adequate or sufficient for that end. Considered as a way of life, however, Pyrrhonian inquiry is rather a therapeutic exercise, in which the method of reasoning for and against a thesis seeks to produce the kind of mental state that may bring about peace of mind (Barnes 1997, p. 62). Of course, the Pyrrhonists continue the inquiry since dogmatism is a rapidly spreading philosophical disease and *ataraxia* a very elusive end. But the Socratic legacy of the Academy entailed a very different attitude for the Academy. The Academics did not seek *ataraxia* but renewed the inquiry every time the Stoics and other dogmatists produced a new piece of doctrine because they thought dogmatism ought to be overcome in order to approach truth. And, for Carneades's school, in particular, suspension of judgment was a stage in the more extensive inquiry process, which consisted in evaluating the degree of persuasiveness of impressions. Now, as the Pyrrhonists knew well, examining impressions while searching for the most persuasive one or as a means to come closer to the truth does not free the skeptic from anxiety. Although finding highly persuasive impressions on which to choose to act may give temporary relief, the fact that these impressions are only approved and not assented to extends the skeptic's state of doubt while failing to provide mental release. But, as de Olaso argued, this does not seem to have been an issue for the Academics:

> The Academic, to the contrary, will claim that to be perturbed is not the worst of evils, and that we do not need to renounce knowledge, even if we have to be contented with an approximate knowledge, which brings about disquiet, either as joy or as suffering. All perturbations are only partially alleviated because we ought not to give up opinion (de Olaso 2018, p. 59; my translation).[52]

Thus, it seems that the Academics – and Cicero, following them – did not even envisage the possibility that skepticism could lead to *ataraxia* through suspension of judgment, in the way Sextus did. According to them, the goal of rational autonomy or, as Maia Neto has put it, of "intellectual integrity" (Maia Neto 1997, p. 202), in both theoretical and practical matters involved an attitude that never gave up doubt, even if it allowed for the practical adoption of persuasive impressions and their corresponding judgments. This approach to skepticism entails important Socratic commitments, such as valuing the search for truth above peace of mind and understanding that ignorance will never be conquered, not even by those called sages.

* * *

By way of conclusion, I wish to draw attention to some of the issues mentioned throughout this chapter. First, the refutation of the Stoic theory of *kataleptic* impressions does not force the skeptic of the New Academy to maintain the principle of inapprehensibility of impressions as a dogma, but only as an uncertain yet sufficiently persuasive principle that allows her to continue the search. Indeed, this sort of skeptic is not compelled to suspend judgment on all matters but may continue investigating – that is, may continue examining impressions and judgments according to their degrees of persuasiveness to find the ones that most approach the truth. Second, although both the Academy and Pyrrhonism share the goal of refuting all sorts of dogmatism, they differ in that the latter has as its final purpose obtaining *ataraxia* through suspension of judgment, while the former has as its aim attaining rational autonomy. Thus, Pyrrhonian inquiry is primarily a therapeutic effort, while the New Academy's is a skeptical investigation to gradually approach

---

[52] Burnyeat agrees: "One may feel that this added element of belief is the very thing that gives meaning and sense to a life, even if it is also the source of trouble and disturbance. Without it, the sceptic's life will be a hollow shell of the existence he enjoyed and was troubled by, prior to his skeptical enlightenment" (Burnyeat 1997, p. 46).

truth. Finally, Carneades's response to the *apraxia* objection entails reducing the scope of *epochê* but not a renunciation of the skeptical outlook. The recommendation of " approving" sufficiently persuasive impressions involves an attitude of critical distance towards opinions, based on the recognition of their temporary and merely practical validity. In other words, while approving persuasive impressions for practical purposes, the Academic skeptic maintains her theoretical doubts on the matter.

This contrast between the school's answers to the *apraxia* objection will be essential to understand, in the remainder of the book, why Academic skepticism was a more suitable position for Enlightened thinkers such as Hume and Kant. The delicate balance between maintaining theoretical doubt and adopting a belief for practical purposes would provide them with a way to resolving metaphysical issues without falling into the extremes of dogmatism and radical skepticism. But before we move to these modern authors, I will discuss in the next chapter Cicero's application of the New Academy's method to two fundamental metaphysical issues – the existence and nature of the gods, and the immortality of the soul. In *De Natura Deorum* (*DND*), Cicero followed Carneades's recommendation of examining the Hellenistic schools' impressions and judgments on the existence and nature of the gods and approving the ones that were most persuasive or probable. However, in the *Tusculan Disputations* (*TD*), he used a more rhetorical argumentation to support opposed claims regarding the nature of the soul—namely Plato's view that the soul is immortal and the Epicureans' opinion that it is mortal. As we will see, the contrast between *DND*'s rather refutative style and the *TD*'s more constructive one, far from showing inconsistency, should be read as indicative of the New Academy's rhetorical roots and affinities, which, unsurprisingly, Cicero expressed in masterful ways.

Bibliography

Alcalá, Román R. 1994. *El escepticismo antiguo, posibilidad del conocimiento y búsqueda de la felicidad*. Córdoba: Universidad de Córdoba.

———. 1996. Enesidemo: la recuperación de la tradición escéptica griega. *Anales del seminario de historia de la filosofía*, Extra 1: 79–96.

Aristotle. 1984. Metaphysics. In *The Complete Works of Aristotle*, ed. Jonathan Barnes, vol. II, 1552–1728. Princeton/New Jersey: Princeton University Press.

Augustine of Hippo. 1951. *Against the Academics*. New York/Ramsey: Newman Press.

Barnes, Jonathan. 1997. The Beliefs of a Pyrrhonist. In *The Original Sceptics: A Controversy*, ed. Myles Burnyeat and Michael Frede, 58–91. Indianapolis: Hackett Publishing Company.

Bett, Richard. 1990. Carneades' Distinction Between Assent and Approval. *The Monist* 73: 3–20.

———. 1997. What does Pyrrhonism Have to do with Pyrrho? *The Society for Ancient Greek Philosophy Newsletter* 251: 1–15.

———. 2000. *Pyrrho, his Antecedents, and his Legacy*. New York: Oxford University Press.

Bolzani, Roberto. 2016. Académicos versus pirrónicos: escepticismo antiguo y filosofía moderna. *Praxis Filosófica, Nueva serie* 43: 245–289.

Brittain, Charles. 2001. *Philo of Larissa: The Last of the Academic Sceptics*. Oxford: Oxford University Press.

Brochard, Victor. 2002. *Les sceptiques grecs*. Paris: Librairie Générale Française.

Burnyeat, Myles. 1997. Can the Sceptic Live His Scepticism? in *The Original Sceptics: A Controversy*, eds. In *Myles Burnyeat and Michael Frede*, 25–57. Indianapolis: Hackett Publishing Company.
Chiesara, Maria Lorenza. 2007. *Historia del escepticismo griego*. Trans. Pedro Bádenas. Madrid: Siruela.
Cicero, Marcus Tullius. 1950. *Tusculanae Disputationes*. Trans. J.E. King. Cambridge/London: Harvard University Press.
———. 2000. *De Natura Deorum / Academica*. Trans. H. Rackham. Cambridge/London: Harvard University Press.
———. 2006. *On Academic Scepticism (Academica)*. Trans. Charles Brittain. Indianapolis/Cambridge: Hackett Publishing Company.
Conche, Marcel. 1994. *Pyrrhon ou l'apparence*. Paris: Presses Universitaires de France.
Correa Motta, Alfonso. 2016. ¿Es posible vivir el escepticismo? In *Convertir la vida en arte: una introducción a la filosofía como forma de vida*, ed. Andrea Lozano and Germán Meléndez, 91–132. Bogotá: Universidad Nacional de Colombia.
Couissin, Pierre. 1983. The Stoicism of the New academy. In *The Skeptical Tradition*, ed. Myles Burnyeat, 31–63. Trans. Jonathan Barnes and Myles Burnyeat. Berkeley: University of California Press, 1929.
De Montaigne, Michel. 2001. *Les Essais*. Paris: Librairie Générale Française.
de Olaso, Ezequiel. 2018. *Escepticismo e Ilustración. La crisis pirrónica de Hume y Rousseau*. Buenos Aires: Centro de Investigaciones Filosóficas.
Descartes, René. 1996. *Oeuvres*, eds. C. Adam, & P. Tannery, Eds. Paris: Vrin.
Diogenes, Laërtius. 1972. *Lives of Eminent Philosophers, Vol I*. Trans. R.D. Hicks. Cambridge/London: Harvard University Press.
Dutton, Blake D. 2016. *Augustine and Academic Skepticism*. Ithaca/London: Cornell University Press.
Eusebius of Caesarea. 2003. *Praeparatio Evangelica (Preparation for the Gospel), Books XIV–XV*. Trans. E. H. Gifford. UKIpswich: http://www.tertullian.org/fathers/eusebius_pe_01_book1.htm.
Fosl, Peter. 2020. *Hume's Scepticism: Pyrrhonian and Academic*. Edinburgh: Edinburgh University Press.
Frede, Michael. 1997. The Sceptic's Two Kinds of Assent and the Question of the Possibility of Knowledge. In *The Original Sceptics: A Controversy*, ed. Myles Burnyeat and Michael Frede, 127–151. Indianapolis: Hackett Publishing Company.
Gellius, Aulus C. 1927. *Noctes Atticae (Attic Nights). Vol. II*. Trans. J.C. Rolfe. Cambridge/London: Harvard University Press.
Glucker, John. 1978. *Antiochus and the late Academy*. Göttingen: Vandenhoeck & Ruprecht.
———. 1995. *Probabile, Veri-simile*, and Related Terms. In *Cicero the Philosopher*, ed. Jonathan Powell, 115–143. Oxford/New York: Clarendon Press.
Görler, Woldemar. 1997. Cicero's Philosophical Stance in the *Lucullus*. In *Assent and Argument: Studies in Cicero's Academic Books*. eds. Brad Inwood y Jaap Mansfeld, 36–57. Leiden: Brill.
Hacking, Ian. 2006. *The Emergence of Probability*. Cambridge: Cambridge University Press.
Hankinson, Robert J. 1995. *The Sceptics*. New York: Routledge.
Hirzel, Rudolf. 1883. *Untersuchungen zu Cicero's philosophischen Schriften (Vol. III)*. Leipzig: S. Hirzel.
Lévy, Carlos. 1978. Scepticisme et dogmatisme dans l'Académie: 'l'ésotérisme' d'Arcésilas. *Revue des Etudes Latines Paris* 56: 335–348.
———. 1992. *Cicero Academicus: Recherches sur les Académiques et sur la philosophie Cicéronniene*. Rome: Ecole Française de Rome.
Maia Neto, Jose R. 1997. Academic Skepticism in Early Modern Philosophy. *Journal of the History of Ideas* 58 (2): 199–220.
———. 2014. *Academic Skepticism in Seventeenth-Century French Philosophy. The Charronian Legacy 1601–1662*. Heidelberg: Springer.
Niiniluoto, Ikka. 2000. Scepticism, Fallibilism, and Verisimilitude. *Acta Philosophica Fennica* 66: 145–170.

Perin, Casey. 2005. Academic Arguments for the Indiscernibility Thesis. *Pacific Philosophical Quarterly* 86: 493–517.

———. 2010. *The Demands of Reason*. Oxford: Clarendon Press.

Photius. 2003. *Bibliothèque*, vol. 3, 186–222. Trans. R. Henry. Paris: Les Belles Lettres.

Plutarch. 1967. Adversus Colotem (Reply to Colotes). In *Moralia Vol. XIV*. B. Trans. B. Einarson and P.H. De Lacy, 153–315. Cambridge: Harvard University Press.

Sedley, David. 1983. The Motivation of Greek Skepticism. In *The Skeptical Tradition*, ed. Myles Burnyeat, 9–29. Berkeley: University of California Press.

Sextus Empiricus. 1933. *Outlines of Pyrrhonism*. Trans. R. G. Bury. Cambridge/London: Harvard University Press.

———. 1996. *The Skeptic Way. Outlines of Pyrrhonism*. Trans. Benson Mates. New York/Oxford: Oxford University Press.

———. 2000. *Outlines of Scepticism*, eds. Julia Annas and Jonathan Barnes. Cambridge: Cambridge University Press.

———. 2006a. *Adversus Mathematicos IX (Against the Physicists; Against the Ethicists)*. Trans. Robert G. Bury. Cambridge: Harvard University Press.

———. 2006b. *Adversus Mathematicos VII (Against the Logicians)*. Trans. Robert G. Bury. Cambridge: Harvard University Press.

Spinelli, Emidio. 2000. Sextus Empiricus, the Neighbouring Philosophies, and the Sceptical Tradition. *Acta philosophica Fennica* 66: 35–62.

Striker, Gisela. 1996. *Essays on Hellenistic Epistemology and Ethics*. Cambridge: Cambridge University Press.

———. 2010. Academics Versus Pyrrhonists Reconsidered. In *The Cambridge Companion to Ancient Scepticism*, ed. R. Bett, 195–207. Cambridge: Cambridge University Press.

Tarrant, Harold. 2007. *Scepticism or Platonism? The Philosophy of the Fourth Academy*. Cambridge: Cambridge University Press.

Thorsrud, Harald. 2002. Cicero on His Academic Predecessors: The Fallibilism of Arcesilaus and Carneades. *Journal of the History of Philosophy* 40 (1): 1–18.

———. 2009. *Ancient Scepticism*. Berkeley/Los Angeles: University of California Press.

Vogt, Katja. 2010. Scepticism and Action. In *The Cambridge Companion to Ancient Scepticism*, ed. Richard Bett, 165–180. Cambridge: Cambridge University Press.

Zeller, Eduard. 1923. *Die Philosophie der Griechen in ihrer geschichtlichen Entwicklung*. Leipzig: O.R. Reisland.

Chapter 3
Academic Skepticism and Metaphysics: On the Gods and the Immortal Soul

Abstract In this chapter, I turn my attention to how, according to Cicero, the New Academy evaluated the Hellenistic schools' metaphysical claims, following Carneades's method. Specifically, I explore the metaphysical discussions on the nature of the gods and the immortality of the soul, portrayed by Cicero in *De Natura Deorum* (*DND*) and the *Tusculan Disputations* (*TD*). I also compare the Academic treatment of these issues with the Pyrrhonian approach, accounted by Sextus Empiricus's *Outlines of Pyrrhonism* (*Pyr.*) and *Adversus Mathematicos* (*M IX*). My aim is twofold: On the one hand, to show how Cicero followed the Carneadean criterion of "the persuasive" (*to pithanon*) to examine metaphysical claims and determine the ones that should be adopted for practical purposes, and, on the other, to discuss how Sextus's skeptical treatment of these matters was compatible with his claim that the Pyrrhonists could have a pious life.

The previous chapter focused on summarizing the central epistemic debates between the Platonic Academy and Stoicism, and outlining the difference between the Academics' and Pyrrhonists' responses to the *apraxia* objection. However, I did not refer to how the Academics applied their method to specific questions. And, one of the most interesting features of their skepticism is that they used their method for examining the persuasiveness of impressions in both empirical and metaphysical problems. In this chapter, then, I turn my attention to how, according to Cicero, the Academy evaluated the Hellenistic schools' metaphysical claims. Specifically, I explore the metaphysical discussions on the nature of the gods and the immortality of the soul, portrayed by Cicero in *De Natura Deorum* (*DND*) and the *Tusculan Disputations* (*TD*). I also compare the Academic treatment of these issues with the Pyrrhonian approach, accounted by Sextus Empiricus's *Outlines of Pyrrhonism* (*Pyr.*) and *Adversus Mathematicos* (*M IX*).[1] My aim is twofold: On the one hand,

[1] The immortality of the soul is discussed in Book I of the *Tusculan Disputations*. The rest of the book is devoted to the psychological explanation of emotions according to the Stoics. Other dialogues in which Cicero discusses metaphysical questions are *De Fato* and *De Adivinatione*. I will refer to the former in Chap. 6 when I summarize Hume's influence on Kant's discussion of necessity and freedom.

C. González Quintero, *Academic Skepticism in Hume and Kant*, Synthese Library 449, https://doi.org/10.1007/978-3-030-89750-5_3

to show how Cicero followed the Carneadean criterion of "the persuasive" (*to pithanon*) to examine metaphysical claims and determine the ones that should be adopted for practical purposes, and, on the other, to discuss how Sextus's skeptical treatment of these matters was compatible with his claim that the Pyrrhonists could have a pious life.

To accomplish this, I will begin with the Hellenistic discussions on the existence and nature of the gods, and the role these schools gave to the "preconception" (*prolêpsis*) of the gods in such disputes. Ancient philosophers thought that all human beings possessed a common conception or, in modern terms, an "idea" or "concept" of the gods, which was the focus of their metaphysical discussions. Therefore, they undertook a detailed examination of this preconception to determine whether the gods existed and had specific attributes, such as providence, wisdom, benevolence, etc. Cicero described the schools' disagreements on this preconception in *De Natura Deorum*. After this, I will continue to describe Cicero's Academic refutation of the Epicurean and Stoic views on the gods, and his approval of the Stoic doctrine as the most persuasive and practically useful at the end of the treatise. From there, I will turn to Cicero's account of the Hellenistic debate on the immortality of the soul in the first book of the *Tusculan Disputations*, where he presents the Platonic and Epicurean theses on the matter. This time, however, Cicero does not seek to *refute* the theses in the traditional Academic way but rather claims that they are both sufficiently persuasive and helpful to overcome the fear of death. To explain how this argumentation coheres with Academic skepticism, I appeal to Cicero's views in the first book of the *Academica* (*Acad. I*) on the evolution of Plato's Academy and its relationship with Aristotelean rhetoric. I suggest that the influence of this rhetorical tradition on the Academy probably explains Cicero's different argumentative approaches, particularly in the case of these metaphysical questions. Finally, I will consider how Pyrrhonism deals with the existence and nature of the gods to show the discrepancies between Pyrrhonism's and the Academy's treatment of such matters. I suggest that Pyrrhonism recommends adopting a kind of theological agnosticism, accompanied by the practical abidance of religious tradition—what in the previous chapter I called a non-dogmatic heteronomy. Conversely, in Cicero's version,[2] the New Academy advocates for the practical adoption of theological beliefs considered sufficiently persuasive and practically useful after the skeptical test. Let us start now by discussing how Cicero applied the New Academy's method to the question of the nature of the gods.

[2]There is a debate about whether Carneades held an agnosticism or followed the criterion of the persuasive impression in these matters too. I will refer to this problem at the end of the first section.

3.1 The Hellenistic Schools' Preconception of the Gods

In his introduction to *De Natura Deorum*, Cicero announced that he would examine the problem of the existence and nature of the gods as an Academic philosopher, that is, skeptically pondering the different opinions on the matter. He continued to say that if, after careful analysis, one of the theses examined were to be established as true, he would happily abandon the Academic's *epochê* and embrace the corresponding doctrine (*DND I*, 13–14). But, of course, the comment was ironical—Cicero did not think certainty on the matter could be attained, only that a sufficiently persuasive or verisimilar view, the closest possible to the truth, could be found. To make the skeptical case stronger, he employed the metaphor of a courtroom. He argued that he would proceed as forensic orators did, presenting and cross-examining the different positions at stake in order to provide a final verdict:

> This being so, I feel disposed to [. . .] attend in court, try the case, and deliver their verdict as to what opinions we are to hold about religion, piety and holiness, about ritual, about honor and loyalty to oaths, about temples, shrines and solemn sacrifices [. . .] for all of these matters ultimately depend upon this question of the nature of the immortal gods (*DND I*, 13–14).

This metaphor is important for at least two reasons: first, because it makes Cicero's intention to blend rhetorical argumentation with skeptical analysis clear from the beginning of the treatise, and, second, because it sets the stage for Kant's understanding of the skeptical method, as we will see in Chap. 6. This is a common way for Cicero to begin his dialogues. In several of his introductions (*TD* I, 22–23; *Fin.*, I, 15–16; *De Divinatione* I, 6–10; *Fato* I, 1), he refers to the wide variety of philosophical views on the issue and the numerous contradictions that arise when they are carefully examined. He also presents the method of the Academy and the criterion of the persuasive impression in diverse ways, one of them, the metaphor of the courtroom and the final verdict. As M. Schofield argues, these introductions played a vital role in Cicero's philosophical writing. In a letter to Atticus (*Ad Att.* III, 16.6), he acknowledged that he enjoyed so much writing prefaces for his works that he even kept a folder with a repertoire of them at his Tusculan villa (Schofield 2009, pp. 63–84). Cicero probably wrote these prefaces intending to present his Academic position before giving voice to the points of view of other schools through his dialogues' characters. For this reason, even if, in subsequent periods of history, philosophers did not have access to the *Academica*, most probably they obtained a general view of the Academics' stance from these introductions.

The doctrines Cicero will analyze in *DND* concern the nature of the gods and are based on a corresponding "impression"—namely the "preconception" or *prolêpsis* of the gods. The Hellenistic schools focused their attention on this common notion to discuss whether the gods existed and, if so, whether they had a specific nature or set of attributes. The term *prolêpsis* is first found in Epicurus, who defined it as a universal concept or idea that served as a criterion for truth analogous to sensations (Schofield 1980, p. 292). We find a definition in Cicero's *DND*:

> Such notions [the preconceptions] Epicurus designates by the word *prolêpsis*, that is, a sort of preconceived mental picture of a thing, without which nothing can be understood or investigated or discussed (*DND* I, 43).

In the same vein, Diogenes Laërtius claims:

> By preconception they [the Epicureans] mean a sort of apprehension or a right opinion or notion, or universal idea stored in the mind; that is, a recollection of an external object often presented – e.g., such and such a thing is a man: for no sooner is the word 'man' uttered than we think of his shape by an act of preconception in which the senses take the lead (*Diog. Laërt.* X, 33).

Based on these descriptions, we can deduce that the Epicureans saw preconceptions as universal and abstract concepts, originating in sensation and memory. So, the preconception of 'man,' in the example provided by Laërtius, results from multiple sense impressions arising from the observation of numerous individual human beings. These sensations are stored in our memory and, once joined by the imagination, produce the preconception or abstract concept of the corresponding object. Because they arise from sense impressions, they share the latter's epistemic status and can function, for the Epicureans, as a criterion of truth (*Diog. Laert.* X, 31). The point seems to be that, being a sub-product of sensations, these preconceptions directly related to the objects they represented. For this reason, they believed that the preconception of the gods was necessarily and directly connected to the corresponding gods and provided accurate information about them. Any traits attributed to the gods that were not included in this preconception were false. In his *Letter to Menoeceus* (*LM*), Epicurus states:

> First of all, you should think of deity as imperishable and blessed being (as delineated in the universal preconception of it common to all men), and you should not attribute to it anything foreign to its immortality or inconsistent with its blessedness. On the contrary, you should hold every doctrine that is capable of safeguarding its blessedness in common with its imperishability. The gods do indeed exist since our knowledge of them is a matter of clear and distinct perception; but they are not like what the masses suppose them to be, because most people do not maintain the pure conception of the gods (*LM*, 123–124).

On the one hand, the preconception of the gods allowed the Epicureans to assert the existence of the gods. Since preconceptions originate in sensations, and sensations are caused by the object they represent, there exists entities corresponding to the Gods—that is, there are Gods. On the other hand, the Epicureans used the preconception to evaluate theological doctrines and common opinions on the nature of the Gods. For them, if we just restricted ourselves to the information engraved in the preconception, we would only ascribe to them immortality and happiness, and not other qualities, such as power over human matters, benevolence, wrath, etc. In sum, the preconception was a criterion to determine the truth of statements concerning the Gods' attributes.

Among the Stoics, Chrysippus seems to have been the first to adopt the Epicurean concept of *prolêpsis* (Schofield 1980, p. 293). He retained the two main characteristics of the concept—its universality and its function as a criterion for truth. Perhaps unsurprisingly, Laërtius reported that Chrysippus's introduction of this concept to the Stoic system produced a doctrinal contradiction, since it entailed accepting not one but two criteria for truth:

> The standard of truth, they [the Stoics] declare to be the apprehending presentation—i.e., that which comes from a real object, according to Chrysippus in the twelfth book of *Physics* and to Antipater and Apollodorus. [...] [But] Chrysippus in the first book of his *Exposition of Doctrine* contradicts himself and declares that sensation and preconception are the only standards, preconception being a general notion which comes by the gift of nature (an innate conception of universals or general concepts) (*Diog. Laërt.* VII, 54).

For Laërtius's, Chryssipus' introduction of this criteria brought about an epistemological contradiction since *kataleptic* impressions are sensible while preconceptions are innate, and, consequently, they should not have the same epistemic status. By adopting the epicurean standard for truth, Chryssipus either made an error or betrayed the empirical nature of the *kataleptic* impression. Cicero, however, offers a response to this problem. According to him, Chryssipus did not see preconceptions as innate but understood them as arising from *kataleptic* impressions. Lucullus explains the Stoic bond between kataleptic impressions and preconceptions thus:

> Such are the things we claim are apprehended by the senses. The next set are just like them, though we don't claim that these are apprehended by the senses themselves, but by the senses in a certain respect – e.g., 'That is white,' 'This is sweet,' 'That is melodious,' 'This is fine-scented,' 'This is rough.' Our apprehension of this set now comes from the mind rather than the senses. Next comes: 'That is a horse,' 'That is a dog.' [...] It's from this set of impressions that our conceptions of things are stamped on our minds, and without them, there can be no understanding, investigation, or argument (*Acad. II.*, 21).

In other words, Lucullus argues that preconceptions, which the Stoics called *ennoia* and Cicero translates to Latin as *notitia*, are abstract concepts that retain the main aspects of the *kataleptic* sense-impressions from which they originate. This is a very close view to the Epicureans', except that, for the Stoics, preconceptions did not come from any sense-impression whatsoever but only from *kataleptic* ones. According to Lucullus' report, in a similar vein to the Epicureans, the Stoics thought they could determine whether a particular judgment was true or false by analyzing the abstract content of the corresponding preconceptions. In other words, they claimed to be able to find whether a judgment corresponded to the content of the preconception, in which case it was true, or added something foreign to it, in which case it was false. As a result, they were able to focus philosophical disputes on one particular notion and achieve a *kataleptic* knowledge of the resulting doctrine (Schofield 1980, p. 297).

Evidently, the Academics attacked this view on the grounds that no preconception could function as a criterion of truth since the impressions at their basis were never apprehensible or *kataleptic*. In this way, they extended doubt from impressions to preconceptions, showing that the latter could be formed out of false impressions or impressions that, even if true, could not be distinguished from false ones. In other words, from the *principle of the inapprehensibility of impressions*, they seem to have derived a *principle of the inapprehensibility of preconceptions*. Cicero explained:

> But suppose there were false conceptions (since you seemed to use the term 'conception' for *ennoia*). Well, if our conceptions were false or stamped on our minds from true impressions that couldn't be discriminated from false impressions, then how would we put them to use? How would we see what was compatible with something or incompatible with it? (*Acad. II.*, 22).

As would be expected, the Academics used their skeptical method to examine preconceptions and the doctrines built on them, in order to determine the latter's corresponding degree of *persuasiveness* or *probability*, which Cicero sets out to discuss in *DND*. I will dive further into how he did that now.

The first book of the treatise deals with the Epicurean doctrine on the Gods. Velleius, the spokesperson for the philosophy of the Garden, begins by affirming the universality of the preconception of the gods:

> [. . .] For he [Epicurus] alone perceived, first, that the gods exist, because nature herself has imprinted a conception of them on the minds of all mankind. For what nation or what tribe of men is there but possesses untaught some 'preconception' of the gods? Such notions Epicurus designates by the word *prolêpsis*, that is, a sort of preconceived mental picture of a thing, without which nothing can be understood or investigated or discussed (*DND* I, 43).

In this paragraph, we see that the first feature of the Epicurean doctrine concerning the gods is the belief that, given the universality of their preconception, the judgment asserting their existence is true. The argument goes something like this: "if all humans possess the preconception of the gods, it is because all, in some way, have perceived the gods. Hence, we can trust that the gods exist." Epicurus claimed that the preconception of the gods came from sense impressions received from the gods themselves, in either visions or dreams (*Lucr.* V, 1193–1206). Thus, their theological beliefs had a kind of empirical foundation since these visions and dreams had, for them, the same epistemological status of sense-impressions.[3]

Moreover, according to Velleius, an anthropomorphic characterization of the gods is also stamped on their preconception: "From nature, all men of all races derive the notion of gods as having a human shape and none other; for in what other shape do they ever appear to anyone, awake or asleep?" (*DND* I, 46). And, finally, the Epicureans believe that, given their preconception, they can also claim with certainty that the Gods are happy and immortal:

> [. . .] Their life is the happiest conceivable, and the one most bountifully furnished with all things. God is entirely inactive and free from all ties of occupation, he toils not neither does he labour, but he takes delight in his own wisdom and virtue, and knows with absolute certainty that he will always enjoy pleasures at one consummate and everlasting (*DND* I, 51).

Velleius's basic idea is that the preconception of the gods guarantees their existence, their anthropomorphic nature, their happiness and immortality, and their absolute disregard for human affairs. Evidently, the last part of the characterization has the purpose of providing a non-religious theology, to which we will refer later. But let us continue with Cicero's discussion.

---

[3]According to Diogenes Laërtius, Epicurus said: "Hence it is from plain facts that we must start when we draw inferences about the unknown. For all our notions are derived from perceptions, either by actual contact or by analogy, or resemblance, or composition, with some slight aid from reasoning." (Diog. Laërt. X, 32.)

After the Epicurean arguments are laid out, Gaius Cotta begins his Academic refutation. Cotta was a Roman Patrician who presided over the city's religious ceremonies. Given his office, he would have been delighted to obtain proof of both the existence and anthropomorphic nature of the gods. However, he was also an Academic, or so tells us Cicero, and thus expressed his scruples on the matter. He announced his intention to refute Velleius's doctrine, with the caveat that he would only do so until he found a more cogent doctrine on the existence and nature of the gods (*DND I*, 61). As we shall see later, this characterization of Cotta as a high priest will be essential to understand Cicero's position at the end of the dialogue. For now, I will summarize the main points of his refutation.

The most crucial argument Cotta offers against the Epicurean theses is that the preconception of the gods does not have a universal character:

> You said that a sufficient reason for our admitting that the gods exist was the fact that all the nations and races of mankind believe it. But this argument is both inconclusive and untrue. In the first place, how do you know what foreign races believe? For my part, I think that there are many nations so uncivilized and barbarous as to have no notion of any gods at all (*DND I*, 62).

For Cotta, the existence of numerous manifestly atheist social groups, particularly among foreign nations, attests that the preconception of the gods is not universal, as the Epicureans thought. Given these cases, either not all human beings possess the alleged preconception or it is a very obscure and weak one for some of them. If we recall Carneades's set of criteria to evaluate an impression, we could say that the existence of atheistic cultures provides impressions that "counter" or "reverse" the Epicurean thesis that the preconception of the gods is universal. Consequently, Cotta determined that the Epicurean arguments were highly implausible or had a very low degree of persuasiveness. To be certain that the preconception of the gods is universal and provides sufficient proof of both their existence and anthropomorphic nature, the Epicureans would have to demonstrate that no atheist society exists and that every human collectivity agrees on the same qualities of the gods. But, this cannot be done. For this reason, the Epicureans should instead affirm that, given a more-or-less accepted notion of the gods, it is somewhat but not highly probable that the gods exist and that they have an anthropomorphic nature. Although Cotta does not use this Carneadean terminology, we can see how his refutation agrees with the New Academy's indications and how Cicero is preserving the spirit of Carneades's method in his discussion. Moreover, Cotta's refutation of Epicureanism was relatively easy since Velleius's arguments, at least in Cicero's account, were not very sharp philosophically.

Things became more complex when Cotta turned to evaluate the Stoic doctrine of the gods, which had more consistent arguments than the Epicurean. In Book III, Cota refuted Balbus, the spokesperson for Stoic theology, but, before doing so, he made a provisional evaluation of the two doctrines. Addressing Velleius, the Epicurean, he stated:

> [. . .] Because I think that your master Epicurus does not put up a very strong fight on the question of the immortal gods; he only does not venture to deny their existence so that he

> may not encounter any ill-feeling or reproach. [. . .] But as for Balbus, I am sure you must have noticed how much he had to say, and how, though lacking in truth, it was yet consistent and systematic. Hence what I have in mind, as I said, is not so much to refute his discourse as to ask for an explanation of the things that I could not quite understand (*DND* III, 3–4).

While there is a measure of irony in Cotta's appraisal,[4] the general intention of the exchange cited above is to show the indisputable superiority of the Stoic doctrine over the Epicurean. In particular, the Stoics' arguments seem to provide more proof for the existence and nature of the gods than the Epicurean assertion of their preconception's universality. In other words, there is greater argumentative variety and coherence in favor of the Stoic theses. Let us look closer at these arguments.

The Stoics claimed that the gods existed and governed the universe according to their wills. They inferred from the argument from the intelligent design of the universe that the preconception of the gods included both their existence and their providence. According to Cicero, the argument was first put forward by Cleanthes of Assos (300 BCE – 232 BCE), Zeno's successor as head of the Stoa (*DND* III, 16–17). In Cicero's version, it is explained as follows:

> When we see something moved by machinery [. . .] we do not doubt that these contrivances are the work of reason; when therefore we behold the whole compass of the heaven moving with revolutions of marvelous velocity and executing with perfect regularity the annual changes of the seasons with absolute safety and security for all things, how can we doubt that all this is effected not merely by reason, but by a reason that is transcendent and divine? (*DND* II, 97)

The argument holds that, just as artifacts created by human minds show order and purpose, the universe, which exhibits such order and purpose to the highest degree, must have been created by an intelligent maker. It has the form of an analogy, based on a supposedly *kataleptic* impression—that of the ordered universe. Based on this argument, the Stoics claimed that the preconception of the gods necessarily involved their providence, namely their having created the world purposefully.[5] For Cotta, this argument was much more persuasive than the Epicureans' because the analogy between an ordered universe and the products of human design had a firmer basis than the argument about the universality of the preconception. For him, the Epicureans wielded a deficient view, disproved by various opposing experiences, whereas the Stoic doctrine presented a highly persuasive thesis, reinforced by the impressions corresponding to three of the four terms of the analogy—namely the ordered universe, the artifacts of human contrivance, and the human mind.

Of course, this does not mean that the Stoic doctrine could weather all criticism. While it was more challenging to refute, Cotta was still able to present forceful arguments against it. He began by questioning the logical inference from the order and regularity of natural phenomena to the mind of a creator:

---

[4]Evidently, Book III of *DND* is a skeptical attack on Stoic doctrine and not merely a clarification of their theses.

[5]For Sextus Empiricus, the Stoics found the origin of the preconceptions of the gods in the empirical verification of the order of the universe (*M IX*. I, 22–28).

> But not all things, Balbus, that have fixed and regular courses are to be accredited to a god rather than to nature. What occurrence do you think could possibly be more regular than the repeated alternation of flow in the Euripus al Chalcis? [...] Why, if all motions and all occurrences that preserve a constant periodic regularity are declared to be divine, pray shall we not be obliged to say that tertian and quartan agues are divine too, for nothing can be more regular than the process of their recurrence [...]? (*DND* III, 24).

Cotta's point in the above passage was that, although we tend to infer that everything revealing order is rational and, therefore, the product of an intelligent cause, this inference is contradicted by many experiences. Many regular things – for instance, malarial fevers – are not the result of rational causes or a purposeful will. Neither are other things, whose regularity is also accompanied by beauty and utility, such as the city of Rome, the product of a divine mind (*DND* III, 21–22). These counter-examples offered cases that opposed or hindered the Stoic inference by refuting the conclusion of the analogy. And Cotta further undermined the design argument's founding impression – namely that the universe is well-ordered – using, among others, the argument on the existence of evil. He claimed that even if the universe seemed to be ordered, beautiful, and useful for humankind, in light of the pervasiveness of evil – especially of moral evil – we could not accept that it was really so. If the gods created the universe, why did they consent to the existence of moral evil?[6] And, if they did consent to the existence of moral evil, how could it be attributed to human beings without also attributing it to the gods? Cotta says:

> Do we then really think that this enormous crop of evil was sown by the immortal gods? For if the gods gave man reason, they gave him malice, for malice is the crafty and covert planning of harm; and likewise also the gods gave him trickery and crime and all the other wickednesses, none of which can be either planned or executed without reasoning (*DND* III, 75).

The existence of evil was one of the Epicurean arguments to prove that the Gods are not concerned about the world. Here, it is used dialectically to undermine the claim that the ordered, beautiful, and useful universe was caused by a provident, divine

[6]In the extant text of *DND*, Cicero does not address directly natural evil. However, Lactantius records the following Academic argument: "But the Academics, arguing against the Stoics, are accustomed to ask why, if God made all things for the sake of men, many things are found even opposed, and hostile, and injurious to us, as well in the sea as on the land. And the Stoics, without any regard to the truth, most foolishly repelled this. For they say that there are many things among natural productions, and reckoned among animals, the utility of which hitherto escapes notice, but that this is discovered in process of the times, as necessity and use have already discovered many things which were unknown in former ages. What utility, then, can be discovered in mice, in beetles, in serpents, which are troublesome and pernicious to man? Is it that some medicine lies concealed in them? If there is any, it will at some time be found out, namely, as a remedy against evils, whereas they complain that it is altogether evil. They say that the viper, when burnt and reduced to ashes, is a remedy for the bite of the same beast. How much better had it been that it should not exist at all, than that a remedy should be required against it drawn from itself?" (Lactantius, *De Ira Dei* XIII, pp. 408.)

nature. Thus, the Stoic doctrine, although more convincing than the Epicurean, is shown to be liable to doubt by Epicurean means. Notwithstanding, Cotta concluded his assessment of the Stoic views on the nature of the Gods by affirming that his intention was not to undermine their system but only to show "how obscure it is and how difficult to explain" (*DND* III, 93). With this statement, he confirms his allegiance to the New Academy's recognition of the difficulty involved in theoretically investigating such matters and the need to continue the inquiry.

However, things do not end here, with a suspension of belief and the promise to resume the search in the future. Instead, the dialogue's characters now embark on the practical question of the social utility of religion.

In the remainder of the dialogue, Balbus asks Cotta to direct the discussion to this matter:

> You have indeed made a slashing attack upon [...] the Stoic doctrine of the divine providence. But as evening is now approaching, you will assign us a day on which to make our answer to your views. For I have to fight against you on behalf of our altars and hearths, of the temples and shrines of the gods, and of the city walls, which you as pontiffs declare to be sacred and are more careful to hedge the city round with religious ceremonies than even with fortifications; and my conscience forbids me to abandon their cause so long as I yet can breathe (*DND* III, 94).

Balbus assures his listeners that the theoretical inquiry is to be continued at another time but insists that they should not put off discussing the issue from a practical standpoint, given the weighty practical consequences of the matter. He reminds his friends that, given the political and moral utility of religious practices, it is not desirable to suspend the city's rituals until the issue is completely settled. For this reason, despite their possibly weak theoretical justification, the metaphysical claims on the existence and nature of the Gods should be re-assessed to determine their practical advisability. Cicero's metaphor of the city, better protected by the walls of religion than by military defense, shows the significance he allots to this part of the examination. I will not discuss the details of Cicero's account of the practical value of religion, wich has the form of a rhetorical peroration, but only mention that he gives a peculiar final 'verdict' on the issue. Cicero, who throughout the dialogue had appeared merely as a witness, now says: "Here the conversation ended, and we parted, Velleius thinking Cotta's discourse to be the truer, while I felt that that of Balbus approximated more nearly to a semblance of truth" (*DND* III, 95).

This final statement has been the object of many interpretations. The main guiding question on the issue is whether Cicero betrayed the skeptical spirit of the Academy by adhering to the Stoic point of view, represented by Balbus, instead of siding with Cotta.[7] Diverse answers have been given to this question, which, according to Pease, can be grouped into three: first, Cicero adhered to the Stoic doctrine

[7] The question is thus expressed by Pease: "Why does Cicero, who in the first part of Book I declares himself an Academic, and who at his entrance into Cotta's house is recognized by his friends as the fellow-schoolman and natural supporter of Cotta, now cast his vote, not with Cotta and Academicism, but with the Stoic speaker and those views which Cotta has been refuting?" (Pease 1913, p. 26). Knuuttila and Sihvola (2000, p. 126) also point to the existing contradiction between Cicero's final adherence to Stoicism and his early expression of preference for Academic skepticism.

simply to save his reputation, that is, not to appear excessively irreligious to his contemporaries (Pease 1913, pp. 27–30). Second, Cicero's eclectic leanings led him to abandon his Academic position and adopt Stoicism. Pease synthesizes this view, held by many nineteenth- and twentieth-century commentators, in a single sentence: "An Academic by profession, he is by sympathies and at heart a stoically inclined eclectic" (Pease 1913, p. 30; Schofield 2009, p. 82). Finally, Pease's own interpretation is that Cicero's position was neutral or impartial, since the dialogue aimed at describing as objectively as possible the theological views of the schools and letting the reader decide for herself which was the most probable claim. Given that Cicero was divided between his theological preferences – sometimes in favor of Cotta's agnosticism, sometimes leaning toward Stoic theism – he finally opted for the position most consistent with his practical and political interests. Pease concludes:

> But Cicero was an intensely practical man, and [...] he seems to have been a little nettled by the charge that the Academy was unpractical [...]. The formal assent to Stoic principles which he gives, in the final sentence of the dialogue is an example alike of the freedom from dogmatic requirements allowed to the Academics, and of the possibility of using such individual liberty for the acceptance of any practical working principle (Pease 1913, p. 37).

I have mixed feelings about this assessment. On the one hand, I think Pease's claim that the dialogue only sought to describe the doctrines of the schools and had no polemical aims is mistaken. On the other, I agree with his view that Cicero's final adherence to Stoicism had political or practical motives. I believe that his verdict should be understood in light of the immediately-preceding passage, in which Balbus introduces the issue of the practical utility of the state religion. The central tenet of Stoic theology, as I said before, was the belief in divine providence, on which religious rituals were justified. Ancient philosophers knew well that without divine providence – namely the belief that God ordered the world according to purposes and had the power to change the course of events at will – the practical aspects of religion became dispensable. The anti-religious tenets of Epicureanism were evident in this respect. If the gods neither planned the universe according to their will nor intervened in human affairs, the religious practices – prayers, sacrifices, omens, etc.– which aimed at obtaining the gods' favor, became unnecessary. Hence, ultimately setting aside the Stoic doctrine would amount *in practice* to accepting the Epicurean one.

Cicero had warned about this practical outcome at the very beginning of the dialogue:

> For there are and have been philosophers who hold that the gods exercise no control over human affairs whatever. But if their opinion is the true one, how can piety, reverence or religion exist? [. . .] But if, on the contrary, the gods have neither the power nor the will to aid us, if they pay no heed to us at all and take no notice of our actions, if they can exert no possible influence upon the lives of men, what ground have we for rendering any sort of worship, honor or prayer to the immortal gods? (*DND* I, 3)

Thus, Cicero did not recommend the agnosticism that would result from suspending judgment on theological issues after Cotta refuted both the Epicurean and Stoic doctrines, because it would entail abolishing Roman religion. Like Cotta, Cicero opposed the views of the Epicureans because their arguments were both theoretically defective and practically anti-religious. But, unlike him, he accepted the Stoic doctrine, finding that it better withstood the skeptical scrutiny and was practically useful. It warranted the continuity of the state religion, whose function was to reinforce social bonds of trust and justice:

> Piety, however, like the rest of the virtues, cannot exist in mere outward show and pretense; and, with piety, reverence and religion must likewise disappear. And when these are gone, life soon becomes a welter of disorder and confusion; and in all probability, the disappearance of piety towards the gods will entail the disappearance of loyalty and social union among men as well, and of justice itself, the queen of all the virtues (*DND* I, 4).

Cicero's verdict in favor of Stoic theology is, then, primarily explained by these practical motives. However, given Cotta's refutation of some of the core Stoic theological tenets, his final acceptance cannot be considered a wholehearted adherence to Stoic doctrine. It would be incorrect to believe that Cicero advocated a return to Stoic dogmatism. With his verdict, he rather suggested that the beliefs in the existence and providence of the gods should be neither held dogmatically nor suspended completely. Let us remember that Carneades's mitigated skepticism established the difference between the mental acts of "assent" and "approval." Assent entailed a dogmatic adherence to an impression or judgment, while approval only a practical acceptance of it, with theoretical reservations (*Acad. II*, 104). In Cicero's judgment, the degree of persuasiveness of the Stoic theological doctrine, together with its practical utility, was sufficient for it to be "approved" for practical reasons—that is, to allow the continuation of the city's religious rituals. For this reason, I think Pease and other commentators are mistaken when they say that Cicero was divided between his Academic and Stoic leanings (Pease 1913, p. 37; Schofield 2009, p. 82). His final acceptance of Stoicism was neither an expression of eclecticism nor the result of political strategy. On the contrary, it was the direct and most consistent consequence of adopting the principles of the New Academy.

There remains, however, a final question to be answered regarding Cicero's dramatic role in the *DND*. Unlike other dialogues, particularly later ones, in which Cicero was both one of the main characters and the representative of the Academy, his role here is that of an observer. Why did Cotta refute the theological doctrines of the schools and not Cicero himself?[8] And, if Cotta represented the Academic point of view, why was it necessary, both from a dramatic and a philosophical perspective, that Cicero appeared as a witness? These questions are important, first, to better

---

[8] One reason for Cicero's preference for Cotta and other historical characters in his dialogues is the fact that they were not alive, and hence, Cicero could give them views that were more philosophical than the historical characters could have had. Based on a letter to Atticus (*Ad Att.* III, 13.19), Schofield argues that Cicero changed this dramatic pattern to include Varro in the first book of the *Academica*, as a courtesy suggested by Atticus, and decided to portray himself in the exchange, following the style of the Aristotelian dialogues (Schofield 2009, pp. 74–75).

understand Cicero's nuanced position and, second, to eventually interpret Hume's recourse to this narrative strategy in the *Dialogues concerning Natural Religion.*

It seems to me that Cicero played this role to disclose his own position *vis-à-vis* the leaders of the Academy. It could be the case that Cotta was envisaged to represent more accurately Arcesilaus's or even Carneades's dialectical arguments. In fact, according to various testimonies, Carneades was more a radical skeptic than a moderate one on matters theological, offering no positive response to the issue of the nature of the gods. As Brochard affirms, he rebutted with particular fervor the Stoics' religious tenets:

> On all points, Carneades takes them on: he denies their ends, calls into question the proofs of the existence of the gods, holds that the idea we make of divinity is contradictory, reduces to the absurd the supporters of popular religion (Brochard 2002, p. 164, my translation).

To explain this attitude, Thorsrud argues that, for Carneades, the criterion of "the persuasive" only functioned as a response to the *apraxia* objection, not as a way to decide on issues such as theology (2009, pp. 59–83). And, if this is the case, then, Cicero's adaptation of the criterion to decide metaphysical disputes could be his own call. I agree with Thorsrud's view that Cicero may have extended the criterion of the persuasive impression to these questions, which Carneades himself treated only with refutative aims. However, I also believe that, in doing this, Cicero sought to be entirely consistent with the teachings of the New Academy and did not become a dogmatist of sorts. Supporting this claim, of course, would require a more detailed historical analysis than I can give here. For the moment, I just want to suggest that Cicero's final acceptance of the Stoic thesis for practical purposes most probably was motivated by his Academic allegiances—by his adoption of the criterion of the persuasive impression—and not by eclectic or Stoic inclinations.

To conclude this section, I want to call attention again to Cicero's metaphor of the courtroom. In *DND*, Cicero debates the Hellenistic doctrines on the existence and nature of the Gods as if in a forensic scene—all the sides' claims have been examined and a final verdict has been given by a judge, Cicero himself. This verdict is that the Stoic doctrine, while not true, is the most persuasive one because its theses are more consistent and less contradicted by others than the Epicurean ones. In other words, they are "persuasive," "tested," and "irreversible," while the Epicurean theses are not persuasive enough. Furthermore, the Stoic characterization of the gods as provident is practically more advisable than the Epicurean denial that they intervene in human affairs, given the moral and political utility of the Roman state religion. Thus, while the theoretical investigation on these matters ought to continue, some of the Stoic claims can be provisionally approved to provide practical orientation. With this final verdict in mind, let us now turn to Cicero's treatment of the metaphysical beliefs in the immortality of the soul, in order to observe whether or not he applies the same Academic method to this question.

3.2 The Immortality of the Soul and the Fear of Death

What takes place in the first book of the *Disputations* has remarkable similarities and differences with *De Natura Deorum*'s line of argument. Cicero states here that, even though it is impossible to attain certainty on the question of the human soul's immortality, he would examine some Hellenistic views on the matter to determine their probability or persuasiveness (*TD* I, 22–23). However, in this dialogue, he takes a different, more constructive approach than we saw in *DND*. He examines the Platonic and Epicurean doctrines about death and the soul by arguing in favor of both views, instead of refuting one and accepting the other as the most persuasive and practically useful. Let us begin our rehearsal of these topics with Cicero's approach to the first view examined—the Platonic thesis in favor of the immortality of the soul.

According to this thesis, stated in Plato's *Phaedo*, death is a good for human beings, since the soul is immortal. Cicero argues that, even though Plato did not conclusively prove the immortality of the soul, he sufficiently supported it, and we should, therefore, deem it as highly persuasive:

> [...] And yet no reason really suggests itself to my mind why the belief of Pythagoras and Plato should not be true. For though Plato produced no reasoned proof – note the tribute I pay to the man – he would crush me by the mere weight of his authority: he has, on the contrary, produced such a number of proofs that it seems he wished to convince others, and beyond doubt, he seems to have convinced himself (*TD* I, 49).

The ironic allusion to Plato's authority in this paragraph is significant. Since one of the New Academy's goals was precisely to free the skeptics from beliefs held from authority (*Acad. II.*, 8–9), Cicero could not sincerely claim that Plato's authority was sufficient for him to adopt the belief in immortality—that is why he jokingly says "note the tribute I pay to the man." Cicero suggests, however, that Plato was more deserving of credit than the leaders of other Hellenistic schools – who sought to secure the blind adherence of their followers – because his arguments were more numerous and consistent, and he earnestly held them. That being said, he proceeds to summarize each of Plato's arguments: his mind-body dualism (*TD* I, 44–47), his theory of knowledge as reminiscence (*TD* I, 57–58), his claim that the soul has a simple and indivisible nature (*TD* I, 66–71), and, finally, his practical *dictum* that philosophy is a preparation for death (*TD* I, 75). I will not delve into de details of this account but only say that, in general terms, Cicero considered Plato's views sufficiently persuasive because they were internally consistent and complemented each other in the best possible way. Moreover, since the aim of the dialogue was to free his interlocutor – a student from whom we have no further characterization – from his fear of death, he maintained that it was practically advisable to believe in the immortality of the soul, because, if death was regarded as a good, the natural fear of it could be counterbalanced.

Next, however, he turned to defend the Epicurean doctrine that the human soul is mortal, and, for this reason, death is neither good nor evil. But, before continuing, let me just acknowledge that having already accepted the Platonic thesis as sufficiently persuasive, it may seem surprising that Cicero did not end the discussion here, with the approval of this claim for practical purposes. Why did he feel he had to continue to examine the contrary thesis? Let me offer some suggestions. First, because the skeptical method requires that pro and con arguments related to a given thesis be scrutinized before deciding anything on the matter; and, second, because Cicero thinks that, even though believing in the immortality of the soul is theoretically justifiable and practically advisable, it is not so easy to maintain such conviction. Therefore, to appease our fear of death, it is necessary to offer yet another argument—the Epicurean claim that given that the soul is mortal, death is neither good nor evil. Cicero puts it this way:

> M: You do not disdain, do you, to hear why, even if this view be true, death is still not reckoned among evils?
>
> S: As you like, but no one will drive me to give up immortality.
>
> M: That I approve, and yet we ought not be over-confident in anything: for we are often influenced by some cleverly drawn conclusion, we waver and change our opinion even in questions that are comparatively clear: much more in this question, for it has an element of obscurity (*TD* I, 78).

Cicero presents the Epicurean argument as a supplementary one, so that, in the event the conviction associated with the Platonic thesis faltered, there would be another view in place to alleviate one's fear of death. Cicero seems to worry that, on matters as difficult as these, the practical approval of a thesis that has withstood skeptical examination could weaken over time. Given that we cannot know with certainty that the soul is immortal, it may be the case that, as we forget the philosophical considerations that led us to accept it in the first place, it may come to appear obscure, implausible, or remote from the truth. This is precisely what happened when the student in the dialogue finished reading the *Phaedo*. He says: "[...] somehow I am sorry to find that I agree [with Plato's doctrines] while reading, yet when I let the book aside and begin to reflect in my own mind upon the immortality of souls, all my previous assent slips away" (*TD* I, 25). Cicero imagines that the same thing may occur after the conversation ends and prepares himself to offer his pupil a supplementary argument to help relieve his fear of death.

It is also plausible that Cicero expounded the Epicurean viewpoint after the Platonic one because the conviction about the soul's immortality may bring about yet another fear, that of the torments of Hades (*TD* I, 111). The Epicurean arguments, particularly the argument that since the soul dies with the body, it is impossible to experience pain when dead, is designed to counteract this fear:

> But suppose, as these thinkers hold, that souls do not survive after death: I see that in that case we are deprived of the hope of a happier life. But what evil does such a view imply? For suppose that the soul perishes like the body, is there, then, any definite sense of pain or sensation at all in the body after death? (*TD* I, 82)

Of course, not everyone would concede that these arguments are sufficient to alleviate our fear of death. To the obvious objection that the evil of death is not the state after it but the very privation of life it entails,[9] Cicero responds that human life is actually full of suffering and thus, "death distances us from evils, not from what is good" (*TD* I, 83). This argument is accompanied by others, which seek to mitigate the fear associated to the process of dying; among them, the Epicurean claim that dying is usually brief and, although painful, more often than not it is a relief compared to the ailments of illness (*TD* I, 82). Cicero also refers to the standard Epicurean view on the symmetry between birth and death, and offers the famous dictum: "Death cannot be where life is; where life is, there is no death" to stress the point that non-existence is not to be feared, because we will not experience it (*TD* I, 91). Resorting to an arsenal of Epicurean topics, then, Cicero sought to persuade his pupil that it was unnecessary to fear death, for either the soul is immortal and death is a good, or the soul is mortal and death is not an evil. At the conclusion of the dialogue, the student enthusiastically expresses his conviction:

> During the first time of what you said the effect was to make me long for death, while the effect of the latter part was sometimes to make me feel not unwilling, sometimes feel untroubled; the net result of all you said, however, is that I do not reckon death among evils (*TD* I, 112).

The book closes with a classical rhetorical peroration. Cicero uses traditional fables from Herodotus, Euripides, Crantor, and others (*TD* I, 113–117), to conclude that we should not be afraid of death since it is part of the human condition: "can that which is necessary for all be wretched for one alone?" (*TD* I, 119) In the end, the pupil says that Cicero's final exhortation had reinforced his practical conviction: "Indeed [. . .] I may tell you that your epilogue has really strengthened me." (*TD* I, 119)

Given the importance Cicero bestows to the Epicurean arguments in this dialogue, we can say that, if in *De Natura Deorum* he had judged the Stoics' theological views as the most persuasive and practically useful, in this dialogue the relationship is inverted. Cicero rapidly dismisses the Stoic thesis on the matter – namely that the soul is not immortal, but it can sometimes survive the body –[10] by arguing that it is more suitable for crows than for humans (*TD* I, 77)[11] and sides with the Epicureans on the view that the soul cannot outlast the body. However, the same cannot be said about the Platonic thesis. This one is, for Cicero, as persuasive as the Epicurean one, but they do not, for this reason, cancel each other out. Instead, they are complementary from a practical point of view – that is, they are helpful to alleviate the fear of death – and, therefore, they should both be used as weapons, so to speak, against a common enemy.

Let us pause at this point to recapitulate Cicero's argumentative strategy in this dialogue. As I said before, in the *Tusculan Disputations*, Cicero did not use the

[9]This is T. Nagel's critique of the Epicurean arguments (Nagel 1979, pp. 1–10). For a contrary view, see E. Tugendhat (2002, pp. 161–181).

[10]In general terms, this is the argument held by Cebes in Plato's *Phaedo* (*Phd.* 87a-88b).

[11]"M: Is there any reason, then, to stop us from sending our friends the Stoics about their business? I mean those who say that souls survive on their departure from the body, but not forever? S. Send them surely [. . .]" (*TD* I, 78).

typical dialectical argumentation, where the adversary's arguments are refuted piecemeal, but rather employed a more constructive reasoning, defending two opposite claims. He first examined Plato's thesis to determine the persuasiveness of the view that the soul is immortal and that death is, therefore, a good. Then, he evaluated the Epicurean doctrine to conclude that its arguments are *likewise* sufficiently persuasive, and we could on their grounds accept that, even if the soul is not immortal, death is not an evil. But he never actually refuted the Platonic or Epicurean arguments, and, what is more, he only offered a brief, barely worth noticing, negative appraisal of the Stoic position. Why did Cicero argue in this way? Again, we may ask ourselves if he lost his Academic disposition in favor of a more dogmatic or eclectic tendency.

In my opinion, Cicero remains consistent with Academic skepticism in this dialogue too. In a nutshell, he claimed that, even though it is impossible to attain metaphysical certainty on the soul's immortality, the Platonic thesis is sufficiently persuasive to be "approved" for practical motives. And, if doubts arise concerning immortality or if this opinion brings about an unbearable fear of Hades, the Epicurean arguments on the mortality of the soul and death as a non-evil may be adopted as a good substitute. This could seem to many as a piece of very inconsistent advice from a philosophical point of view. But it is neither theoretically contradictory nor inconsistent from a practical point of view, and for a skeptic who accepts from the beginning the impossibility of attaining knowledge on the matter, particularly since the tenacity of our fear of death is such that we may need to resort to as many arguments as we can to alleviate it, irrespectively of their mutual coherence. The Academic philosopher understands that, because the truth of the matter is never to be obtained, the criteria to assess these beliefs must be different—in this case, their persuasiveness and utility to deter our fear of death.

Cicero displays in this exchange the constructive or positive face of the New Academy's skepticism. Even though the Academy recommended *prima facie* presenting arguments for and against a thesis to determine their degree of persuasiveness and to approve for practical purposes the claim that better withstood the skeptical test, sometimes the equal persuasiveness of both thesis and antithesis, and the practical ends sought for, suggested approving both of them, even if they opposed each other. This way of proceeding, however, does not entail falling back into dogmatism since no thesis is held as true, but both are regarded as merely persuasive and practically advisable. Notwithstanding, this constructive aspect of Academic skepticism cannot be sufficiently understood unless we consider for a moment the relationship between the Academy and the classical rhetorical tradition. I will turn to this discussion now.

3.3 Aristotelean Rhetoric and Academic Skepticism

While the relationship between skepticism and the art of rhetoric has not yet been explored to the extent it deserves, it is undeniable that the two traditions historically converged in the Platonic Academy (Michel 2003; Brittain 2001). As we have seen,

Cicero's argumentative strategy in the *Tusculan Disputations*, while retaining the Academic method of examination, had rhetorical aims—namely to convince his student that death is good or, at least, not evil and, hence, should not be feared. With this purpose in mind, Cicero presented the arguments supporting the Platonic and Epicurean theses, which he considered the most persuasive ones on the matter, and finished the dialogue with a rhetorical epilogue amplifying his discourse on the natural character of our mortal condition. Someone could argue, however, that this example of joined skeptical exposition and rhetorical expertise does not authorize us to claim that the skepticism of the Academy shared with the rhetorical tradition their methods and theoretical assumptions. After all, Cicero was both a philosopher and an accomplished orator, but this double expertise does not necessarily extend to the philosophers of the Academy before him or to the Academy's outlook as a school. Notwithstanding the possibility of this view, I believe that the connection between both traditions, in fact, ran very deep in the Academy.

To begin with, we should entertain some historical considerations. In ancient doxography, not only Cicero but also other members of the Academy were portrayed as excellent orators. One example is provided by Laërtius, who described Arcesilaus as both an expert dialectician and persuasive orator:

> Very fertile in invention, he could meet objection acutely or bring the course of discussion back to the point at issue, and fit it to every occasion. In persuasiveness he had no equal, and this all the more brought pupils to the school, although they were in terror of his pungent wit. But they willingly put up with that; for his goodness was extraordinary, and he inspired his pupils with hopes (*Diog. Laert.* IV, 37).[12]

The testimony describes how Arcesilaus's dialectical and rhetorical capabilities complemented each other. When referring to his sharpness to invent objections and his ability to re-direct the conversation toward the nucleus of the discussion, Laërtius undoubtedly highlights Arcesilaus's skill in skeptical refutation. But, when he suggests that Arcesilaus was, likewise, extremely persuasive, so that young students were eager to attend his school despite their fear of being refuted, he is rather portraying Arcesilaus's rhetorical ability. It seems that the scholarch was a pleasant orator, who probably used the middle style and excelled at Aristotelian *ethos*—i.e., the capacity to persuade by projecting a virtuous and trustworthy character (*Rh.* 1356a 1–10).

In turn, Carneades's rhetorical ability is mainly accounted for in the testimonies of his famous participation in the 155 BCE Greek embassy to Rome, together with the Stoic Diogenes of Babylon and the Peripatetic Critolaus. The legend goes that he gave consecutive speeches in favor and against justice, and that his speeches on both sides were so eloquent that Cato the Elder was enraged and asked the Senate to

---

[12]Cicero also considered Arcesilaus a highly-accomplished dialectician: "Arcesilaus' position wasn't widely held at first, despite the prominence he achieved by the sharpness of his intellect and his curiously attractive manner of argument" (*Acad. II.*, 16). Eusebius of Caesarea offers a more critical testimony but also expresses his admiration for Arcesilaus's discursive capacity (*Praep. evang.*, XIV, 5–6).

respond promptly to the requests made by the delegation so that Carneades would leave Rome as soon as possible (*De Or.* II, 155, *TD.* IV, 5).[13] Whether true or not, the legend emphasizes both his skepticism and his rhetorical competence. Carneades also seems to have possessed a vigorous discursive style. Again, according to Laërtius:

> Such was his predominance in philosophy that even the rhetoricians would dismiss their classes and repair to him to hear him lecture. His voice was extremely powerful so that the keeper of the gymnasium was sent to him and requested him not to shout so loud. To which he replied, "Then give me something by which to regulate my voice." Thereupon, by a happy hit, the man replied in the words, "You have a regulator in your audience" (*Diog. Laërt.* IV, 63).

Given this portrayal, we can say that if Arcesilaus's oratory followed Aristotelean *ethos*, Carneades's rhetoric could have been an instance of Aristotelean *pathos*, aiming at the arousal of passions (*Rh.* 1356a 10–15). Likewise, in Ciceronian terms, his style would be grandiloquent or pathetic (*Or.*, 97–99).[14] Cicero agrees with Laërtius's view that Arcesilaus's "mastery of arguments and force of eloquence" was significant, but Carneades's was still greater (*Acad. II.*, 60). Furthermore, in *De Oratore* (*DO*), he describes Carneades as "a speaker who, for spirited and copious oratory, surpassed, it was said, all other men" (*DO* I, 45), and he refers, again, to the force and effectiveness of Carneades's discourse by saying: "[. . .] the extraordinary power and diversity of his oratory would be extremely to our liking; since in the debates of his, he supported no contention without proving it, and attacked none which he did not overthrow" (*DO* II, 161). In sum, Carneades was, for Cicero, an example of the "perfect orator," characterized by a mixture of great discursive power and incomparable philosophical sharpness—the ideal rhetorician that he himself sought to become.

Among Carneades's successors at the Academy, his disciple, Charmadas, seems to have been as eloquent as his master, and Cicero also says that Philo of Larissa used to combine philosophical teaching with rhetorical exertion (*Or.*, 51; *DO* I, 84 y III, 110; Brittain 2001, pp. 312–342). All these testimonies indicate that the teachings of the rhetorical tradition played an essential role in the training of the heads of the Middle and New Academies up to Cicero himself.

Another insight into this historical connection is given in the first book of the *Academica*, where Cicero argues that the Academy directly inherited the Aristotelean rhetorical canon. In this book, Varro – the character who represents Antiochus's position – described the general development of the Hellenistic schools and claimed that the Old Academy and the Lyceum shared the same doctrinal corpus: "Following

[13] For a critical assessment of this historical legend, *cf.* J.G.F. Powell 2013, pp. 219–247. Powell argues convincingly that the legend may be a construction of Cicero's, who invented it to give a plausible setting to his *De Republica.*

[14] Cicero describes the grand style as "[. . .] magnificent, opulent, stately and ornate; he [who uses it] undoubtedly has the greatest power. [. . .] This eloquence has the power to sway men's minds and move them in every possible way. Now it storms the feeling, now it creeps in; it implants new ideas and uproots the old" (*Or.*, 97).

Plato's complex and eloquent lead, a single and concordant system of philosophy developed under two names: the philosophy of the Academics and the Peripatetics. Despite their difference in name, they agreed in their doctrine" (*Acad.* I, 17).

The corpus that Academics and Peripatetics shared was the traditional threefold system, encompassing ethics, natural philosophy, and dialectics. But, they differed from other Hellenistic schools in that they taught dialectics and rhetoric together. As Varro reports: "This was their [the Academics' and Peripatetics'] teaching of the whole method of dialectic, i.e., speech used in formal argument. Its counterpart, as it were, was the ability to use rhetoric, i.e., the development of continuous speech adapted for persuasion" (*Acad.* I, 32). This paragraph's phrasing – particularly the use of the term "counterpart" – shows Cicero's familiarity with the Aristotelean formulation of rhetoric as the "antistrophe" or counterpart of dialectics (*Rh.* 1354a). According to this view, both dialectic and rhetoric were arts of arguing about non-specialized issues and only differed in their scope of application—dialectics being directed toward the evaluation of opinion; rhetoric to the discussion of political and judiciary matters (*Rh.* 1354a 1–10).

Still more important for us is that, through Varro, Cicero claimed that this method of teaching dialectics and rhetoric together was naturally embraced by the Academy: "[. . .] we [the Academics] must obey the precepts of the dialecticians and the orators as if they were laws, since our school thinks that dialectic and rhetoric are virtues" (*Acad. I,* 5). Therefore, if Varro's words bear witness to the historical truth, the Old Academy adopted the Aristotelean view of rhetoric as a counterpart of dialectics. Nonetheless, it should be noted that Varro was the spokesman for Antiochus's views in the first book of the *Academica*, hence it is still possible that, in the missing part of the dialogue, Cicero himself had given a response endorsing a different view of the relation between the two schools. However, I do not believe this is the case for at least two reasons. First, in a letter to Atticus, where Cicero appears to be very pleased with himself for having introduced Varro as a character in the *Academica*, he argues: "Nothing could have been more appropriate for expounding the system of philosophy in which he seems to be especially interested, and for introducing a part which prevents me from seeming to give my own cause the superiority." (*Ad Att. III*, 13, 19). These words suggest that, while wanting to detach himself from the views of Antiochus, Cicero did not wish to refute Varro too forcefully in this book. Most likely, he differed from Varro only in his exposition of Antiochus's epistemological doctrine but not in his characterization of the Academy's historical development (Schofield 2009, p. 81).

Second, Cicero repeatedly supported the same view in other works. *De Oratore*, for instance, is explicit in this respect. Cicero laments there the split of "brain and tongue" – that is, of philosophy and rhetoric – effected first by Socrates and later by the Socratic schools, particularly, the Stoa:

> For, as I said before, the older masters down to Socrates used to combine with their theory of rhetoric the whole study and science of everything that concerns morals and conduct, and ethics and politics; it was subsequently, as I have explained, that the two groups of scholars were separated from one another by Socrates and then, similarly, by all the Socratic schools, and the philosophers looked down on eloquence and the orators on wisdom (*DO* III, 72–73).

For Cicero, a more natural approach was the original sophistic teaching of philosophy and rhetoric as one and the same art. After the Socratic split of "brain and tongue," he continues, the Academy and the Lyceum sought a reunification of the disciplines and returned to their conjoined teaching. Such reunification, Cicero believed, would take place in the ideal figure of the "perfect orator;" an ideal incarnated in "either the modern dialectic of Carneades or the earlier method of Aristotle." (*DO* III, 71–80). Therefore, the view that the Lyceum and the Academy shared a unified system of rhetoric and dialectic is indeed Cicero's and supports his conviction that the original link between the two traditions should be restored in the figure of the ideal orator.

But the coincidences between the skeptical and the rhetorical traditions, the Academy and the Lyceum, are not only historical or methodological. Philosophical commonalities between the teachings of Academic skepticism and Aristotelian rhetoric further suggest their mutual influence and development. To begin with, we have Carneades's choice of the "persuasive" (*to pithanon*) as a criterion to assess impressions and judgments. This decision places rhetorical invention at the center of the skeptical method.[15] As Couissin affirms:

> Their [the Academic's] dialectic leads to the persuasive, neither more nor less than Aristotle's does; basically, indeed, it is nothing more than rhetoric, whose object is the *pithanon*. [...] The *pithanon* is his [Carneades'] end, in the sense that he is setting out to convince his audience now of this thesis, now of the contrary thesis – not to prove one or the other of these theses but to win approval for it. [. . .] Everything is inapprehensible, but everything is, in a sense, persuasive (Couissin 1983, p. 50).[16]

The criterion of the persuasive impression seems to come directly from the heart of the rhetorical tradition. This suggests that the New Academy took Aristotle's view that rhetoric naturally complements dialectics seriously, and chose to unite skeptical dialectics with rhetorical argumentation to examine the plausibility of a given philosophical thesis.[17] Although the two traditions seem to have had different means and ends – dialectics provided arguments pro and against a view to refute it or show its inherent contradiction, while rhetoric employed long discourses containing many arguments in support of a specific thesis to persuade the audience – they complement each other by joining the dialectical effort of examining both sides of a

[15] Cicero defines invention as, "[. . .] the action of thinking true and plausible things, which make a cause probable" (*Inv.* I, 7).

[16] Brittain also highlights the rhetorical origin of Carneades's criterion: "[. . .] Philostratus, in *The Lives of the Sophists*, considers Carneades to have been among the group of philosophical precursors who were termed 'sophists' for their eloquence [...] It is likely that Carneades himself was aware of the parallel – it was presumably one reason for his choice of the *pithanon* as his criterion for life. It is at any rate not inconsistent with his understanding of the (scant) success of philosophy that he should consider it to be merely a special form of rhetoric (or persuasion)" (Brittain 2001, p. 324).

[17] Also, the type of dialogue that Cicero develops seems to come directly from Aristotle and combine dialectical reasoning with the longer and richer rhetorical exposition (Schofield 2009, p. 70).

question with the rhetorical one of determining the arguments that are most persuasive. And, indeed, I believe they were combined in the New Academy to obtain the skepticism of Carneades and some of his followers, up to Cicero.

To better explain this claim, allow me to elaborate on both discursive practices by using a metaphor. Skeptical dialectic is like peeling an onion. Each layer of the onion is a belief or opinion. Taken together, all the opinions – the onion as a whole – constitute a corpus of belief, a doctrine held as true. The skeptics peel away each layer of the onion until nothing remains. They refute each claim by demonstrating either that it is intrinsically contradictory (as Socrates did) or that an opposed one is equally valid (in the style of Pyrrhonism). As a result, they arrive, as it were, at the empty core of the onion, and suspend judgment.

Rhetorical activity, on the other hand, is like the formation of an onion. A particularly weak or uncertain claim is strengthened by inventing persuasive arguments in its favor. Each layer of the onion is one of these arguments. Just as an onion grows bigger and stronger with additional layers, so too does an opinion strengthen as the number of persuasive arguments in its favor increases. And, just as an onion can grow strong and have a pleasant flavor without a solid core, so too can a claim be strengthened without having certainty as a founding principle. The New Academy combined the two procedures—on the one hand, they refuted dialectically; on the other, they built rhetorically. Or, to put it in terms of our metaphor, the New Academy peeled away the layers of the onion to reveal the fundamental insufficiency of the doctrines examined, while simultaneously constructing new beliefs through a process of choosing the most persuasive ones and "approving" them for practical purposes. The result, of course, was not a theoretical system such as the one represented by the original onion – in other words, it was not strictly speaking a doctrine – but it neither was a disjointed collection of arguments. On the contrary, it was a new set of articulated beliefs with enough plausibility or persuasive force to direct individual or social action.

Let us now take the metaphysical discussions explored in the first two sections of this chapter as examples of this kind of argumentation. First, in *DND*, Cotta presents arguments against the Garden's and the Stoa's theological doctrines to show their weaknesses. However, at the end of the dialogue, Cicero, as a spectator, pronounces a verdict—that the Epicurean case is easily undermined by the skeptical objections, while the Stoic doctrine better withstands the skeptical charge give that it is more solidly constructed and has a practical utility. Thus, he recommends approving the latter, albeit non-dogmatically and only to orient action. Second, in the first book of the *Disputations*, Cicero offers arguments in favor of both the Platonic thesis on the immortality of the soul and the Epicurean one about its mortality. Despite the two dialogues' argumentative differences, the same outcome can be observed. *DND* does not end with a recommendation to suspend judgment, which would have been expected, had Cicero employed skeptical dialectics without rhetorical elements. In this case, the skeptical reaching of the empty core of the onion would have meant the embracement of a broad-scope *epochê*. But this exercise was rather complemented by the determination of the persuasiveness of the claims, and the non-dogmatic and acceptance of, as it were, a less robust onion, stripped of rancid or deteriorated layers. By the same token, nothing in the *Disputations* points to a typical dogmatic reasoning in favor of the thesis of death as a good or a non-evil. Cicero's constructive strategy does not have a solid grounding – principles taken as true – but rather moves freely from one argument to the other, selecting the ones that better suit his practical purpose of alleviating the fear of death. The two dialogues, thus, follow the New Academy's rule, according to which the result of examining impressions and their derived judgments must always be the non-dogmatic adoption of highly persuasive opinions for practical purposes.

In conclusion, although much more is still to be explored about the confluence of skepticism and rhetoric in the Platonic Academy, I believe it is safe to claim that the Lyceum's rhetorical teachings inherited by the Academy helped Carneades develop his moderate kind of skepticism. Not only is the criterion of "the persuasive" taken from this tradition, but his proposed method of argumentation points to a reunification of dialectical refutation and rhetorical persuasion, in the fashion Cicero envisaged in *De Oratore*. Had Carneades and his followers up to Cicero not been trained in rhetoric and not considered dialectic and rhetoric as counterparts, most probably they would not have produced the skeptical innovation for which the New Academy is renowned.

With this general account of how Cicero applied the method of the Academy to metaphysical questions in mind, we should now move to the Pyrrhonian treatment of these matters, in order to appreciate how the latter school refuted the Hellenistic theses on the existence and nature of the gods and embraced what I would like to call a "pious agnosticism."

3.4 Pyrrhonism's Pious Agnosticism

Just as Cicero did in *De Natura Deorum*, Sextus Empiricus begins his discussion about the nature of the gods in the *Outlines of Pyrrhonism* by declaring that his skeptical reasoning only seeks to undermine the arrogance of the dogmatists, not to endanger religious practice. He states:

> Since, then, the majority has declared that God is a most efficient Cause, let us begin by enquiring about God, first premising that although, following the ordinary view, we affirm undogmatically that the gods exist and reverence the gods and ascribe to them foreknowledge, yet as against the rashness of the Dogmatists we argue as follows (*Pyr.* III, 2).

The Pyrronists, Sextus recalls, accept undogmatically both that the gods exist and are provident, and act piously in their everyday lives. At first glance, then, it appears as if both skeptical schools, The Academy and Pyrrhonism, had the same stance about the gods—neither of them completely denied their existence and providential nature, and neither recommended discontinuing religious practices.[18] Nonetheless, there are essential differences in their approaches and results that we should bring to the fore.

[18] Some of the topics used by Sextus in his discussion have a Carneadean origin (*M IX*. I, 182–190).

First, Sextus's most synthetic theological argumentation in the *Outlines* refutes various Hellenistic doctrines by showing their disagreement rather than presenting specific arguments against each one. To begin with, Sextus affirms that the preconception of the gods contains no evidence whatsoever of their essence, form, and situation in space and time:

> Since [...] some of the Dogmatists assert that he [God] is corporeal, others that he is incorporeal, and some that he has human form, others not, and some that he exists in space, others not; and of those who assert that he is in space some put him inside the world, others outside; how shall we be able to reach a conception of God when we have no agreement about his substance or his form or his place of abode? (*Pyr.* III, 2–3).

As we can appreciate from this passage, Sextus generally took the schools' dissensus about the attributes of the Gods to show that we cannot know their essence. The Stoics, for example, contended that the gods are not anthropomorphic and that they either exist in the world or are the world itself. In contrast, the Epicureans claimed that the gods are similar to humans and live in a world outside the one humans inhabit. For Sextus, this opposition suggested that human beings have different preconceptions of the gods, and, hence, these preconceptions cannot be used as a criterion of truth.

In the most crucial point of the dispute – divine providence – Sextus also placed the Stoics' and Epicureans' doctrines in opposition to undermine both:

> And apart from this, let them tell us what a "blessed" thing is – whether it is that which energizes according to virtue and foreknows what is subject to itself, or that which is void of energy and neither performs any work itself nor provides work for another. For indeed, about this they also disagree interminably and thus render "the blessed" something we cannot conceive, and therefore God also (*Pyr.* III, 5).

The Epicureans insisted that since the happiness of the gods entails their absolute imperturbability, they are *not* concerned with human affairs, while the Stoic thesis held that it is their providence, particularly their caring for the world, that makes them happy. Given this contradiction, Sextus concluded, neither imperturbability nor providence could be an attribute of their nature.

Sextus also refuted the Stoic views on divine providence using a version of the Epicurean argument on the existence of evil. The argument is long and convoluted, yet magnificently crafted:

> He who affirms that God exists either declares that he has, or that he has not, forethought for the things in the universe, and on the former case that such forethought is for all things or some things. But if he had forethoughts for all, there would have been nothing bad and no badness in the world; yet all things, they say, are full of badness; hence it shall not be said that God forethinks all things. If, again, he forethinks some, why does he forethink these things and not those? For either he has both the will and the power to forethink all things, or else he has the will but not the power, or the power but not the will, or neither the will nor the power. But if he had had both the will and the power, he would have had forethought for all things; but for the reasons stated above he does not forethink all; therefore he has not both the will and the power to forethink all. And if he has the will but not the power, he is less strong than the cause which renders him unable to forethink what he does not forethink; but it is contrary to our notion of God that he should be weaker than anything. And if, again, he has the power but not the will to have forethought for all, he will be held to be malignant; while if he has neither the will nor the power, he is both malignant and weak – an impious thing to say about God (*Pyr.* III, 9–11).

In brief, if God is provident (has forethought), he has created all things in the world or some of them with a purpose. If the former is the case, then there should be no evil. But there is evil, so not all things are created with a purpose. Then, the latter must be the case. But since providence entails that God has the will and the power to create the world as he thinks fit, the fact that God has made some things without a purpose should lead us to conclude that God either has not the will or power to do this. But if he has the will and not the power, he is weak; while if he has the power and not the will, he is evil. And if he does not have both, then he is not God. In all of these cases, he is not provident and has not arranged the work as he sees fit.

The argument is undoubtedly powerful, but Sextus does not stop there. Immediately after this Epicurean attack on the Stoics, he turns around and refutes the Epicureans with the Stoic argument that if the gods did not care for the world, we would not know anything about their nature, because they would not have manifested themselves to us in dreams or any other sort of vision, and, therefore, we would not have a preconception of the gods (*Pyr.* III, 11). As we see, both the Epicureans and the Stoics give Sextus powerful weapons to refute their respective doctrines.

Now, in *Against the Physicists* (*M IX*), Sextus offers a longer and more developed refutation of the Hellenistic theological doctrines and goes beyond confronting the different schools' views to producing arguments of his own. He begins by arranging the arguments about the origin of the preconception of the gods into three groups: (i). these preconceptions came from ancient legislators who spread them among their peoples; (ii). they came from very virtuous ancestors who were regarded as gods by their descendants; or (iii). they came from visions and dreams of much bigger human beings than us, which, as a result of this difference in size, were called "gods." Against the first explanation, he argues that ancient legislators could not have gathered all the primitive peoples together, so a preconception with such an origin cannot be universal (*M IX.*, I, 33). Then, he claims that the second possibility neither provides a reasonable explanation because before the descendants attributed divinity to their forbearers, they must already have had this very notion. Besides this, the argument does not offer any facts supporting the idea that the worship of these gods, and not of others, survived through generations (*M IX.*, I, 34–35). And he refutes the final argument by noting the obvious point that having dreams or visions of giants is not the same as having visions of gods since, besides a great size, other attributes should be contained in the preconception of the gods for these entities to actually be gods and not merely giants (*M IX.* I, 43–44). Based on the above, Sextus concludes that there is no evidence whatsoever of the origin of the preconception of the gods.

The second part of the discussion addresses the existence of the gods (*M IX.* I, 50–58). Here, Sextus uses wide-ranging arguments, from the pre-Socratics' to the Stoics'. Some arguments *in favor* of the existence of the gods include the Epicurean claim that their preconception is universal, the Aristotelean doctrine of a "prime mover," the Stoic argument from the design of the universe, and the also Stoic argument about the animal nature of the world (*M IX.* I, 60–136). To attack these arguments, Sextus argues, first, that just as having a notion of Hades taken from myths and poetry does not prove that it exists, having a preconception of the gods does not guarantee their existence (*M IX*. I, 66–74). Second, against the Stoic thesis that the world is a living being or an animal, he uses the Carneadian argument that if the world is an animal, God must also be an animal. But, if this is the case, God can perceive and, hence, some things cause him pleasure and pain, but if things cause him pain, they must harm him and, thus, he is a mortal animal. However, a mortal animal is not a god, and, consequently, God does not exist (*M IX*. I, 142–146). Finally, Sextus also questions the argument about the moral perfection of God, maintaining that if God is morally perfect, he possesses each and every virtue, and if he possesses each and every virtue, then he has the virtues of moderation and resistance to adversity. But, since one only develops these virtues if one faces things difficult to withstand, then, some things are difficult for God to withstand. And if there are such things, God is susceptible to change and harm, and is, therefore, mortal. But God cannot be mortal; hence he does not exist (*M IX*. I, 152–157). There are also other variations on this *sorites*, which deal with diverse attributes of God—if God is rational, he learns; but if he learns, he is susceptible to change, etc. Or, if God is virtuous, he deliberates; but if he deliberates, there are things he considers prejudicial; and if there are things he considers prejudicial, he is, again, susceptible to change, etc. Sextus rehearses many of these (and similar arguments) to conclude that God does not exist.

When Sextus finishes recounting the arguments against the existence of the gods, he lists Carneades's *sorites*, which consists of a number of variations on the following argument:

> [. . .] if Zeus is a god, Poseidon too, being his brother, will be a god. But if Poseidon is a god, Achelous too will be a god; and if Achelous, the Nile too; and if the Nile, every river; and if every river, streams too would be gods; and if streams, mountain run-offs. But streams are not; therefore, neither is Zeus a god. But if there were gods, Zeus would be a god. Therefore there are not gods (*M IX*. I, 182–190).

The argument claims that if Zeus is a god, then something similar to Zeus, namely, Poseidon, must also be a god; and if this is the case, something similar to Poseidon, that is, the Achelous, must also be a god, and so on, until something that cannot be a god, for example, a creek, is posited. This argument is a sort of *reductio ad absurdum* based on a chain of accepted similarities. At the end of the section, Sextus claims that, given the equal validity of the arguments for and against the existence of the gods, the skeptical attitude of suspending judgment on the matter is the most suitable one (*M IX*. I, 191).

Now, a question arises about the practical bearings of this skeptical refutation. If Sextus openly encourages suspension of judgment on the existence of the gods – that is, agnosticism – how can he continue to defend his initial position, namely that the Pyrrhonian skeptic is pious and holds, albeit undogmatically, the existence and providence of the gods? The problem we face here is yet another version of the famous *apraxia* objection. If the Pyrrhonists do not believe that the gods exist and are provident, it is difficult to see how they can act piously. But, as we have seen in Chap. 1, Sextus did offer a response to this objection, according to which, by "adhering to the appearances," the Pyrrhonists "live in accordance with the normal rules of life, undogmatically, seeing that we cannot remain wholly inactive" and they follow, among other observances, the "tradition of customs and laws, whereby we regard piety in the conduct of life as good, but impiety as evil" (*Pyr.* I, 24).

Several commentators have inquired about the type of piety a Pyrrhonian would profess. Thorsrud examines this problem in the context of the common-sense objection that participating in religious rituals without believing that God exists and is provident would be insincere: "Why would the skeptic say that the gods exist if he does not believe they exist? And why would he bother to participate in religious ceremonies? One answer immediately suggests itself: the skeptic is insincere in his statements about the gods and in his pious observances" (Thorsrud 2011, p. 91). Although the objection strikes us as obvious, it is misleading. As Thorsrud argues, skeptical psychology is somewhat more complex than this standard view seems to consider. His thesis is that, while the Pyrrhonists do not hold any kind of belief in God on which to justify their religious practices rationally, they do have the kind of "affective states" required to undertake them sincerely. Just as, following appearances, they eat when hungry or drink when thirsty, and they do not if they do not have these inclinations, so too, when feeling piety or devotion, they perform religious practices, and when not, they avoid them. But, given their upbringing in a specific community, with particular laws, customs, and attitudes toward religion, they tend to have the kind of affective states that are necessary to engage in the community's religious rituals. Thus, they do feel devotion and piety, and engage in religious practices sincerely:[19]

> Insofar as the skeptic's evaluative impressions are the product of habituation or enculturation, they will not be contrary to the established traditions of his community. So, we may suppose that, for the most part, when the skeptic abides by his city's laws and customs, it is because it seems good to do so (Thorsrud 2011, p. 106).

In other words, education and habituation spontaneously incline the Pyrrhonists towards harboring feelings of religious devotion. Consequently, the only difference between ordinary believers and Pyrrhonists is that the latter do not associate their affective dispositions with specific beliefs, while the former do and find it a necessary component of their faith. In other words, the Pyrrhonists' piety has no rational grounds—it is entirely founded on subjective impressions and affective dispositions. Unlike the dogmatists, who base their religious practice on the belief that God exists and is provident, and unlike the Academics, who approve these judgments as sufficiently persuasive and practically useful but still doubt them theoretically, the Pyrrhonists act piously only because, having been educated in the religious customs of their community, possess the appropriate affective dispositions to do so.

Let us remember that, for Sextus, this attitude is the most conducive to *ataraxia* or peace of mind. Suspension of judgment alleviates the distress associated with any sort of conviction, including religious ones. Unlike the dogmatists, the Pyrrhonists

---

[19] Of course, if they were born and raised in an secular society, they would not have feelings of piety. But, then, this would not be a problem at all.

do not agonize over issues such as whether the gods exist and are provident, whether they are kind-hearted and almighty, whether their goodness and power can be reconciled with the existence of evil, etc. Along with the constant frustration for not being able to attain truth in these matters, these questions are motives of permanent anguish for the dogmatist. But the Pyrrhonist does not feel this distress. As Thorsrud claims:

> It may be the case that the gods appear to the skeptic at one time provident and at another time malicious. But again, that would pose no particular problem since he does not experience these impressions in competition with one another. Having no such contradictions to resolve is in fact an essential component of the skeptics' tranquility (Thorsrud 2011, p. 109).

We can affirm that the Pyrrhonists are pious by convention or, as I said in the previous chapter, by exercising a sort of non-dogmatic practical heteronomy. They follow their affective dispositions towards religion and the traditional practices of their community, without holding the corresponding beliefs and without being insincere or hypocritical, given that they possess the appropriate impulses and affective states to perform the actions of any ordinary believer.

By contrast, the skeptic of the New Academy is pious on the grounds of a non-dogmatic approval of particular beliefs for practical purposes. These skeptics have tested their belief in God from every angle and conclude that it is persuasive enough to orient their practice. Their religious behavior is, in most respects, the same as the dogmatists', but their attitude towards religion is similar to the Pyrrhonists' in that it is not zealous but reserved and open to criticism. However, since their aim is not to attain *ataraxia* but practical autonomy, unlike the Pyrrhonists, they adopt a particular view on the matter, which is, nonetheless, cautious, and never definitive.

3.5 Literary Excursus: Caesar's Skepticism in "*The Ides of March*"

To illustrate the contrast between the theological positions of the New Academy and Pyrrhonism, I find it helpful to finish this chapter by discussing a section of the novel "*The Ides of March*," by the U.S. playwright Thornton Wilder (1961). The work recreates the diary and correspondence of the emperor Julius Caesar during the last years of his life. In the entry entitled "On a possible abolition of all religious observances," Caesar writes a letter to his friend Lucius Mamilius Turrinus about his recent thoughts on the state religion and its rituals. He relates how, disappointed by the inefficiency of the priestly bureaucracy, he decided to pen an edict revoking the state religion:

> Last night I sat down and wrote the edict abolishing the College of Augurs and declared that henceforward no days were to be regarded as unlucky. [...] I disbanded the College of Vestal Virgins; I married the daughters of our first houses, and they gave sons and daughters to Rome. I closed the doors of the temples, of all the temples except those of Jupiter. I tumbled the gods back into the gulf of ignorance and fear from which they came and into that treacherous half-world where the fancy invents consolatory lies. And finally, the moment

came when I pushed aside what I had done and started to begin again with the announcement that Jupiter himself had never existed; that man was alone in a world in which no voices were heard than his own, a world neither friendly nor unfriendly save as he made it so. And having reread what I wrote, I destroyed it (Wilder 1961, p. 39).

Caesar claimed to have had substantive reasons for abolishing the Roman religion. He firmly believed that the gods were created only by human ignorance and fear, and that the assertion of their existence was nothing but a consolatory lie. In his view, the world was exclusively human and neither good nor bad; it was just the way humans decided it to be. With his decree, he would dismantle all the institutions of the state religion—shut down the temples, prohibit prophetic practices, dismiss all ecclesial authorities. As we learn from the diary entry, his atheistic conviction was not a sudden whim but had a history of philosophical reflection that went back to his youth (Wilder 1961, pp. 39–40). However, when put to the test of abolishing religious institutions, he began to doubt and decided to shred the edict:

In myself, I was not certain that I was certain. [...] In drawing up the edict last night and in foreseeing the consequences that would ensue, I was driven to examine myself most strictly. [...] I must be certain that in no corner of my being, there lingers the recognition that there is the possibility of a mind in and behind the universe which influences our minds and shapes our actions. If I acknowledge the possibility of one such mystery, all the other mysteries come flooding back: there are the Gods who have taught us what is excellent and who are watching us; there are our souls which are infused in us at birth and which outlive our death; there are the rewards and punishments which furnish a meaning to our slightest action (Wilder 1961, pp. 39–41).

Caesar's doubts sprang from the recognition that his belief in the inexistence of the gods and the mortality of the soul could still be erroneous. After all, it could be the case that the gods existed and cared about the world, and that the human soul was immortal. And this mere possibility made it impossible for him to act as planned.

At this point, it may be worth asking ourselves if the attitude of this hesitant Caesar is in keeping with Pyrrhonism or with Academic skepticism. Where should we place his skeptical hesitation? Caesar himself provided a clue. In the same entry, he made clear that he destroyed the edict, not because he considered that the state religion served a practical purpose, as Cicero did, but rather because he felt an internal drive not to do it:

I destroyed it not for Cicero's reasons – not because the absence of a state religion would drive superstition into clandestine forms and still baser practices (that is already taking place); not because so sweeping a measure would disrupt the social order and leave the people in despair and dismay like sheep in a snowstorm. [...] No, it was not the possible repercussions of the move which arrested my hand and will; it was something in and of myself (Wilder 1961, p. 39).

As we can see in this passage, Caesar distanced himself from Cicero's views on the social utility of religion. He did not consider the state religion useful, but instead a cheap and unhealthy superstition. Therefore, Cicero's approval of religious beliefs for practical motives was not a reasonable answer to Caesar.

Did he act as a Pyrrhonian, then? To my mind, this is the case. His honest recognition that there was the slightest chance that the gods existed and the soul was immortal was enough for him to not act against religion. As a good Pyrrhonian who opposes thesis and antithesis of equal validity and suspends judgment on the matter, Caesar abandoned his dogmatic atheism and embraced agnosticism as soon as he acknowledged the possibility of the contrary thesis. In practical matters, he also followed Sextus's advice—he did not enact his initial impulse to revoke religious practices but instead decided to allow religion to continue according to tradition.

This literary example shows us how skepticism regarding metaphysical matters can function in a double direction—it can attack both religious and irreligious dogmatism.[20] Caesar's initial atheistic position is dogmatic in the technical sense of the term. For a long time, the emperor held his beliefs in the non-existence of the gods and the mortality of the soul as true and certain. Wilder even makes him affirm that he "took a vow never to [. . .] doubt" his atheism (Wilder 1961, p. 40). The force of his conviction was, therefore, equal to that of the theistic dogmatist who holds the belief in the existence of God and the immortality of the soul with alleged certainty. And, both the atheist and the theist may find themselves, like Caesar, overwhelmed by the most trifling possibility that the opposite view is true. The pressing question, of course, is how to act given these doubts. For the ancient schools of skepticism, there were two options: either to suspend judgment and follow, without believing, the precepts of tradition, or to consider that one of the opposed beliefs may, after all, be held as the most persuasive and the most useful one, and adopt it to act while maintaining theoretical doubts on the matter.

The example also helps illustrate how arguments about the utility of religion can be opposed by arguments of equal weight about its harmfulness. Thus, whoever, in the final balance, adopts atheism because it is a persuasive enough position and more useful for society than theism would also be following the orientations of the New Academy, provided that she does it non-dogmatically. In other words, the method of the New Academy does not commit us to accept Cicero's particular view about religion. Insofar as Academic skepticism only provides a set of criteria for evaluating beliefs and deciding which ones to approve, the result may be either the provisional adoption of religious beliefs or of non-religious ones. And as we will see in the following chapters, Hume's and Kant's adoption of Academic skepticism gave way to divergent positions regarding religion. Hume's position was that the arguments of natural theology were sufficiently persuasive or probable to support the belief in the existence of God but not strong enough to support many of his alleged attributes, particularly providence. Besides this, he believed that religion's harmfulness was amply proved by history, and hence, theism should not be recommended practically. As a result, the only theological belief worth adopting for him was a *non-dogmatic, anti-religious deism*.

[20] I do not wish to leave the reader with the impression that I am using this example to defend skepticism as a means to refuting or opposing atheism. As an agnostic myself, I deplore the use of skepticism to advance any kind of religious dogmatism.

In contrast to this position, in his Transcendental Dialectic, Kant argued that since both the thesis of theism and the same logical validity, neither amounted to knowledge. However, he thought that the grounds of this opposition should be revealed and some religious beliefs could be maintained for practical purposes—that is, he approved a sort of *non-dogmatic practical or moral theism*. Thus, the New Academy's method may produce diverse results precisely because it is a procedure for examining beliefs and not a *doctrine* or a set of justified beliefs. Its usefulness rests on the fact that it supports an autonomous but non-dogmatic way of thinking, particularly in metaphysical matters where no certainty is ever to be attained.

* * *

To summarize this chapter, it is worth first mentioning how the Hellenistic schools examined the preconception (*prolêpisis*) of the Gods to determine their existence and attributes or nature. Cicero contended in *De Natura Deorum* that the Epicurean arguments in both respects were the weakest, while the Stoic ones proved persuasive and practically advisable, given that they supported the state religion's institutions and practices. However, Cicero neither recommended adopting Stoicism dogmatically nor sustained a sort of eclecticism. Instead, his final verdict that the Stoic doctrine was the closest to the truth derives from his Academic commitments.

In the first book of the *Tusculan Disputations*, Cicero showed a more constructive than refutative aspect of Academic skepticism, as he argued in favor of both the doctrines of Plato and the Epicureans on the immortality of the soul. Even if they express opposite theses – the former claims that the soul is immortal and death is something good, while the latter argues that the soul is mortal, and death is neither good nor evil – Cicero considered both to be persuasive enough and practically useful to deter our fear of death. The complementarity of both approaches suggests that the Aristotelian view of rhetoric as a counterpart of dialectic helped shape the method of the New Academy. Finally, the Pyrrhonian treatment of theological matters shows very substantial differences from the Academic approach. More importantly, the sort of piety that Sextus advocated was based on affective dispositions acquired by education and not from a cautious but autonomous adoption of religious beliefs. A good illustration of this position is Caesar's decision in Wilder's *Ides of March* to allow the continuation of Rome's religious practices and institutions while theoretically embracing agnosticism.

Bibliography

Brittain, Charles. 2001. *Philo of Larissa: The Last of the Academic Sceptics*. Oxford: Oxford University Press.

Brochard, Victor. 2002. *Les sceptiques grecs*. Paris: Librairie Générale Française.

Cicero, Marcus Tullius. 1950. *Tusculanae Disputationes* (*Tusculan Disputations*). Trans. J.E. King. Cambridge/London: Harvard University Press.

———. 1962. *Orator* (*Orator*) Trans. H.M. Hubbell. Cambridge/London: Harvard University Press.

———. 1968. *De Inventione* (*On Invention*) Trans. H.M. Hubbell. Cambridge/London: Harvard University Press.
———. 1999. *Ad Atticus* (*Letters to Atticus*). Trans. D.R. Shackleton Bailey. Cambridge/London: Harvard University Press.
———. 2000. *De Natura Deorum* (*On the Nature of the Gods*). Trans. H. Rackham. Cambridge/London: Harvard University Press.
———. 2001a. *De Fato* (*On Fate*). Trans. H. Rackham. Cambridge/London: Harvard University.
———. 2001b. *De Finibus* (*On Moral Ends*). Trans. Julia Annas. Cambridge: Cambridge University Press.
———. 2001c. *De Oratore* (*On the Orator*) Books I–II. Trans. E.W. Sutton and H. Rackham. Cambridge/London: Harvard University Press.
———. 2006a. *Academica* (*On Academic Scepticism*). Trans. Charles Brittain. Indianapolis/Cambridge: Hackett Publishing Company.
———. 2006b. *De Divinatione* (*On Divination*). Trans. D. Wardle. Oxford/New York: Oxford University Press.
Couissin, Pierre. 1983. The Stoicism of the New Academy. In *The Skeptical Tradition*, ed. Myles Burnyeat. Trans. Jonathan Barnes and Myles Burnyeat, 31–63. Berkeley: University of California Press.
Diogenes Laërtius 1972. *Lives of Eminent Philosophers, Vol I.* Trans. R.D. Hicks, Cambridge/London: Harvard University Press.
Epicurus. 2012. Letter to Menoeceus. In *The Art of Happiness*. Trans. George K. Strodache, 155–173. London: Penguin Books.
Eusebius of Caesarea. 2003. *Praeparatio Evangelica* (*Preparation for the Gospel*), Books XIV–XV. Trans. E.H. Gifford. UKIpswich: http://www.tertullian.org/fathers/eusebius_pe_01_book1.htm
Knuuttila, Simo, and Juha Sihvola. 2000. Ancient Scepticism and Philosophy of Religion. *Acta Philosophica Fennica* 66: 125–144.
Lactantius. 1965. The Wrath of God. In *The Minor Works*. Trans. M.F. McDonald, 59–116. Washington, DC: The Catholic University of America Press.
Lucretius. 1880. *On the nature of things*. Trans. John Selby Watson. London: G. Bell.
Michel, Alain. 2003. *Les rapports de la rhétorique et de la philosophie dans l'oeuvre de Cicéron: recherches sur les fondements philosophiques de l'art de persuader*. Leuven: Peeters.
Nagel, Thomas. 1979. *Mortal Questions*. Cambridge: Cambridge University Press.
Pease, Arthur Stanley. 1913. The Conclusion of Cicero's De Natura Deorum. *Transactions and Proceedings of the American Philological Association* 44: 25–37.
Plato. 1997. Phaedo. In *Plato, Complete Works*, ed. John M. Cooper. Trans. G.M.A. Grube, 49–100. Indianapolis/Cambridge: Hackett Publishing Company.
Powell, Jonathan G.F. 2013. The Embassy of the Three Philosophers to Rome in 155 BC. In *Hellenistic Oratory: Continuity and Change*, ed. C. Kremmydas and K. Tempest, 219–247. Oxford: Oxford University Press.
Schofield, Malcolm. 1980. Preconception, Argument, and God. In *Doubt and Dogmatism: Studies in Hellenistic Epistemology*, ed. Malcolm Schofield, Myles Burnyeat, and Jonathan Barnes, 283–308. New York: Oxford University Press.
———. 2009. Ciceronian Dialogue. In *The End of Dialogue in Antiquity*, ed. Simon Goldhill, 63–84. Cambridge/New York: Cambridge University Press.
Sextus Empiricus. 2000. *Pyrrhoniae Hypotyposes* (*Outlines of Scepticism*), eds. Julia Annas and Jonathan Barnes. Cambridge: Cambridge University Press.
———. 2006. *Adversus Mathematicos IX* (*Against the Physicists; Against the Ethicists*). Trans. Robert G. Bury. Cambridge: Harvard University Press.
Thorsrud, Harald. 2009. *Ancient Scepticism*. Berkeley/Los Angeles: University of California Press.
———. 2011. Sextus Empiricus on Skeptical Piety. In *New Essays on Ancient Pyrrhonism*, ed. D.E. Machuca, 91–111. Leiden: Brill.
Tugendhat, Ernst. 2002. *Problemas*. Trans. Román Cuartango. Barcelona: Gedisa.
Wilder, Thornton. 1961. *The Ides of March*. London: Penguin Books.

Chapter 4
Hume's Academic Skepticism: A Normative Theory of Belief

Abstract The close link between Hume and the skepticism of the New Academy was contested for a long time by an interpretative tradition that saw Hume as a modern and accomplished Pyrrhonist. Even though Hume acknowledged his debt to Academic skepticism and was very critical of Pyrrhonism, several commentators preferred to interpret Hume's thought within the framework of Pyrrhonian skepticism. In this chapter, I first explore the general interpretation of Hume as a Pyrrhonist, detailing the account that exerted the most significant influence, that is, Popkin's article of 1951 "David Hume: His Pyrrhonism and his Critique of Pyrrhonism." Then, I describe Hume's critique of Pyrrhonian skepticism in the *Treatise on Human Nature* (*THN*). From there, I discuss Hume's praise of Academic skepticism in *An Enquiry Concerning Human Understanding* (*EHU*) and why he believed that this kind of "mitigated skepticism" helps overcome the negative consequences of Pyrrhonism. The second half of the chapter describes the normative aspects of Hume's theory of belief in connection with the Carneadian set of criteria to determine the degrees of persuasiveness of an impression. To this effect, I focus mainly on Hume's typology of beliefs into "knowledge," "proofs," and "probabilities," and on his "rules for judging of causal inferences." I defend the thesis that these aspects of his theory are largely Academic. Particularly, I show how the rules for causal inferences are consistently Academic, despite the prevalent interpretation that they have Newtonian roots, by arguing that Hume's allusions to the "sage" in these passages of the *Treatise* clearly indicate this connection.

The close link between Hume and the skepticism of the New Academy was contested for a long time by an interpretative tradition that saw Hume as a modern and accomplished Pyrrhonist. Even though Hume acknowledged his debt to Academic skepticism and was very critical of Pyrrhonism, several commentators preferred to interpret Hume's thought within the framework of Pyrrhonian skepticism

C. González Quintero, *Academic Skepticism in Hume and Kant*, Synthese Library 449, https://doi.org/10.1007/978-3-030-89750-5_4

(Aguirre 2010, pp. 13–40; Popkin 1951; Fosl 2020).[1] The origin of this misunderstanding is a crucial article for the history of skepticism—namely Richard Popkin's "David Hume: His Pyrrhonism and his Critique of Pyrrhonism," published in 1951 (Popkin 1951, pp. 385–407). Popkin claimed that notwithstanding his strong critique of this tradition, Hume's skepticism constitutes the most consistent form of modern Pyrrhonism.[2] His argument rested on two premises: first, that Hume's portrayal of Pyrrhonism as an extreme form of skepticism was inaccurate because Pyrrhonism does not recommend a complete suspension of judgment but allows for the adoption of beliefs that are necessary to act; and second, that since the latter view is very similar to Hume's naturalism, his skepticism should be considered an authentic form of Pyrrhonism (Fosl 1994).

Briefly, my assessment of this interpretation is that it accurately portrays Hume's mistakes in his description of Pyrrhonism, but does not give us a real assessment of Hume's kind of skepticism. I think that Popkin and other commentators since him (de Olaso 2018, pp. 79–80; Norton 1982, pp. 266–267; Penelhum 1983, p. 124; Smith 2007, p. 118) rightly claimed that Hume's critique of Pyrrhonism overlooks Sextus's response to the *apraxia* objection. They also accurately indicated a similarity between the Pyrrhonian practical observance of "yielding to appearances" and Hume's naturalistic claim that some beliefs essential for acting are adopted irrationally, almost instinctively. Hence, their arguments contained a grain of truth, which perhaps revealed a Pyrrhonian facet of Hume's thought. However, this grain of truth obscured the fundamental nature of Humean skepticism, which, as he himself claimed, was Academic. Thus, I contend, along with other commentators (Calvo 2012, pp. 388–391; de Olaso 2018, pp. 77–80; Penelhum 1979, pp. 253–278; Stroud 1991, pp. 271–291), that despite the apparent similarities between Pyrrhonism and Humean naturalism, Hume's skepticism is thoroughly Academic. Some of these scholars have convincingly criticized Popkin's argument, while others have merely acknowledged that the most influential Ancient school for Humean skepticism was the New Academy (Fosl 1994, pp. 103–120; Immerwhar 1979; Livingston 1998, pp. 161–168; Norton 1982; Passmore 1968, p. 150; O'Connor 2001;

[1] Interestingly enough, for some scholars, the contrary is the case. Fosl argues, for example, that linking Hume with Academic skepticism is the mainstream interpretation while arguing for a Pyrrhonian influence is a less accepted view: "While Hume is commonly today accepted as some kind of Academic sceptic, the claim advanced here, that Hume is also deeply Pyrrhonian, has remained decidedly more contested" (Fosl 2020, p. 2). I believe that, while many commentators do acknowledge Hume's debt to Cicero and the Academy, when it comes to a detailed analysis of Hume's skepticism the prevalent arguments situate his thought in the Pyrrhonian tradition. Perhaps this is due to a more widespread reading of Sextus's than of Cicero's works among contemporary scholars. The fact remains, in my view, that the "Pyrrhonian Hume" is still a very important interpretative model among Hume scholars.

[2] In Popkin's words: "What I wish to show, as the philosophical significance of Hume's criticisms of Pyrrhonism, is that it is the only 'consistent' version of the original sceptical theory, more consistent than even the formulation in Sextus Empiricus" (Popkin 1951, p. 403).

Olshewsky 1991; Smith 2007, pp. 105–126).[3] Before delving into my own interpretation of this influence, let me briefly survey some of these accounts.

An essential antecedent for recognizing Hume's Academic roots is J. Laird's study at the beginning of the twentieth century (1932). In this work, Laird clearly acknowledges Cicero's influence on Hume's skepticism. He claims:

> In the *Academic Questions*, Hume must have heard much about Philo the skeptic, of Socratic humanism, and the alleged "Socratic doubt." [. . .] That Cicero, speaking as an "opinator," not as a "sapiens," assented perforce from the violence of impressions [. . .] That even Carneades had admitted that man might know nothing and yet be guided by opinion or by probability (Laird 1932, p. 181).[4]

In the second half of the twentieth- and over the course of the twenty-first century, several voices have corroborated this interpretation. Among them, we find D. F. Norton, who recounts all the skeptical traditions identifiable in Hume's works and concludes that Academic skepticism lies at the heart of his theory of belief (Norton 1982, pp. 278–279); E. de Olaso, for whom, Hume's "moderate skepticism corresponds [. . .] to Academic philosophy" (de Olaso 2018, p. 70); D. O'Connor, who affirms that, despite the common interpretation of Hume as Pyrrhonist, "a better ancestral home for Humean skepticism is in Cicero's conception" (O'Connor 2001, p. 13);[5] A. Calvo, who understands Hume's Academic skepticism as the best expression of his "true philosophy" (Calvo 2012, pp. 329–452); P. J. Smith, who finds remarkable similarities between Carneades's and Hume's characterization of belief (Smith 2007, pp. 105–126), and, more recently, R. Fosl, who claims that Hume's thought is "[. . .] in many ways [. . .] the most powerful rendering of the Academic's philosophical achievement ever produced" (Fosl, 2020, p. 39).

---

[3] And still others have argued that both schools deeply influenced Hume (de Olaso 2018, p. 78; Fosl 2020; Smith 2007, pp. 105–126). Robert Fosl argues in favor of the thesis that Hume's form of skepticism is "radical" and "hybrid," insofar as it is composed of both Pyrrhonian and Academic strands (2020, p. 2). However, he seems to lean towards a Pyrrhonian Hume, since he states that, notwithstanding his disavowal of Pyrrhonism, "many dimensions of Hume's thought bear remarkable similarities to [Pyrrhonism's] most important formulations—so much so that it would be neglectful and misleading to describe Humean skepticism *as anything but deeply Pyrrhonian*" (p. 79, my emphasis).

[4] Popkin considers that his argument is supported by Laird's and he cites the following line: "For Hume remained a complete Pyrrhonian regarding all ultimate principles" (Laird 1932, p. 180; Popkin 1951, p. 385). Indeed, taken out of context, it would seem that Laird does link Hume's skepticism with Pyrrhonism. However, this line appears scarcely a page before Laird's explicit recognition of the Ciceronian sources of Humean skepticism. To explain this misunderstanding it is important to take into account the eighteenth- and nineteenth-century tendency to use the term "Pyrrhonian" to describe all kinds of skeptics and "Pyrrhonism" to refer to all varieties of skepticism. This tendency may have somehow survived into the first half of the twentieth century, when Laird was writing. I think that Popkin may have been misguided by Laird's terminological ambiguity.

[5] And he continues stating: "[. . .] The Ciceronian model of skepticism that Hume uses for his own position is a useful reminder that, while indeed a modernist, he also has deep intellectual and cultural roots in the ancient world, especially, ancient Rome" (O'Connor 2001, p. 14).

Building on these and other accounts, in this chapter and the next one, I will expand and deepen the discussion on the influence of Academic skepticism on Hume's theory of belief. In particular, I argue that Hume's Academic skepticism is most recognizable in the *normative* aspects of his theory of belief (sometimes I will refer to them as a normative theory of belief *simpliciter*), which were overlooked for a long time given the prominence of the interpretations that emphasized his naturalism, but have been more amply discussed in the last decades (Calvo 2012, pp. 453–507; de Pierris 2001, pp. 351–383; Falkenstein 1997, pp. 29–72; Hickerson 2013, pp. 1133–1153; Loeb 2002). These normative aspects, often referred to as Humean "probabilism," provide a set of criteria for the rational assessment of beliefs which are closely related to the skepticism of the New Academy (Norton 1982, p. 278; Smith 2007, pp. 105–126; Calvo 2012, pp. 453–507).

However, understanding the influence of Academic skepticism on Hume requires analyzing these connections in a more detailed way than has been done up to now. While some scholars have discussed Hume's normative theory of belief and others have acknowledged Hume's relation to Cicero and the Academy, the accounts that connect the two features of his thought are scarce. Most probably, this is due to the fact that commentators who focus on Hume's normativity usually come from the analytic tradition, while those that stress Hume's debt to the Academy instead do research in the history of philosophy. That professional philosophy is so compartmentalized nowadays does not help construct a more cohesive view of an author's epistemic views and historical influences. I attempt to do that, first, by describing Hume's views on both kinds of skepticism and the Academic underpinnings of his theory of belief, and, second, by closely examining his relation to Cicero and the application of his theory of belief in the *Dialogues concerning Natural Religion* (*DNR*).

To begin this discussion, in this chapter, I explore the general interpretation of Hume as a Pyrrhonist, detailing the account that exerted the most significant influence, that is, Popkin's article of 1951. Then, I describe Hume's critique of Pyrrhonian skepticism in the *Treatise on Human Nature* (*THN*). From there, I discuss Hume's praise of Academic skepticism in *An Enquiry Concerning Human Understanding* (*EHU*) and why he believed that this kind of "mitigated skepticism" helps overcome the negative consequences of Pyrrhonism. The second half of the chapter will describe the normative aspects of Hume's theory of belief in connection with the Carneadian set of criteria to determine the degrees of persuasiveness of an impression. To this effect, I focus mainly on Hume's typology of beliefs into "knowledge," "proofs," and "probabilities," and on his "rules for judging of causal inferences." I defend the thesis that these aspects of his theory are largely Academic. Particularly, I show how the rules for causal inferences are consistently Academic, despite the prevalent interpretation that they have Newtonian roots, by arguing that Hume's allusions to the "sage" in these passages of the *Treatise* clearly indicate this connection.

4.1 The Pyrrhonian Hume: Between Doubt and Dogmatism

Given the great weight that the interpretation of Hume as a Pyrrhonian skeptic has had from the second half of the twentieth century until now, I find it necessary to survey, once again, Popkin's claims on the issue, which inaugurated this line of interpretation. In the fashion of the Ancient skeptics, who began their arguments with a refutation of a central piece of dogmatic philosophy, in this section, I indulge in a sort of dialectical exercise to show how Popkin's authoritative account is flawed. Succeeding at this, I should be able to undermine – like Descartes envisioned to do – the foundations of a common interpretation without having to address each one of the accounts subsequently built on its basis.[6] I must insist, however, that this does not mean that Hume's skepticism does not have any Pyrrhonian features. Indeed, I am absolutely willing to accept a more nuanced claim – namely that it involves *some* Pyrrhonian aspects – since, evidently, Hume not only read Cicero but also had access to Sextus Empiricus's works and was deeply involved with the thought of some modern Pyrrhonists (Fosl 1998, pp. 261–278; Perinetti 2018, pp. 45–74). Nonetheless, I believe that the best characterization of Hume's skepticism is that it is Academic, and I am convinced that there is still much to be said in this respect.

Popkin begins his analysis with Hume's critical description of Pyrrhonian suspension of judgment, which he assesses as inaccurate. Among other places, Hume's critique appears in the following fragment from the first *Enquiry*:

> But the Pyrrhonian cannot expect that his philosophy will have any constant influence on the mind: Or if it had, that its influence would be beneficial to society. On the contrary, he must acknowledge, if he will acknowledge anything, that all human life must perish, were his principles universally and steadily to prevail. All discourse, all action would immediately cease; and men remain in a total lethargy, till the necessities of nature, unsatisfied, put an end to their miserable existence (*EHU*, 160).

If adopted, Hume says, the Pyrrhonian principle of suspending belief would entail the end of the skeptics' life. Of course, this argument is yet another formulation of the famous *apraxia* objection, which, as we have seen, dates back to Aristotle, Diogenes Laërtius, and Plutarch, among others.[7] Popkin maintained that Hume's objection is unwarranted since it is based on the erroneous presumption that the Pyrrhonists suspended judgment in all matters and were, therefore, unable to act, while Pyrrhonism, in fact, envisaged a less-demanding suspension of judgment, which allowed for the adoption of beliefs that are necessary for everyday life. Popkin explained:

> In this brief sketch of the doctrines of Pyrrhonian scepticism, as Hume conceived them in his *Treatise*, *Enquiries*, and *Dialogues Concerning Natural Religion*, it is worth noting two ways in which the theory, so stated, differs from the standard formulation of it in Sextus Empiricus. First of all, Hume's version is more dogmatic than that of Sextus. The latter

[6]Recall Descartes: "[. . .] because undermining the foundations will cause whatever has been built upon them to crumble of its own accord, I will attack straightway those principles which supported everything I once believed" (AT.IX, 9).

[7]Curiously enough, the argument is known by most contemporary analytic philosophers as "Hume's argument against skepticism."

> attempted to list a series of arguments pro and con on many questions, and then suspended judgment on them, instead of dogmatically holding that all questions are unanswerable. Secondly, Hume's rendering of Pyrrhonism omits any reference to the basis that the Pyrrhonian offers for deciding practical questions once the suspensive attitude has been adopted (Popkin 1951, p. 36).

Indeed, as Popkin claims, and as we have seen in the first chapter, Sextus answered to the *apraxia* objection by laying out a set of practical criteria for the skeptics: the so-called "fourfold observances," which include following the "appearances," acting on the grounds of natural impulses, abiding by the customs and norms of their society, and adopting technical precepts to master an art (*Pyr.* I, 23–24). Despite Sextus's insistence that these observances allowed the skeptics to lead a normal life, Hume tends to describe Pyrrhonism as a sort of radical skepticism, in which suspension of judgment confines the skeptic to total inactivity. For this reason, Popkin concluded that Hume's Pyrrhonism was rather an extreme and inexact formulation of Sextus's skepticism and that we should not take his criticism at face value.

After this, Popkin proceeded to link Hume's skepticism to Pyrrhonism by expanding on the naturalistic elements of the former's epistemology. Before examining his arguments, however, it is worth providing some context about what is usually understood by Hume's naturalism and why this is an essential aspect of the more circumscribed interpretation of Hume as a Pyrrhonian skeptic. As it is well known, in 1905, N. K. Smith published two articles entitled "The Naturalism of Hume" (I and II), where he outlined for the first time the now widely accepted thesis that Hume held a naturalistic theory of belief. Smith explains this thesis thus:

> We cannot by means of reason explain any of the ultimate characteristics of our experience – the origin of our sensations, the true 'secret' nature of causal connection, apprehension of external reality, appreciation of beauty, judgment of an action as good or bad. And the alternative is not scepticism but the practical test of human validity. Certain beliefs or judgments [. . .] can be shown to be 'natural,' 'inevitable,' 'indispensable,' and are thus removed beyond the reach of our sceptical doubt (Smith 1905, p. 152).[8]

According to Smith, Hume thought that the majority of beliefs on which practical life rests – for example, the beliefs in the existence of an external world, causality, the unity of the self, moral good and evil, etc. – lack rational justification, and are adopted unreflectively, almost instinctively or mechanically. After Smith, scholars have referred to these beliefs as "natural beliefs." By "natural beliefs," it is usually understood the sort of fundamental beliefs that are essential for conducting everyday life and are, for this reason, immune to skeptical refutation. Hume's view about these beliefs is that, even if philosophical reflection may lead us to question their rational justification and to skeptically conclude that we should suspend them, in fact, as soon as we wanted to determine ourselves to act, we would have to immediately, and inadvertently, adopt them again, because otherwise we would not be able to lead a normal life. The belief in the existence of the external world is one of such natural beliefs, perhaps the most important one since without it, we could not orient ourselves among objects and

[8]This thesis is expanded in Smith (1941).

people around us. By the same token, the belief in causality is indispensable to have a continuous experience of the world; and the belief in the unity of the self, to be able to set goals for our lives and choose the most appropriate means to achieve them. We permanently assent to these fundamental beliefs in our everyday lives, for, without them, we would indeed be confined to inactivity.

Fosl claims that Smith's interpretation of Hume as a naturalist was crucial to change the previous view of Hume as a radical skeptic:

> [It] reversed preceding negative readings [of Hume as a radical skeptic], arguing that, for Hume, natural belief trumps skepticism and establishes epistemically capable scientific inquiry on the indefeasible ground of nature. One might call this the 'naturalistic anti-sceptical' reading (Fosl 2020, p. 5).

Indeed, the adoption of Smith's thesis led subsequent interpreters to soften the radical or destructive aspects of Hume's skepticism and sometimes even to argue in favor of a Humean metaphysical or scientific realism. This, of course, has caused heated disputes among Hume scholars who claim, on the one hand, that his skepticism and his naturalism are incompatible and, on the other, that their compatibility can be worked out. For some of the scholars in the first category, Hume should be regarded either as a scientific or metaphysical realist, on account of his naturalism, or as an anti-realist, given his skepticism.[9] Far from wanting to take part in this dispute, I suggest that Hume's theory of belief has *both* naturalistic (or descriptive) and normative (or prescriptive) aspects, and the latter are the expression of his Academic skepticism.

Now, returning to Popkin's thesis, he used Smith's naturalistic interpretation to argue that Hume's conception of "natural belief" is similar to Sextus's response to the *apraxia* objection and, for this reason, Hume should be seen as a modern Pyrrhonist. Let us remember that Sextus's answer to the objection involves the practical criteria of following "appearances," natural impulses, laws and customs, and technical rules to conduct the skeptic's life. In the same way, according to Popkin, Hume argued that the skeptic should follow natural beliefs to act:

> Hume maintained that, in order to exist in this world, to act as human beings do, one must hold to some opinions. Our actions are not based upon a rational philosophy, but on irrational natural instincts and mechanisms, which allow us to persevere. This is true of all men, including the Pyrrhonian sceptics. What we believe, in order to act, is not a matter of rational choice, and did we not believe these matters, we would perish (Popkin 1951, pp. 401–402).

For Popkin, then, Hume's Pyrrhonism is based on his naturalism concerning belief. Interestingly, while Smith's naturalism was adopted mainly to stress the non-skeptical features of Hume's thought, Popkin used it precisely to identify Hume with ancient Pyrrhonism. To support his interpretation, Popkin pointed to the numerous passages where Hume's criticism of Pyrrhonism is followed by a *naturalistic qualification*—his frequent assertion that skeptical doubt is always defeated by our natural inclination to believe. For example, in the first *Enquiry*, immediately after claiming that human life would be at risk if Pyrrhonian principles

[9] For some good accounts of this debate see: Calvo 2012, pp. 385–386, and Fosl 2020, pp. 5–13.

were adopted, Hume added his well-known observation that, fortunately, nature would not allow such a state of affairs:

> It is true; so fatal an event is very little to be dreaded. Nature is always too strong for principle. And though a Pyrrhonian may throw himself or others into a momentary amazement and confusion by his profound reasonings; the first and most trivial event in life will put to flight all his doubts and scruples, and leave him the same, in every point of action and speculation, with the philosophers of every other sect, or with those who never concerned themselves in any philosophical researches (*EHU*, 160).[10]

Since nature precludes a permanent and complete suspension of judgment, the dangers associated with it actually do not affect the skeptic's life. All philosophical questioning is naturally limited by our instinctive tendency to adopt beliefs on which we base everyday action. Consequently, whenever the philosopher pauses her speculations and returns to the exigencies of common life, she instinctively gives up *epochê*. For Popkin, in conclusion, Hume's view that our most fundamental beliefs are adopted in this natural and irreflective way should lead us to identify him as a genuine Pyrrhonist, despite his critique of Pyrrhonism.

Popkin's article was well-received, and his interpretation has been highly regarded, even today. But we should not allow this notoriety to distract us from serious flaws in his line of argument, which I will now discuss. Although it is correct that Hume's naturalism has many similarities with Sextus's response to the *apraxia* objection, I find, nevertheless, that they are insufficient to conclude that Hume is a "genuine Pyrrhonist."

Undoubtedly, Hume's claim that "nature is always too strong a principle" for the suspension of judgment to be permanent reveals him as a philosopher conscious of the inevitability of belief and its grounding of human action. It also shows his awareness of the irrational motives that support some of our most cherished beliefs, and his dismay at the futility of the effort involved in suspending judgment. But it should not be taken as an explicit assertion of his skeptical affinities. Indeed, while the naturalistic aspects of his theory of belief may be one of the most important features of Hume's epistemology, it cannot be said that his *naturalism* is the key expression of his *skepticism*. Let me explain why.

To begin with, it is crucial to notice that Hume's naturalistic adoption of belief is not comparable to Sextus's response to the *apraxia* objection. As we saw in Chap. 2, the Pyrrhonian practical advice to follow appearances to orient daily life does not involve their *assenting* to them, that is, their holding them as beliefs.[11] On the contrary, the Pyrrhonists keep a critical distance from these subjective impressions, expressed in the *dictum* that they follow appearances "undogmatically" (*Pyr.* I, 23–24). This critical distance, however, is not involved in Humean naturalism, which refers instead to the way common-sense beliefs are usually adopted—that is, to the way human beings typically give in to their dogmatic inclinations and

---

[10] See *TNH* I, 269 and *DRN*, 31–32, for a similar qualification.

[11] Some commentators, however, think that this is the fact. In the first chapter, I discuss some of their views. See Burnyeat 1997, pp. 25–57; Frede 1997, pp. 127–151; Perin 2010, among others.

sustain all sorts of beliefs (Smith 2007, p. 119). But Popkin seems to overlook this difference, claiming that the authentic Pyrrhonist is *at once* skeptical and dogmatic:[12]

> The true Pyrrhonist is both a dogmatist and a sceptic. In being entirely the product of nature, he welds his schizophrenic personality and philosophy together. He believes whatever nature leads him to believe, no more and no less. He is compelled to believe, and in accepting the compulsion, he is exhibiting his scepticism. He is led to philosophize by certain natural inclinations, and through them to come to certain conclusions, and in so doing, he is again exhibiting his scepticism (Popkin 1951, p. 406).

In my view, when Popkin refers here to "skepticism" and "dogmatism," he must be speaking figuratively about the natural movement of human reason between doubt and belief. Otherwise, his claim does not hold. In the figurative sense, evidently, all human beings are both "skeptical" and "dogmatic," since at some point we all doubt our opinions, but we also dispel those doubts or change our previous beliefs for new ones when we find good reasons to do so. However, this does not make us *skeptics* or *dogmatists*. In fact, in the strict sense of the term, one is only a skeptic when one is determined to doubt, be it for attaining peace of mind or for adopting a critical point of view about one's or other people's beliefs. And one is only a dogmatist, again in the literal sense of the term, when one holds that some beliefs can be considered knowledge on the grounds of their compliance with a particular criterion of truth. If we accept, as Popkin seems to do, that Hume was at times a skeptic and at others a dogmatist,[13] we would have to conclude that he was, in reality, neither a skeptical nor a dogmatic philosopher, but rather a normal human being—and, perhaps, like many of us, an inconsistent philosopher.[14] However, Popkin insists on this point: "Hence, if one is really Pyrrhonian, as Hume was, one will be as dogmatic and as opinionated as one is naturally inclined to be" (Popkin 1951, p. 406; *cf.* de Olaso 2018, p. 23). It just strikes me as contradictory that being dogmatic or holding opinions is precisely what makes one a "true skeptic." So I must reject Popkin's

---

[12] As Fosl argues, Popkin thinks that Hume's avowal of this natural tendency to dogmatism is precisely what defines his Pyrrhonism: "In Popkin's view, even positive assent is, for Hume, a type of dogmatic assent, and it is moreover, a kind of dogmatism that Hume's properly consistent Pyrrhonism endorses. [. . .] Popkin, that is, reads Hume's endorsement of dogmatic natural belief not as a rejection of Pyrrhonism (as Kemp Smith et al. do) but instead as a critique of Pyrrhonism from within a properly Pyrrhonian framework" (Fosl 2020, pp. 11–12).

[13] "If we follow nature, as the Pyrrhonists said that we should, we believe whatever we have to. Hence, though we have found no refutation of skeptical reasonings, we are both skeptics and dogmatists; skeptics when we examine questions theoretically, dogmatists when we are constrained to believe" (Popkin 1951, p. 70).

[14] Although in a different context, Fosl also argues that the view that Hume is both a skeptic and a dogmatist is erroneous: "Situating Hume as perhaps the most profound inheritor of the skeptical traditions allows readers to see that there are not two (or more) Hume's, that in Hume there is not a fundamental sceptic who is also a dogmatic natural scientist. Hume is not schizoid or what Michael Williams calls "bi-perspectival." He is no kind of dogmatic realist – skeptical or anti-sceptical – but neither is he a dogmatic anti-realist. Hume is a radical skeptic through and through." (Fosl 2020, p. 11). While I agree with Fosl's general line of argument, I take distance from his last claim, namely that Hume is a radical skeptic. I rather believe Hume when he affirms that he is a mitigated one.

assertions on the grounds of an inadvertent *reductio ad absurdum* within his own argument.

But the most critical flaw in Popkin's reading is that he thinks that Hume's naturalism commits him to conflate the natural adoption of beliefs with a philosophical one. For this reason, he argues that Hume's philosophical skepticism would be no more than a mood that could easily be overcome by another one—that of common-sense dogmatism: "Hume, the naturalist who subverts all reason to emotion was just Hume the Pyrrhonist in his dogmatic mood. Hume, the reasoner, was Hume the Pyrrhonist in his sceptical mood" (Popkin 1951, p. 407, n. 86). I do not believe that, for Hume, the epistemic "schizophrenia," as Popkin puts it,[15] of the ordinary person is a desirable state for a philosopher. In this sense, I wholly concur with Stroud's critique of Popkin's argument:

> Someone who was "sceptical" only some of the time, and "dogmatic" the rest of the time, would not really be following a sceptical way of life. He would simply be a creature of nature. And every creature is a creature of nature. But not everyone leads a sceptical way of life (Stroud 1991, p. 281).

Like Stroud, I think that while Hume's naturalism *describes* the non-rational or instinctive adoption of belief, his skepticism *prescribes* the ideal of philosophical activity. Although, for Hume, natural dogmatism saves us from extreme skepticism by forcing us to adopt beliefs necessary for daily life, this does not mean that we should embrace *all* our beliefs in this natural and unreflective way. Hume's skepticism is much more than a simple philosophical mood to be alternated with an irreflective, everyday dogmatic disposition. As Calvo argues, Hume does not think philosophy should be either thoroughly Pyrrhonian or thoroughly dogmatic, or alternatively both, but it should actually overcome both extremes:

> To think that being a philosopher means being a radical skeptic is to overlook that Hume's comprehension of the nature of philosophy and true philosophy has been transformed after his crisis. Equally, to believe that in everyday life, the philosopher simply stops being a philosopher and forgets his reflections implies denying the Humean idea that the philosopher ought to be a critical participant of the lifeworld through his resistance to common superstition (Calvo 2012, p. 390, my translation).

In other words, Popkin's alleged fluctuation of skepticism and dogmatism overlooks Hume's own characterization of the philosopher as a moderate skeptic who neither abandons philosophy while participating in everyday activities, nor limits his philosophical speculations to abstruse topics outside the realm of common life. This mitigated skeptic maintains a critical attitude regarding belief—a voluntarily adopted stance, involving the permanent disposition to philosophically evaluate beliefs based on a set of criteria. This set of criteria constitutes Hume's *normative* theory of belief, as opposed, but complementary, to his *naturalism*.

In this sense, it is essential to understand Hume's philosophical and skeptical endeavor for what it is—an Enlightenment project. If we attribute nothing other than

[15] "Perhaps it was Hume's view that any 'honest' philosopher would be this sort of schizophrenic Pyrrhonist." (Popkin 1951, p. 407, n. 86).

naturalism to Hume, we will interpret him as an anti-Enlightenment philosopher—namely a modern irrationalist, who endorses our spontaneous disposition to believe or doubt whatever our natural faculties inclined us to, and who perhaps explains the equilibrium between both tendencies as an eventual balance of opposite forces (Fogelin 2009, p. 158).[16] But following this path requires overlooking Hume's critical efforts on matters as important as the justification of religious belief and superstition (Calvo 2012, pp. 373–377). Hume did not recommend naturalism, only recognized and described it. Although our natural inclination to believe has the advantage of checking skeptical doubt when it is too radical and risks leaving us inactive, its associated disadvantage is that it can make us take for true all kinds of irrational opinions, founded on our wild imagination and ardent passions. In Norton's words, "[. . .] however often Hume may say that we have certain natural propensities to believe this or that, he does not go so far as to say that what we must naturally *believe* must be *true*." (Norton 1982, p. 201).

P. Smith also suggests that the unavoidability of a belief is not a guarantee of its truth:

> While many skeptics judged that natural beliefs would hinder the triumph of skepticism and that naturalism would better characterize Hume's position, Humean science shows that the beliefs nature imposes on us are unavoidable errors and fictions. And, does the recognition that an error is inevitable and natural constitute a solution for it? (Smith 2011, p. 83, my translation).

And Calvo insists on Hume's critical attitude concerning belief:

> Hume recommends maintaining skepticism both in everyday life and in philosophical investigation; he sees skepticism as a normative attitude that can be generalized so that its effects in every realm of life become visible. In common life, it does not entail the suspension of customary beliefs but only of the natural attitude's dogmatism, which assumes that if we believe something, it is because things are in such a way objectively (Calvo 2012, p. 365; my translation).

Along with these scholars, I believe that the return to nature and custom stressed by some commentators of Hume should be taken with a grain of salt. Humean philosophy, even with its emphasis on the passions and its doctrine of belief as a feeling (I will come later to this aspect of his thought), is not an irrationalist avowal of spontaneous belief, nor needs to rely on custom's gradual and, so to speak, dialectical movement between natural doubt and natural assent, to progressively refine and discard the prejudices of every era (Livingston 1998, pp. 144–145). Instead, philosophy has a *normative* task for Hume—correcting false beliefs and showing the relative solidity or weakness of most of our beliefs' justifications. We should engage in philosophical investigation to recognize how both our natural beliefs, with their

---

[16]Fogelin says: "It is important to recognize that the mitigated skepticism of the *Enquiry* is not the result of reasoning. Reason, left to itself, provides no response to Pyrrhonian arguments. Mitigated skepticism results from the interaction of two forces: a departure from common instinctive belief, counteracted by the persistence of common instinctive belief." (Fogelin 2009, p. 158). A similar line of thought can be found in Livingston 1998, p. 156. As I argue in this and next chapter, this view completely overlooks the normative aspects of Hume's theory of belief, which are at the core of his Academic and mitigated skepticism.

unavoidability and instinctive character, and other beliefs that we tend to take for true based on prejudices, education, or the passions, lack sufficient rational justification. While the former are unavoidable, the latter are and should be liable to rational criticism and correction. As Norton argues:

> It is true that natural instincts or natural beliefs are given a central role by Hume.[. . .] But it does not follow from these facts that reason –reflective reason, consciously chosen analysis, conscientiously undertaken critical examination– is without function and effect, or that the interludes of reason are entirely insignificant when compared to the interludes of backgammon (Norton 1982, p. 235).

Of course, we unavoidably, almost instinctively, hold natural beliefs. Still, these beliefs are very few in number compared to the stockpile of possible beliefs a person may adopt during her life. Norton's distinction between "reflexive" and "reflective reasoning" may be helpful to understand this point. Natural beliefs are part of our "reflexive reasoning," insofar as they are "involuntary or unwilled." But, "reflective reasoning" is "characterized by deliberation or voluntary, willful activity" (Norton 1982, p. 209), which involves critically assessing our various beliefs and modifying them when it is necessary. Precisely this is what Hume intended to do with his critical treatment of traditional metaphysics. In religious matters, particularly, our natural inclination to believe easily leads to superstition and fanaticism. De Olaso vividly portrays it when he says: "In the end, the dilemma Hume faces is whether to prefer superstition or philosophy, fantasy or understanding, prejudice or judgment" (de Olaso 2018, p. 45). Hume was genuinely concerned about our natural tendency to believe everything that our imagination vividly suggests, and, for this reason, he recommended philosophical activity as a way of avoiding gullibility, superstition, and religious enthusiasm (Calvo 2012, pp. 374–375; Stroud 1991, p. 284).

In metaphysical matters especially, Hume found that it was necessary to examine our assertions before deciding whether or not to accept them. Thus, for example, to hold the theses of natural theology instead of those of revealed religion is more advisable, if not completely satisfactory, while maintaining the latter is more desirable than blindly adopting popular superstition. A merely naturalistic (and irrationalist) Hume would be unable to defend this gradation. He would regard all beliefs as equally justified or equally unjustified, and would find it unnecessary to provide any kind of criteria for deciding which ones to discard and which ones to adopt in everyday life.

In conclusion, naturalism *per se* does not give us a complete picture of Hume's project (Falkenstein 1997, p. 30). Neither when commentators interpret this aspect of his thought as the best expression of Hume's skepticism (*à la* Popkin), nor when they find in it reasons to understand Hume as an irrationalist or anti-Enlightened supporter of custom and tradition, are they giving us an accurate and complete view of Hume's theory of belief. Instead, Hume's naturalism should be given its due value by recognizing the normative aspects of his theory of belief, which aimed at overcoming prejudice, particularly in religious matters, and is deeply Academic. But before turning to explain this aspect of Hume's thought, let us review for a moment his critique of Pyrrhonism, which involves not only the famous Humean appropriation of the *apraxia* objection, but also his complaint that this sort of skepticism brings about the severe disadvantages of psychological melancholy and social ostracism. In this regard, Hume's remarks will help us see how Academic skepticism is a fundamental complement for his naturalism.

4.2 Humean Consequences of Pyrrhonism: *Apraxia*, Melancholy, and Ostracism

Before I begin with Hume's critique of Pyrrhonism, I should say a word about the convenience of interpreting it as his sincere opinion on the skeptical school. Commentators who regard Hume as Pyrrhonist or retrieve the Pyrrhonian features of Hume's skepticism argue that his critique of Pyrrhonism is not entirely sincere (Fosl 2020, pp. 77–80) or is based on a misunderstanding of this sort of skepticism (Popkin 1951, pp. 385–407; de Olaso 2018; Fosl 2020, p. 77). Fosl, for example, argues in favor of Hume's voluntarily hiding Pyrrhonian references in his works, as a strategy to avoid the censure attached to this kind of skepticism:

> There are [...] also good reasons to think that Hume suppressed in his publications a Pyrrhonian self-understanding of his work. [...] Hume enjoyed access to Pyrrhonian texts and used them [...], and considerations of audience and reputation would have led any prudent and ambitious author of Hume's time to resist characterization of his or her work as Pyrrhonian. In either case (accident or suppression), as a matter of philosophical interpretation, even if not biographical fact, Hume's skeptical philosophy is reasonably and properly read as profoundly Pyrrhonian (Fosl 2020, p. 79).[17]

While I accept that Hume's skepticism shares some Pyrrhonian features[18] – the most important one, in my view, is explaining causal inferences on the grounds of habit and custom – I have a difficult time believing that Hume's repeated-*ad-nauseam* criticism of Pyrrhonism is only an esoteric device to hide his real skeptical allegiances. Let me explain why.

Usually, Hume's critique of Pyrrhonism has three steps: first, he asserts that the radical or universal doubt of Pyrrhonists would destroy human life; second, he offers his naturalistic qualification, according to which, this total suspension should not be feared since it is impossible to maintain it (the famous line, "nature is always too

[17] Fosl also thinks that Hume may have followed Huet's, Augustine's, and Cicero's strategy of hiding his own skeptical allegiances for prudential reasons (Fosl 2020, p. 63).

[18] I do agree with Fosl's assessment in this point: "There is [...] substantially more to Hume's Pyrrhonism that this common reading suggests. The Pyrrhonian tradition comprises an expansive and complex network of ideas and practices, developed over a long period of time, and many dimensions of Hume's thought bear remarkable similarities to its most important formulations" (Fosl 2020, p. 79). His conclusion, though, that this is "so much so that it would be neglectful and misleading to describe Humean skepticism as *anything but deeply Pyrrhonian*" (Fosl 2020, p. 79) I find excessive, and to some extent, an expression of the interpretative prejudice that Popkin popularized.

strong for principle"); and finally, he claims that the proper function of skepticism is not to destroy ordinary life but to bring about moderation in judgment. The first two steps serve the purpose of distancing himself from radical skepticism and universal doubt, while somehow defending Pyrrhonism from the charge that it brings about a total disruption of human life. Indeed, Hume's insistence that this danger is actually unreal may be seen as a veiled defense of Pyrrhonism. However, such a veiled defense should not be conflated with a veiled avowal of Pyrrhonism since the third step of the criticism is there to show his adherence to another version of skepticism, a mitigated one, that espouses intellectual modesty while setting a clear limit to suspension of belief.

A variation to this structure, which nonetheless contains the same elements, can be seen in the *Letter for a Gentleman*, where Hume is defending himself from the accusations of radical skepticism and irreligion. At first sight, he seems to be defending Pyrrhonism, but soon afterward, we realize that even though he sympathizes with this sort of skepticism, he rejects it on account of its universal suspension of belief. He asserts:

> As to the scepticism with which the author is charged, I must observe, that the doctrines of the Pyrrhonians or Sceptics have been regarded in all ages as principles of mere curiosity, or a kind of *jeux d'esprit*, without any influence in a man's steady principles of conduct in life. In reality, a philosopher who affects to doubt of the maxims of common reason, and even of his senses, declares sufficiently that he is not in earnest, and that he intends not to advance an opinion which he would recommend as standards of judgment and action. All he means by these scruples is to abate the pride of mere human reasoners, by showing them, that even with regard to the principles which seem the clearest and which are necessitated from the strongest instincts of nature to embrace, they are not able to attain a full consistency and absolute certainty. *Modesty,* then, and *humility* with regard to the operations of our moral faculties, is the result of scepticism, not a universal doubt, which is impossible for any man to support, and which the first and most trivial accident in life must immediately disconcert and destroy (*LG*, pp. 154–155).

We can see how in this passage Hume distinguishes universal doubt, for him the main characteristic of Pyrrhonism, from the intellectual modesty of mitigated skeptics. He does seem to imply that Pyrrhonism is harmless, but he does not side with it, overtly or tacitly. Moreover, two paragraphs after this, Hume presents Socrates and Cicero as representatives of "the highest degree of skepticism" (*LG*, p. 155), which most probably means the *proper kind* of skepticism, namely, Academic skepticism. As a result, I think that we should take Hume's criticism of Pyrrhonism as a sincere one, even if he is sympathetic to it. He may have thought that Pyrrhonism was an inadequate sort of skepticism, although not as dangerous and grievous as the anti-skeptics portrayed it. The reasons he did not recommend it were, first, the all-too-common view that it confines the skeptic to inactivity; second, that it leads to melancholy; and, third, that it ostracizes the skeptic. I will now refer to these three aspects of his critique.

Evidently, Hume's critique of Pyrrhonism directly relates to the tradition of *apraxia* objections that various ancient philosophers made to the skeptics (de Olaso 2018, p. 67, n.7; Norton 1982, pp. 266–267).[19] At the beginning of the *Dialogues concerning Natural Religion*, for example, Cleanthes, the character who defends the thesis of natural theism, seeks to discredit his skeptical opponent, Philo, by putting forward the objection:

> Whether your scepticism be as absolute and sincere as you pretend, we shall learn by and by, when the company breaks up: We shall then see, whether you go out at the door or the window; and whether you really doubt if your body has gravity or can be injured by its fall; according to popular opinion, derived from our fallacious senses and more fallacious experience (*DNR*, 132).

The paragraph shows Hume's familiarity with Aristotle's and Plutarch's versions of the objection. Let us recall Aristotle's words:

> For why does a man walk to Megara and not stay at home, thinking he ought to walk? Why does he not walk early some morning into a well or over a precipice, if one happens to be in his way? Why do we observe him guarding against this, evidently not thinking that falling in is alike good and not good? Evidently, he judges one thing to be better and another worse. And if this is so, he must judge one thing to be man and another to be not-man, one thing to be sweet and another to be not-sweet (*Metaph.* 1008b 10–15).

Hume seems to have taken from Aristotle the example of a person who, deciding whether to walk in one direction or another, cannot suspend judgment about the common opinion that falling into a precipice is a bad thing because it can produce injuries. Besides, he appears to have retrieved from Plutarch his sarcastic comment about the convenience of leaving a room by the door. Plutarch, let us recall, says: "But how comes it that the man who suspends judgment does not go dashing off to a mountain instead of to a bath, or why does he not get up and walk to the wall instead of the door when he wishes to go out to the market-place?" (*Adv. Col.* 1122E). Given the similarities between Hume's and these ancient versions of the objection, there is no doubt that Hume knew them well and even enjoyed paraphrasing them whenever the occasion allowed it.

Moreover, to this acquaintance with the objection in ancient sources, we can add his familiarity with their modern adaptations, such as Pierre Bayle's. Bayle mentions it when referring to the Stoic Epictetus's opinion about the Pyrrhonists:

> Even though Epictetus thought highly of Pyrrho, he felt such unusual scorn for the Pyrrhonians that he could not abide them. One day he said to a Pyrrhonian who was trying to show that feelings were always deceptive, "Who among you, wanting to go to the oven, went to the mill instead?" He also often said, "If I were the servant of those Pyrrhonians, it would give me pleasure to make them suffer. If they said to me, "Epictetus, pour oil into the bath," I would pour brine on their heads. And if they asked me for the tea, I would bring them the vinegar. And if they complained, I would say they were wrong and would convince them

[19] Norton claims that the source of Hume' sarcastic attitude and characterizations of Pyrrhonism can be found in Diogenes Laërtius: "[...] Hume, along with a fair number of other philosophers both before and after him, seems to have given considerable credit to the view of Pyrrho found in Diogenes Laertius. According to this account, Pyrrho doubted to such an extent that he did not turn aside in the face of rushing chariots, mad dogs or imminent precipices [...] and that he was indifferent about everything" (Norton 1982, pp. 266–267).

> that the vinegar was actually tea, or I would make them give up their opinion" (Bayle 2005, pp. 111–112; my translation).

Bayle also retrieves Aristotle's and Plutarch's topics. And his allusion to the skeptics' confusing vinegar with tea recalls Laertius and Eusebius's anecdote of Lacydes being deceived by his slaves (*Praep. Evang.* XIV, 7; *Diog. Laert.* IV, 59) to which I referred in the first chapter of this book.

In the seventeenth and eighteenth century, the objection seems to have been a common-place argument used by anti-skeptics and Christian apologists to insist on the dangers of Pyrrhonism and other kinds of skepticism. It appears, for example, in J.P. de Crousaz's widely read *Examen du pyrrhonisme ancien et modern*, to which, according to Perinetti (2018, pp. 59–61), Hume had access at *La Flèche*. De Crousaz says:

> [. . .] their delight in contradicting, embarrassing others, and holding paradoxes grows more and more alluring with every objection [. . .]. But when the dispute ends, they return to their natural sentiments and act and think like the rest of humans (de Crousaz 1733, p. 3, my translation).

De Crousaz's view is not so much that the skeptics cannot act because they suspend judgment, but that, in reality, they do not maintain their suspension and go back to their natural beliefs immediately after they finish their game of refuting philosophical doctrines. We should pay attention here to the terminology. Indeed, de Crousaz's allusion to "natural sentiments" may have inspired Hume's idea that nature saves us from skeptical doubt. If this is the case, then Hume's naturalism is not so much an expression of his Pyrrhonism, as Popkin believed, but rather a concession to modern anti-skeptical discourses.

Given these accounts of the *apraxia* objection in both ancient and modern sources, with which Hume was familiar, I believe we can confidently assert that his rehearsal of the objection was not his most original position concerning skepticism, as has been usually held by commentators. Notwithstanding the pleasure he must have taken in re-creating the irony of the objection, I believe he rather used it as a standard rhetorical device. The cursory way in which Hume refers to it, usually followed by the naturalistic qualification that we should not worry about this danger because nature saves us from it, makes me think that he did not take the objection too seriously or, at least, that this was not his most heartfelt reproach to Pyrrhonism. But this does not mean, again, that Hume should be considered an adherent a modern Pyrrhonist. Instead, I believe that he held other important objections against this sort of skepticism. In this I concur with L. E. Loeb, who argues that Hume's most crucial critique of Pyrrhonism is related to the deep emotional distress that this skepticism produces, which he famously refers to as "melancholy and delirium" (*THN* I, 269). According to Loeb:

> This is a deeper objection to the Pyrrhonian than Hume's first criticism, the familiar Humean point that there are beliefs that cannot be suspended. Even if belief could be suspended, tranquility could not be achieved through equipollence. [. . .] In Hume's view, this gains nothing for the skeptic in respect to tranquility (Loeb 2002, p. 9).

Let us examine this objection in some detail. In the *Dialogues concerning Natural Religion*, Hume argues that the Pyrrhonian opposition of thesis and antithesis involves exercising a sort of violence against oneself that leads to a state of profound anguish:

> And for what reason impose on himself such a violence? This is a point in which it will be impossible for him ever to satisfy himself, consistent with his sceptical principles: So that upon the whole nothing could be more ridiculous than the principles of the ancient Pyrrhonians; if in reality they endeavored, as is pretended, to extend throughout the same scepticism, which they had learned from the declamations of their school, and which they ought to have confined to them (*DNR*, pp. 132–133).

Curiously, here Hume appears to be unaware of the fact that, for the Pyrrhonists, the point of suspending judgment was precisely attaining peace of mind. The question, "for what reason impose on himself such a violence?" either shows Hume's ignorance of this Pyrrhonian objective or intends to stress his view that this sort of skepticism produces a significant amount of emotional distress. In my opinion, given Hume's knowledge of ancient skeptical sources, the latter must be the case.[20] But since here and in other passages, he insists that the result of radical skepticism is "momentary amazement and irresolution and confusion" (*EHU*. 12.1, SB 155 n 1),[21] it seems that, as most modern philosophers, Hume thinks that peace of mind cannot be attained through suspension of judgment but only through a state of cognitive stability, which involves the adoption of some beliefs (Loeb 2002, pp. 6–37; González 2020, pp. 78–100). A good portrayal of Hume and other modern philosophers' attitude towards suspension of judgment is found in Peirce's renowned description of the mental distress doubt produces and how it naturally leads to "fixating belief" (Peirce 1992, p. 109). Perhaps the ancient Pyrrhonists were able to attain *ataraxia* through suspension of judgment, but the scientific *ethos* of modern philosophers entailed their regarding suspension and doubt as a cause of mental distress. They were, thus, inclined to adopt justified beliefs after their theoretical investigations. As De Olaso rightly claimed: "[...] Hume does not aspire to mental quietude in the Pyrrhonian manner, but he wants the sort of mental repose that comes from probabilistic science, not from *docta ignorantia*" (De Olaso 2018, p. 67). Hume, thus, thought that this Pyrrhonian "self-imposed violence" could only lead to philosophical melancholy, which he famously described in the Conclusion to the first book of the *Treatise*:

> My memory of past errors and perplexities makes me diffident for the future. The wretched condition, weakness, and disorder of the faculties I must employ in my enquiries, increase my apprehensions. And the impossibility of amending or correcting these faculties, reduces me almost to despair, and makes me resolve to perish on the barren rock on which I am at

[20] Perinetti and Fosl have recently shown that Hume had access to Sextus Empiricus's works at *La Flèche*'s library, and have dispelled the doubts scholars, including Fols himself, had about Hume's direct knowledge of Sextus and other Pyrrhonian sources. See: Perinetti 2018, pp. 57–59, Fosl 1998, pp. 261–78, 2020, pp. 155–158.

[21] Although this quote is not about Pyrrhonism but Berkeleyan skepticism, the context indicates that Hume was considering his skepticism as a radical form, akin to Pyrrhonism.

> present, rather than venture myself upon that boundless ocean, which runs out into immensity. This sudden view of my danger strikes me with melancholy; and as 'tis usual for that passion, above all others, to indulge itself, I cannot forbear feeding my despair, with all those desponding reflections, which the present subject furnishes me with such abundance (*THN I*, 264).

Notwithstanding the emotional force of this passage, commentators debate on whether Hume's depiction of his skeptical crisis is sincere or not.[22] I agree with Calvo (2012, pp. 355–356) that the vivacity of the description indicates a real and profound crisis. The fact that Hume suffered the so-called 'disease of the learned' while writing the *Treatise* also seems to confirm his honesty about his skeptical crisis.

In brief, the objection claims that skeptical doubts lead to melancholy and delirium. Instead of resulting in suspension of judgment and *ataraxia*, the dialectic of theses and antitheses produces a mental "combat" in which each view is alternately assented to and no belief wins the battle. In Loeb's words:

> We might imagine that opposing arguments of equal strength invariably cancel each other out, counterbalance one another. [...] Hume maintains, to the contrary, that opposing arguments often struggle with each other in psychological combat, first one prevailing and then the other. The metaphors of combat are Hume's. [...] Hume maintains that in contradictory beliefs, we have a "combat" and "struggle" between "enemies" who seek to "destroy" the "opposition" by which they are "attacked." Though we try "sacrificing the one to the other," the opposing beliefs can prove equally strong, so that neither "will quit the field" (Loeb 2002, p. 8).

The combat is described as a dispute between two sources of belief. On the one hand, the imagination leads us to assent to lively but rationally contradictory claims. On the other, the understanding makes us take for true the most abstract notions, wholly removed from common sense. Sometimes the skeptic follows the lead of her imagination, and then, she falls into logical contradictions. At other times, she trusts the suggestions of her understanding but, again, arrives at absurd conclusions, opposed to the teachings of experience. This movement from one faculty to the other and from one judgment to its opposite ultimately produces deep psychological suffering. Hume continues:

> This question is very difficult and reduces us to a very dangerous dilemma, whichever we may answer it. For if we assent to every trivial suggestion of the fancy, beside that these suggestions are often contrary to each other, they lead us into such errors, absurdities, and obscurities, that we must at last become ashamed of our credulity. [...] But, on the other hand, if the consideration of these instances makes us take a resolution to reject all the trivial suggestions of the fancy, and adhere to the understanding, that is, to the general and more established properties of the imagination, even this resolution, if steadily executed, would be dangerous and attended with the most fatal consequences. For I have already shown that the understanding, when it acts alone, and according to its most general principles, entirely subverts itself and leaves not the lowest degree of evidence in any proposition, either in philosophy or common life (*THN*. I, pp. 267–268).

[22] For a good account of the different positions on this debate, see: Calvo 2012, pp. 355–356.

For Hume, this is what happens in metaphysical and religious issues, when assenting to the imagination's suggestions would amount to sustaining superstitious beliefs, while accepting the inferences of reason would lead to maintaining all sorts of theological subtleties (Calvo 2012, p. 353). The philosopher cannot indulge in any of these outcomes, lest her reflection becomes inadvertently dogmatic. But neither can she persevere in doubting them because this would take her to a state of permanent duplicity. She refrains from assenting to one judgment now and its opposite next, but cannot detach herself from her natural inclination to believe. This state gradually weakens her spirits until she falls into melancholy and delirium, while waiting for the suspension of judgment to materialize.

Given that this skeptical mental combat never ends in the final relaxation of the mind that the Pyrrhonists sought, Hume thinks that a radical skeptic, far from attaining quietude, lives in a permanent state of mental instability. Thus, while *apraxia* is not much of a worry, since the skeptic has natural beliefs that spare her from inactivity, melancholy and delirium are real threats for those who indulge in the Pyrrhonian dialectical method of antinomy.

P. Smith has argued that Hume thought that most antinomical arguments, except those of natural religion, could be solved (Smith 2011, p. 71). I agree with him that Hume thought he could find the solution to some antinomical issues, such as the debate about the (in)finite divisibility of space and time.[23] However, I think that Hume did find these arguments profoundly distressing, and his version of a natural opposition of reason with itself, described in the Conclusion of the First Book of the *Treatise,* made him reject radical or Pyrrhonian skepticism and opt for the milder Academic outlook.[24]

An objection to this view, however, may come from the fact that the origin of this criticism of Pyrrhonism, as Perinneti has shown, can also be found in de Crousaz. In his *Logic*, he distinguishes the Academic skeptics from the Pyrrhonists, and within the latter, those that are "obstinate" but inconstant in their suspension of judgment, from those that "suffer a pathology characterized by excessive timidity and weak temper." (Perinetti 2018, p. 61).[25] These individuals, says de Crousaz, need to "be treated in the same way the mad and melancholic are treated" (de Crousaz, *Logique*, 3:1004, cited by Perinetti 2018, p. 61), that is, with common distractions and

[23]I will further develop this point in Chap. 6, when I describe Hume's influence on Kant's antinomical problems.

[24]In an attempt to stress the Pyrrhonian features of Hume's skepticism, Fols argues that although it is true that for Hume the opposition of claims leads to mental distress, "the Humean sceptic achieves *ataraxia* not only through *isosthenia* and *epochê* but also through Apelletic gain and reflective observance of the Fourfold of appearances in common life." (Fosl 2020, p. 108). I disagree with this assessment, first, because, if my analysis is right, the Humean skeptic does not reach *epochê* but is in a permanent state of duplicity (holding beliefs and doubting them alternatively) and second, because the fourfold observances of Pyrrhonism are not adopted in order to bring about *ataraxia*, but merely to provide a criterion not to remain inactive while suspending judgment.

[25]According to Perinetti, "it is likely that Crousaz's remarks inspired Hume's own association between radical skepticism and melancholy" (2018, p. 61).

pleasures that progressively take them to a more balanced state of mind where they can assent to the truth (Perinetti 2018, p. 61). It appears, indeed, that de Crousaz's characterization of Pyrrhonism was very influential for Hume. The question arises, then, why, if I suggested before that Hume's *apraxia* objection and naturalistic qualification should be seen as a rhetorical concession rather than a sincere objection, should we not treat this charge in the same way, and claim that Hume's melancholy and delirium were just a way for him to pay lip service to the dogmatists of his time?

I do not believe this is the case for two reasons. First, because Hume does not refer to this objection as hastily as he does to the *apraxia* charge, which he leaves behind immediately, with the claim that nature saves us from inactivity. He rather elaborates on the causes of melancholy and describes the state of mind in a very vivid way. Second, because the duality of reason – namely the state in which the understanding provides some inferences while the imagination affords opposite ones – resulting in melancholy, is a common topic throughout Hume's works, to which he only gives a solution when he discusses Academic skepticism. Thus, I think that the charge that Pyrrhonism brings about melancholy or, at least, an uneasy and imbalanced state of mind is not a mere rhetorical topic but a sincere concern.

Finally, there is one more aspect of Hume's critique of Pyrrhonian skepticism that we should consider. For Hume, Pyrrhonian doubt not only produces a state of profound psychological helplessness but also leads to ostracism. This feature of Hume's lamentation in the Conclusion of the first book of the *Treatise* should be taken very seriously, because it attests to his own intellectual frustration and constant fear of censure:

> I am first affrighted and confounded with that forlorn solitude, in which I am placed in my philosophy, and fancy myself some strange uncouth monster, who not being able to mingle and unite in society, has been expelled all human commerce and left utterly abandoned and disconsolate. [. . .] I have exposed myself to the enmity of all metaphysicians, logicians, mathematicians, and even theologians; and can I wonder at the insults I must suffer? I have declared my disapproval of their systems; and can I be surprised, if they should express a hatred of mine and of my person? When I look abroad, I foresee on every side, dispute, contradiction, anger, calumny, and detraction. When I turn my eye inward, I find nothing but doubt and ignorance (*THN* I, 264).

Hume refers here to the social alienation the skeptic suffers, having been banished from his scientific community (Sanfélix 1994, p. 57). He says that he misses the company of the "metaphysicians, logicians, mathematicians and even [of the] theologians" he harshly criticized. With dismay, he acknowledges that his skeptical arguments brought about "dispute, contradiction, anger, calumny, and detraction." As Calvo has argued, Hume's metaphor of the monster seems to convey his anguish genuinely:

> It arises from the lonely perspective in which philosophical reflection has left Hume, since, on the one hand, it has separated him from the world of custom and habitual conversation, and, on the other, it has alienated him from philosophical and theological orthodox systems [. . .] (Calvo 2012, p. 350, my translation).

The metaphor is, in fact, so eloquent that it should call our attention. Added to the "melancholy charge," we obtain a somewhat unfavorable picture of Pyrrhonism, which may not lead to *apraxia*, but does result in anxiety and ostracism. Probably, Hume's concern about the latter is what motivated him to write the *Letter to a Gentleman* as an attempt to regain a place in his academic community. By explaining the harmlessness of his philosophical principles and calling for toleration and freedom of speech (LG, p. 162), he sought to re-establish communication with the intellectual circle of Enlightened Great Britain. But, aside from the biographical aspects of this criticism, the risk of ostracism is a significant charge against radical skepticism because it harms both the skeptic and the scientific community at large. Indeed, without the voice of the skeptic, the community loses the point of view that encourages self-criticism, and this leads to a vicious circle—the skeptic's isolation leaves her in a state of inescapable melancholy, leading to more and more irresoluble doubts, while the scientific community becomes more and more inflexible and incapable of self-examination, that is, increasingly dogmatic. Hume realized that, to avoid these risks, Pyrrhonism should be overcome by a mitigated sort of skepticism, Academic skepticism, which retained the function of combatting dogmatism and superstition, without bringing about the skeptic's isolation and melancholy. As Calvo argues, this skepticism conveys significant social benefits: "[. . .] in the role of interlocutor, [the mitigated skeptic] contributes to undermining prejudice and bigoted, dogmatic views, which are very ingrained in ordinary people and scientific objectivism." And, thus, the community at large becomes able to "refine its manner of believing, passing from certainty to a sound, open criticism of conviction." (Calvo 2012, p. 426, my translation).

In conclusion, for Hume, melancholy and ostracism pose a real threat to the skeptic and her community at large. If *apraxia* is, as Hume claims, actually not a very important threat, the latter two charges should guard us against embracing Pyrrhonian radical skepticism. This is why Hume turned to Academic skepticism.

4.3 Hume's Praise of Academic Skepticism: Epistemic Humility and Moderate Passions

With poignant irony, Ezequiel de Olaso once wrote, "it seems redundant to affirm that Hume said he was Academic and, then, pile up arguments to prove it" (de Olaso 2018, p. 77).[26] However, this is precisely what I intend to do in this section, by re-visiting Hume's praise of Academic skepticism in the *Enquiry*. Hume not only expressed his reservations about Pyrrhonism but spoke in highly favorable terms about Academic skepticism, and even though commentators have already drawn attention to Hume's portrayal of Academic skepticism (Calvo 2012, pp. 403–430; de

[26]De Olaso accepts that the reason we need to do this is the influence of Popkin's interpretation on Humean scholarship.

Olaso 2018, pp. 69–80; Norton 1982, pp. 277–279; Stroud 1991, pp. 286–287), I believe we should consider the issue again, to obtain a more clear view of the reasons why he opted for Academic skepticism.

Hume primarily mentions this school in the fifth section of the first *Enquiry*, where he offers a "skeptical solution" to the problem of causality. The title of the section should already call our attention—he declares to offer a skeptical solution to the problem and not just a skeptical refutation or attack. The section begins by discussing the philosophers' tendency to justify their prejudices via elaborate arguments. Instead of using philosophy to correct inveterate passions and vices, they employ it to justify their temperamental inclinations. In contrast to these common tendencies, Hume highlights the intellectual honesty of Academic philosophy:

> There is, however, one species of philosophy which seems little liable to this inconvenience, and that because it strikes in with no disorderly passion of the human mind, nor can mingle itself with any natural affection or propensity; and that is the Academic or Sceptical philosophy (*EHU* V, pp. 40–41).

This initial description conveys the idea that Academic skepticism nourishes and is nourished by moderate emotions and, hence, is helpful to avoid psychological distress. Given Hume's lamentation of the effects of radical skepticism in the *Treatise,* this aspect of Academic philosophy – which he now equates with "skeptical philosophy" – should not be overlooked. Academic skepticism does not lead to melancholy or any other kind of emotional disturbance, but rather promotes a relatively serene state of mind. Academic skeptics live a balanced life, unlike radical skeptics, who are besieged by melancholy, or superstitious people, who suffer from a constant fear of divine punishments. Their skepticism helps them cultivate a mental equilibrium, with no inclination to either excessive doubt or dogmatic assent. As Stroud maintains, Hume saw Academic skepticism as a philosophy that leads to a "good life":

> Hume recommends the pursuit of the sceptical or Academic philosophy as the best or perhaps the only way to achieve this most natural and therefore most blissful human condition. [. . .] With philosophy as our guide, we will be immune to the disturbances of religion and other forms of superstition. We will have achieved a happy "determination" in which no side of our nature "draws too much" (Stroud 1991, p. 287).

Hume was aware of the Hellenistic view that philosophy is an intellectual endeavor as much as a way of life, aiming at shaping the individual's character in ways that lead to virtue and happiness. Norton also stresses this aspect of Hume's Academic allegiances when he claims that Hume's moderate skepticism brings about "practical and beneficial results" which not only involve clear advances in philosophical inquiry, but also a deep transformation of the philosopher's disposition (Norton 1982, p. 219). Then, for Hume, Academic skepticism was not only an epistemic stance but also a way of directing one's life.

Beyond this, as is usually remarked by commentators, Hume thought that Academic skepticism kept inquiry within the limits of human affairs, maintaining the philosopher away from fanciful or abstruse doctrines:

> The Academics always talk of doubts and suspense of judgment, of danger in hasty determinations, of confining to very narrow bounds the enquiries of the understanding, and of renouncing all speculations, which lie not within the limits of common life and practice. Nothing, therefore, can be more contrary than such a philosophy to the supine indolence of the mind, its rash arrogance, its lofty pretensions, and its superstitious credulity. Every passion is mortified by it, except the love of truth; and that passion never is, nor can be, carried to too high a degree (*EHU*, 41).

The moderate passions of Academic skepticism help restrain speculative or metaphysical impulses and keep reflection in the practical realm. Abstruse speculation should be checked, although not wholly abandoned (Norton 1982, pp. 211–214), and a more productive focusing on human affairs ought to be the goal of philosophy. However, this limitation to the necessities of ordinary life should not be seen as an invitation to accept widely held opinions uncritically. Instead, it should be understood as a way to keep philosophy away from both the arrogance of dogmatism and the indolence of common prejudice (Calvo 2012, pp. 408–409). From this passage, Hume's final reference to the "love for truth" professed by this skeptical school is also worth highlighting. As we know, Cicero repeatedly claimed that Academic skepticism seeks the truth above all (*Acad. II.* 7). Hume acknowledges that this skepticism is not a negative dogmatism, as Sextus portrayed it (*Pyr.* I, 1–4), but rather a philosophy that sought to find truth or, at least, what best approximates it. For this reason, he expressed surprise about the accusations of Academic skepticism for being "libertine, profane, and irreligious" (*EHU* V, 41). Despite his – perhaps exaggerated – astonishment, the fact is that the skepticism of the Academy was widely considered an enemy of religion and traditional metaphysics. In this respect, Pierre Galland's commentary, mentioned by C. Schmitt in his study of the *Academica*'s reception in the Renaissance, is illustrative. Galland says:

> The Academy tries to root out straightaway from the minds of men not only every conviction of sanctity and piety, but everything else as well; and, as did the Titans, wages a ruinous war with the immortal gods. For what can be established and certain concerning God or anything else, for one who is always accustomed to refute the opinion of others, who destroys all confidence in the senses, who disturbs reason, who believes in no evident facts whatever, who in desperation leaves all things uncertain, who abolishes all perception and assent, and hence the stability and constancy of every virtue? (Cited by Schmitt 1972, p. 95).

Many Christian apologists of the sixteenth and seventeenth centuries shared Galland's opinion. And, as we will see in the next chapter, it is perfectly natural that Christian theology would feel apprehension for the Academy, since Cicero's *De Natura Deoum* offered a severe critique of theism, particularly of natural theism. For this reason, the claim that Hume may have chosen to wear the mask of Academic skepticism to avoid the censure attached to Pyrrhonism[27] is not entirely plausible—

[27] According to Fosl, "recent archival research has confirmed that many of Hume's contemporary readers found his work, even in its pruned presentation, to be deeply threatening and irreligious. Perhaps the ease with which this was discerned in part explains Hume's subsequently emphasizing his Academicism in the first *Enquiry*." (2020, p. 162).

the public at large, and the censors, in particular, showed as much apprehension towards Academic skepticism as towards Pyrrhonism.

Hume's second explicit mention of Academic skepticism in the *Enquiry* appears in the final section, entitled "Of the Academical or Skeptical Philosophy," where he defines Academic skepticism as a corrective to Pyrrhonian doubt:

> There is, indeed, a more *mitigated* skepticism, or *Academical* philosophy, which may be both durable and useful, and which may, in part, be the result of this *Pyrrhonism*, or *excessive* scepticism, when its undistinguished doubts are, in some measure, corrected by common sense and reflection (*EHU* XII, 161).

In this passage, it becomes clear that Hume regards Pyrrhonism as an extreme form of skepticism, whose excessive tendency to doubt may be overcome and transformed into more productive ends by the mitigated skepticism of the Academy (Jaffro 2011, pp. 53–69). According to Hume, the skepticism of the Academy curbs Pyrrhonism's radical doubt by applying "common sense and reflection" to their method. Although he does not say much more about how this skepticism achieves this end, we can anticipate that its means involve the skeptical examination of belief and the non-dogmatic adoption of the claims that withstand it and are practically useful. I want to caution, however, that Hume's taking into account the claims of "common sense and reflection" in this passage should not be understood as a return to natural belief or the beliefs sanctioned by tradition. As L. Jaffro has suggested: "mitigated skepticism is not a return to ordinary life. It is not about forgetting the previous doubts but rather about remembering and being aware of them" (Jaffro 2011, p. 63). Even though natural beliefs do check extreme skepticism, they are not the only means to control excessive doubt. Beyond this, we should notice that, for Hume, "common sense" is not an irreflective assent to opinions but rather the ability to judge based on acquired experience. In other words, Humean common sense does not entail assenting uncritically or unreflectively to particular opinions but instead involves carefully choosing one's beliefs on account of what cumulative experience recommends.[28] Moreover, as the passage above indicates, both common sense and "reflection" play the role of controlling excessive doubt. Thus, Hume appears to be telling us that the Academy corrects Pyrrhonian suspension of belief through experience and reflection, not by following our automatic or spontaneous inclinations to believe (Calvo 2012, pp. 456–472; Hickerson 2013, pp. 1133–1153; Jaffro 2011, p. 63; Norton 1982, pp. 192–238; Sanfélix 1994, pp. 63–65). In the next section, I will explain in more detail the role of reflection in the process of correcting beliefs.

To conclude with Hume's praise of Academic skepticism, we can say that he sees it as a mechanism to control prejudice, superstition, and speculative obstinacy. It neutralizes both the arrogance of the dogmatist and the naivety of the ignoramus. It redirects inquiry to the pressing problems of human life by demanding philosophers to refrain from their excesses in metaphysical speculation. However, it does not promote an uncritical adherence to natural and traditional beliefs, and provides the means to overcome the tendency to blindly accept the claims of authorities. In sum,

---

[28] See Norton's discussion of the difference between common-sense philosophers, such as Turnbull et al. (1982, pp. 152–208).

this kind of skepticism constitutes, for Hume, a healthy habit of thought, which, promoted by moderate passions, helps cultivate all the virtues, particularly the intellectual ones. This habit of thought supplies "[. . .] a degree of doubt, and caution, and modesty, which, in all kinds of scrutiny and decision, ought forever to accompany a just reasoner" (*EHU XII*, 162).

Let us turn now to observe whether or not Hume's description of Academic skepticism fits our general understanding of it. Maia Neto has claimed that five main features characterize Academic skepticism:

1. A permanent search for truth, coupled with the principle of inapprehensibility of impressions, that is, with the denial of certainty.
2. A middle-ground position between dogmatic belief and total suspension of belief.
3. The avoidance of dogmatism, particularly concerning metaphysical knowledge.
4. The limitation of philosophical endeavor to affairs that are within the human reach.
5. The virtue of "intellectual integrity," defined as the conscious avoidance of rashness of thought.[29]

From this description, we should highlight the practical consequences of what Maia Neto calls the Academic's "intellectual integrity." As I argued before, the Academic sage not only sought to avoid prejudice and rashness of thought but, choosing to act on the basis of opinions that were highly probable or persuasive, endorsed practical autonomy, which allowed for the attribution of responsibility. De Olaso has also claimed that the most crucial difference between Pyrrhonism and Academics is that the former follow appearances and the customs of their society, while the latter subject these customs to skeptical examination and only approve the ones that are sufficiently persuasive or probable. In this way, the Academic:

> [. . .] can act independently from customs—even against them. And, because of this, the Academic can assess particular customs as prejudices and propose alternatives of action. For the Pyrrhonian, these possibilities are precluded from the beginning. Moreover, the Academic takes responsibility (obtained through evidence) for his acts. [. . .] For this reason, Academicism was always a supporter of *libertas philosophandi*, and always strove to obtain it.[. . .] The academic is the subject of a moral system that includes, as one of its main characteristics, personal responsibility (de Olaso 2018, p. 65, my translation).

I believe that Maia Neto's and de Olaso's characterizations stress those aspects of Academic skepticism that were more attractive to Hume. Perhaps they only missed Hume's emphasis on the sort of moderate passions that Academic skepticism helps cultivate and on the fact that the adoption of belief it suggests involves a *reflective* attitude. This latter feature is what I understand as Hume's Academic *normative*

---

[29] In this respect, Maia Neto claims: "The Academic skeptic holds that not possession of truth but avoidance of errors is what distinguishes the sage. Rashness, that is, coming too quickly to a belief before its full examination, is the evil that the Academic sage most abhors" (Maia Neto 1997, pp. 204–207). Fosl argues that this description is a good framework to situate Hume's academicism. See: Fosl 2020, pp. 69–70. I agree, but I think we should add to this framework the Academics' practical autonomy, which de Olaso also brings to the fore.

theory of belief, which complements his naturalism while doing justice to his Enlightened commitments (Calvo 2012, pp. 456–472; Falkenstein 1997, pp. 29–72; Hickerson 2013, pp. 1133–1153; Loeb 2002, p. 114; Norton 1982, pp. 192–238).

Before proceeding to the next section, however, I must acknowledge a final possible objection to the thesis that Hume's skepticism is Academic. There is, indeed, a difference in emphasis between the treatment of Pyrrhonism and Academic skepticism in the *Treatise* and the *Enquiry*. By insisting on returning to instinctive belief after engaging in arduous philosophical activity, the young Hume of the *Treatise* appears to consider natural belief as the only way out of skeptical doubt.[30] Again, in the Conclusion to the first book, Hume offered the most extensive treatment of how natural belief helps overcome radical skepticism:

> Most fortunately it happens, that since reason is incapable of dispelling these clouds, nature herself suffices to that purpose and cures me of this philosophical melancholy and delirium, either by relaxing this bent of mind, or by some avocation, and lively impression of my senses, which obliterate all these chimeras. I dine, I play a game of backgammon, I converse, and am merry with my friends; and when after three or four hours of amusement, I wou'd return to these speculations, they appear so cold, and strain'd, and ridiculous, that I cannot find in my heart to enter into them any farther (*TNH* I, 269).

The predominance of this argument in the *Treatise* may have triggered Popkin's interpretation, according to which the 'schizophrenic' alternation of skepticism and dogmatism – of doubt and natural belief – is the very destiny of the philosopher. More recent commentators, who also link Hume's skepticism with Pyrrhonism, also retrieve this naturalistic topic, although making a more balanced interpretation of Hume's works. Fosl, for example, finds that Sextus's fourfold practical observances "[...] describe the central features of Humean thought." (Fosl 2020, p. 87). Thus, for him too, naturalism is a key to understanding Hume's Pyrrhonian allegiances, even though he suggests that Hume's "acceptance of common life" is rather reflective (Fosl 2020, p. 110). Given this, I believe that a more precise assessment of Hume's Academic allegiances requires regarding the *Treatise* and the first *Enquiry* as showing slightly but significantly different attitudes towards skepticism. Some commentators have pointed to an evolution in Hume's thought that would begin with a sort of "Pyrrhonian crisis" while composing the *Treatise*, and end in the adoption of Academic skepticism during his mature years, particularly while writing the *Enquiry* (de Olaso 2018, p. 47; Penelhum 1979, p. 259, 1983, p. 124; Calvo 2012, pp. 411–413). Penelhum, for example, maintains that, given natural beliefs' insufficiency to overcome skeptical melancholy, Hume eventually decided to soften his skepticism and adopt the mitigated sort of the Academy:

> [...] a recognition of the hazardousness of philosophy leads him [Hume] in the first *Enquiry* to a more formal attempt to limit skepticism, by limiting the subject matter of philosophical

[30] Popkin's interpretation of Humean skepticism is undoubtedly based on a more extensive treatment of the *Treatise* than of the *Enquiry*. There are approximately twice as many references to the former than to the latter in his article.

> reflection as well as the psychological occasion to it. I refer, of course, to the introduction there of the concept of a mitigated skepticism, which Hume recommends to us as a mode of thought whose virtues can be made obvious if we have learned humility about the powers of our reason from indulgence in the Pyrrhonian variety. I do not agree with Popkin that this later version of skepticism is the same as the earlier (Penelhum 1979, p. 259, 1983, p. 124).

Pehelhum claims that we should take close notice of the appearance of the term "mitigated skepticism" in the *Enquiry*. Hume's new and open references to a "mitigated skepticism" or "Academic philosophy" is, for him, no small indicator that the mature Hume may have adopted this type of skepticism to avoid the psychological suffering that he described in the Conclusion to the first book of the *Treatise*.[31] In turn, Calvo suggests that, in the *Enquiry*, Hume is finally able to comfortably call himself a skeptic once he has gained the conviction that his mitigated skepticism does not hinder the investigation into human nature but instead allows to "distinguish between justified and unjustified beliefs both in science and in everyday life" (Calvo 2012, pp. 412–413). It seems, then, that Academic skepticism is, for Hume, a sort of acquired "second nature," characterized by emotional moderation and cautious judgment—an attitude gained after painful effort and distress, as the Pyrrhonian crisis of the *Treatise* suggests (Penelhum 1983, p. 122). What characterizes Hume as an Academic skeptic is his adoption of a more mature skepticism, consisting in applying voluntarily-adopted habits of thought to determine when it is reasonable to doubt and when to give way to undogmatic belief (de Olaso 2018, pp. 70–77).

However, I also distance myself from these interpretations, in that this change of perspective does not necessarily mean that Hume was Pyrrhonian in the *Treatise* and Academic in the *Enquiry*. As I will shortly show, since the most essential aspects of Hume's Academic normativity are developed in the *Treatise*, his theory of belief seems to involve an unconscious mixture of naturalism and Academic skepticism—indeed, a precarious balance, or better, a conflict, that only in the *Enquiry* becomes a conscious endorsement of Academic skepticism.

I would like to indulge in one last observation before I close this section. For some, arguing that Hume holds an Academic normative theory of belief may seem to claim that he endorses an unnatural, rationalistic method that, by strict compliance to some rules of thought, may subvert the practices of everyday life. However, nothing could be further from Academic skepticism than this way of proceeding. Academic skepticism actually involves the carefree attitude that Hume depicted as ideal for the philosopher:

---

[31] This interpretation is also shared by J. Immerwahr and de Olaso, who think that, although already present in the *Treatise*, Academic skepticism has much more weight in the *Enquiry*. Immerwahr claims: "The Academic skepticism of the *Enquiry* can now be seen as a solution to the dilemma posed in the conclusion of *Treatise I*. In the *Enquiry*, Hume offers a version of skepticism that can be used to attack both metaphysics and superstition, while leaving science and the moral subjects as appropriate subjects of reasoning having a similar foundation" (Immerwhar 1979, p. 237). Similarly, de Olaso argues: "In fact, Hume's was a skeptical crisis that did not end in the Pyrrhonian but rather in the Academic fashion. Far from having taken the possibilities of Pyrrhonism to the extreme, Hume built the most ambitious Academic system of philosophy" (de Olaso 2018, p. 47). Also Fosl argues in a similar vein (2020, p. 158).

> The conduct of a man who studies philosophy in this careless manner, is more truly sceptical than that of one, who feeling in himself an inclination to it, is yet so overwhelmed with doubts and scruples as totally to reject it. A true sceptic will be diffident of his philosophical doubts, as well as of his philosophical conviction; and will never refuse any innocent satisfaction, which offers itself, upon account either of them (*THN* I, 273).

Undertaking philosophy in this way means overcoming both Pyrrhonism and the natural inclination to dogmatically give assent to all sorts of beliefs. Those who are "diffident of [their] philosophical doubts, as well as of [their] philosophical conviction," position themselves on a middle ground between the dogmatist and the radical skeptic, both of which tend to take themselves all too seriously—the dogmatist by adopting an artificial gravity, the Pyrrhonist by falling into self-inflicted melancholy (Calvo 2012, pp. 364–365). The Academic skeptic, in contrast, takes herself seriously enough to undertake inquiry with constancy, but jovially enough to avoid being disappointed by the natural limits of her faculties. She does not philosophize bitterly and frees herself from philosophy happily while playing games with friends; nor does she extend her doubt to every possible belief held by humanity, as a cynic, misanthrope, or profoundly depressed person would do. She adopts beliefs when the occasion requires it, looking for what is more persuasive or probable. And she understands that dogmatism is a threat to sound reason, while radical skepticism, to a healthy emotional life. For this reason, she is cautious to always adopt beliefs in an undogmatic and diffident way (Fosl 2020, p. 31), allowing herself to ironically express her critical distance from both her own and other people's opinions. In fact, Hume's skeptical inclination to irony is one of the philosophical and stylistic features he adopts from Cicero and a mark of his Academic allegiances. Both philosophers found it difficult to make philosophy without a tincture of healthy sarcasm, and both considered themselves to be very far removed from the Stoic sage's gravity and self-assurance.

As a conclusion, then, I suggest that we consider Hume's Academic skepticism as a set of epistemic and practical attitudes that help overcome the damaging consequences of extreme Pyrrhonism—melancholy is replaced by a light-hearted diffidence; *apraxia* by the ability to act on examined but non-dogmatically adopted beliefs; and ostracism by the willingness to incorporate the skeptic's perspective into the philosophical and scientific conversation.

Now, of course, Hume not only praises Academic skepticism but adopts it in what I have previously referred to as his *normative* theory of belief. I shall now explore how he does it.

4.4 Hume's Normative Theory of Belief: Types of Judgments and Rules for Causal Inferences

To properly understand how Hume incorporated the skepticism of the New Academy into his theory of belief, let us return to the distinction made earlier between the *descriptive* and *normative* aspects of his theory of belief. The naturalistic or *descriptive* aspects hold that there are some non-rationally justified beliefs essential for human life that we adopt almost instinctively in order to act. While these beliefs, the so-called natural beliefs, may be temporarily suspended in the course of philosophical examination, they are necessarily re-adopted whenever we prepare to act. Some examples include the belief in the existence of the external world, causality, and personal identity.

In contrast to these descriptive aspects or descriptive theory *simpliciter*, we have a *normative* theory that claims that some beliefs should be adopted only to the extent that they can be *rationally justified* (Calvo 2012, pp. 453–507; de Pierris 2001, pp. 351–383, Loeb 2002; Falkenstein 1997, pp. 29–72; Hickerson 2013, pp. 1133–1153; Norton 1982, p. 235; Smith 2011, pp. 82–83; Stroud 1991, pp. 271–291). These beliefs, in other words, should be subjected to skeptical scrutiny, and only those that resist the test can be accepted, albeit non-dogmatically. Such a theory may be found in Hume' *Treatise* and first *Enquiry,* particularly in the former's section "Of Knowledge and Probability" (*THN* I, 69–179), and the latter's "Skeptical solution of these doubts" and "Of probability" (*EHU*, 40–55). The theory provides the criteria to evaluate claims, especially metaphysical ones, according to their degree of probability. It consists of a typology of judgments according to their degree of assuredness, into "knowledge," "proofs," and "probabilities," and a set of rules to be taken into account when dealing with the latter and making causal inferences. Although some commentators have already discussed these sections of the *Treatise* and the first *Enquiry* in relation to Hume's Academic allegiances (Calvo 2012, pp. 453–507; Fosl 2020, pp. 311–337; Norton 1982, p. 278; Smith 2007, pp. 105–126), I think they have not analyzed them with the degree of detail necessary to fully understand Hume's Academic skepticism. Norton, for example, only cursorily mentions how Hume's typology of judgments corresponds to Carneades's criteria to evaluate impressions,[32] while Fosl provides a general overview of the connection but does not sufficiently delve into it. The latter affirms:

> [. . .] Hume departs from Pyrrhonian passivity by crafting Academic technologies of doxastic management that skeptical philosophers can use consciously, actively, and affirmatively to "methodize and correct" beliefs. These Humean *technai*, guided by the self-understanding of skeptical reflection, can be used to discipline beliefs, to limit them, to attenuate them, and to temper and tranquillise them emotionally. Hume's rules for judging of causes and effects, along with various principles he articulates [. . .], are clear examples of this practice. Hume follows Carneades who [. . .] developed techniques for actively discriminating phenomena that are: 1). merely persuasive, from those that are both, 2). persuasive and undiverted, as well as from those that are 3). persuasive, undiverted, and also thoroughly explored or scrutinized, especially by considering the assessment of other inquirers (Fosl 2020, p. 323).

Fosl's analysis is accurate but, unfortunately, it lacks the degree of detail required to fully understand the link between Hume's normativity and Carneades's

[32] "Hume's own constructive but undogmatic probabilism bears some resemblance to the theory proposed by Carneades, or at least one might consider the extent to which Hume's knowledge, proofs, and probabilities correspond to the three levels of probability which Zeller and others have attributed to the earlier philosopher." (Norton 1982, p. 278).

method. Thus, in this section, I attempt to fill this gap by further examining Hume's probabilism and directly relating it with Carneades's skeptical system.[33]

First, it is easy to see how Hume's notion of "belief" has Hellenistic roots. He defines belief as "a lively idea related to or associated with a present impression" (*THN* I, 96). To exemplify it, Hume opposes a belief to a judgment that we consider fictitious. Why, for instance, do we believe that the city of Rome existed while we do not that the city of Atlantis ever took place? The difference, he argues, lies in the way we conceive the ideas corresponding to each city. Since the former idea is presented to the mind with more force and vivacity than the latter, we believe that Rome existed but not that Atlantis did (*THN* I, 108). Presumably, what Hume had in mind is that we have various and vivid historical testimonies –impressions or empirical evidence – of Rome, while in the case of the Atlantis, we lack a variety of them and the ones we have are not conclusive. Thus, believing in the existence of Rome depends on how its corresponding idea is presented to our mind in connection with the impression or set of impressions that originated it. Hume says:

> But belief is somewhat more than a simple idea. 'Tis a particular manner of forming an idea: And as the same idea can only be vary'd by a variation of its degrees of force and vivacity; it follows upon the whole, that belief is a lively idea produc'd by a relation to a present impression [. . .] (*THN* I, 97).[34]

P. Smith has accurately shown how Hume's conception of belief, rooted in the force or vivacity of an idea, follows Carneades's rationale. As he argues, "Carneades and Hume understand belief in terms of the impressions' intensity: it is the force and vivacity of a representation which makes us accept it, take it for true" (Smith 2007, p. 112; my translation). Not only Carneades's view but the more general Hellenistic notion of "assent" seems to be present in Hume's characterization of belief. Let us recall the Stoic epistemological view that a *kataleptic* impression was, first, presented to the mind, and then, grasped or assented to, given its clarity or evidence, just as a hand first receives an object and then securely clutches it (*Acad. I*, 41). The result of this process of assenting to a clear impression resulted, for them, in knowledge. And, even though the Academics denied *kataleptic* impressions, they accepted the general description of this mental process. For them, assenting to more or less clear impressions produced belief or opinion, which they represented with a half-opened, half-closed hand. Hume seems to agree with this Hellenistic understanding of assent when he affirms that the force and vivacity of an impression are transferred to its corresponding idea in such a way that the mind *believes* it. Believing is caused, then, by the quality of the impression or associated idea, and is conceived as an independent act of the mind, just as Stoics and Academics saw it. Hume describes this act of the mind as a *feeling*, that is, the way we are *affected* by the force and

[33] My analysis is based on the *Treatise*, but I will point to the corresponding passages of the *Enquiry* in footnotes.

[34] In the *Enquiry*, the definition appears thus: "I say, then, that belief is nothing but a more vivid, lively, forcible, firm, steady conception [*conception*] of an object, than what the imagination alone is ever able to attain." (*EHU*, 49).

vivacity of the impression or idea.[35] This definition has led commentators to once again find in Hume an irrational or, at least, a non-rationalistic foundation of belief. However, Hume's idea that belief is a feeling does not need to be understood in any mystified way. It suffices to acknowledge that believing is a possible *effect* of how ideas are presented to the mind, that is, of their force and vivacity. In other words, believing, for Hume, is just what the Stoics and Academics understood as *assenting* to impressions.

Now, for Hume, something else is needed to explain how this force and vivacity is transferred from an impression to an idea and from an idea to other ideas connected in judgment. So, he elaborated on how belief is produced and maintained through the principles of association of ideas—namely, resemblance, contiguity, and cause and effect. These principles transfer the force and vivacity of the initial impression to the idea directly related to it. In this way, we believe or assent to an idea based on a forceful and lively impression and connected to that impression through one of the three principles of association. For example, when I look at a picture of my friend on Instagram, if the impressions I obtain are sufficiently similar to the idea of him I already have, I *believe* that it is indeed my friend in the photo and that he is, let us say, on Copacabana beach. Thanks to the principle of *resemblance,* the vivacity of the impression is transferred to the idea in my memory, making me assent to the idea of my friend being in that enviable place.

The same occurs with *contiguity*. When I am in my hometown, for example, the vivacity of the impressions is transferred by the principle of contiguity to the idea I have of it so that I naturally assent to everything that happens there. It is not the same when I am traveling, and the distance causes my hometown's events to be more weakly present in my imagination. In that case, we have a sensation of estrangement and difficulty imagining facts and circumstances at home, which does not happen when we are there.[36]

Finally, for Hume, the principle of cause and effect confers the greatest force and vivacity to ideas so that, in general, when we connect impressions and ideas through cause and effect, we assent to or *believe* them. The example he uses to illustrate this principle is quite curious and, as Perinetti argues, may be linked to his experiences with the Jesuits at *La Flèche* (Perinetti 2018, p. 52). The relics of the saints, says Hume, allow devout believers of that "strange superstition" that is Catholicism to reaffirm their faith (*THN* I, 99). The argument seems to be that the liveliness of the impression from a piece of clothing or an object owned by the saint produces in the religionist a reaffirmation of her belief in the existence of such saint and his or her

---

[35] In Fosl's words: "To believe is to feel. It is to experience an impression or an idea with more or less of the press of nature, with simply more or less degree of 'force and vivacity'" (Fosl 2020, p. 318).

[36] "When I am a few miles from home, whatever relates to it touches me more nearly than when I am two hundred leagues distant; though even at that distance the reflecting on anything in the neighborhood of my friends or family naturally produces an idea of them. But as in this latter case, both the objects of the mind are ideas; notwithstanding there is an easy transition between them; that transition alone is not able to give a superior vivacity to any of the ideas, for want of some immediate impression" (*EHU* V, 52).

sanctity. Hume concludes: "This phenomenon clearly proves that a present impression with a relation of causation may enliven any idea, and consequently produce belief or assent, according to the precedent definition of it" (*THN* I, 101).[37]

As we can see, then, Hume's explanation of belief builds on the Hellenistic description of the mental process involved: we have an impression, whose degree of clarity or perspicuity leads us to assent to it. As a result of this assent, we have knowledge (if the impression is *kataleptic*) or opinion (if it is not). Hume's argument shares this general scheme and adds a number of mechanisms of association through which our imagination transfer the qualities of the impressions in question to their related ideas, and those of the resulting ideas to yet other ideas in judgment. These mechanisms make us believe or assent to a whole net of associated ideas and judgments to have whole experiences and clusters of beliefs. And finally, on the basis of these mechanisms, common-sense views and philosophical doctrines are formed and maintained.

However, this explanation of the internal mechanisms of belief does not necessarily imply a positive evaluation of their functioning. Hume recognizes that they can be a source of superstition, enthusiasm, and even madness. These mechanisms do not always yield beliefs justified from a rational point of view, particularly when they refer to metaphysical entities. For example, beliefs in testimonies of "apparitions, enchantments, and prodigies" (*THN* I, 113), says Hume, although based on the association principle of cause and effect, which transfers to the ideas involved the force or vivacity of the testimonies heard, have no rational grounds, even though many people believes them.

Hume also underscores other sources of false belief. Education is a very significant one, given that ideas are usually invigorated and reinforced through learning processes of repetition. Indeed, opinions reiterated time and again through one's childhood take on such a degree of forcefulness that, often, regardless of the efforts we make as adults to change them, we continue to maintain them inadvertently. Hume explains:

> I am persuaded that upon examination, we shall find more than one-half of those opinions that prevail among mankind, to be owing to education, and the principles, which are thus implicitly embrac'd, overbalance those, which are owing either to abstract reasoning or experience. As liars, by the frequent repetition of their lies, come at last to remember them; so the judgment or rather the imagination, by the like means may have ideas so strongly imprinted on it, and conceive them in so full a light that they may operate upon the mind in the same manner with those, which the senses, memory, and reason present to us (*THN* I, 117).

Finally, Hume observes that the passions can also endow an idea with the sufficient force and vivacity to make us assent to it. This mechanism is rather complex since the passions are usually interrelated so that, when a particular one enlivens the ideas involved in a certain judgment, they make us assent to it and hold a belief, which in

[37] The same explanation of the three principles of association of ideas as causes of belief and the same examples appear in the *Enquiry* (*EHU III*, 50–55).

turn produces other related passions, which, again, bring to the mind other ideas and make us believe in them, and so on. In other words, beliefs and passions reinforce each other:

> As belief is almost absolutely requisite to the exciting of our passions, so the passions in their turn are very favorable to belief; and not only such facts as convey agreeable emotions, but very often such as give pain, do upon that account become the more readily the objects of faith and opinion. A coward, whose fears are easily awakened, readily assents to every account of danger he meets with; as a person of a sorrowful and melancholy disposition is very credulous of everything that nourishes his prevailing passion (*TNH* I, 120).

In this way, superstitious people not only tend to believe the testimonies of extraordinary events, but such events reinforce their superstition and lead them to adopt other attitudes, such as fanaticism and enthusiasm. To correct this tendency, it is necessary to examine our beliefs skeptically and attach only its due value to each of them. Therefore, having described the mental mechanisms of belief, Hume devotes the second part of his theory to offering a set of criteria to determine whether we are justified or not in holding certain beliefs and to what extent. These are the normative aspects of Hume's theory of belief.

Hume begins by giving the reader a typology of beliefs into three categories: 'knowledge," "proofs," and "probabilities" (*THN I*, 124). He defines them according to the degree of assuredness or certainty we grant to each one:

> By knowledge, I mean the assurance arising from the comparison of ideas. By proofs, those arguments which are derived from the relation of cause and effect, and which are entirely free from doubt and uncertainty. By probability, that evidence which is still attended with uncertainty (*THN* I, 124).[38]

The first type, knowledge, involves "relations of ideas," that is, analytical and demonstrative propositions. These beliefs are the most certain because their logical necessity grants them the greatest degree of force and vivacity. The second type, proofs, designate judgments of experience or "matters of fact" – i.e., causal inferences – whose ideas have a very high degree of force and vivacity, since they have been observed in all past experiences. Hume's examples of proofs include: "all men are mortal," "the sun will rise tomorrow," and "bread is nutritious." In all past experiences, without exception, the sun has risen, men have died, and bread has been nutritious. Although we can assert proofs with great confidence, they are not certain because they do not involve necessity, as knowledge does—that is, we can still conceive as possible a contrary state of affairs. Finally, the third type of belief, *probabilities,* comprises judgments based on ideas with lesser force and vivacity since they have not been observed to obtain in all past experiences. Examples are: "parents love their children," "friends help friends," etc. Given that exceptions to these claims have been observed in past experiences, they involve various degrees of uncertainty. Thus, the difference between these three types of judgments depends on the forcefulness and liveliness of their related impressions and ideas, which determines the type of assent we accord them.

---

[38] The classification of judgments into knowledge, proofs, and probabilities appears in a footnote in section VI of the *Enquiry* (*EHU* VI, 56).

As commentators have acknowledged, this typology directly relates to Academic skepticism (Calvo 2012, pp. 453–507; Fosl 2020, pp. 311–337; Norton 1982, p. 278; Smith 2007, pp. 105–126). In my view, even though largely Academic, the typology also includes a concession to Stoicism. As we have seen in Chap. 1, the Stoics considered that only judgments based on *kataleptic* impressions were to be assented to, given that only these impressions provided knowledge. All other judgments ought to be rejected because they only constituted opinion and, hence, were, for the most part, false. Hume seems to have accepted the Stoic view that there are impressions and ideas with such a force and vivacity that their judgments are held with the greatest degree of assurance or certainty, and should be considered "knowledge." But Hume did not think, like the Stoics, that all the remaining judgments, founded on less forceful and lively ideas, belonged to the same category, namely opinion. Neither did he think that they should be rejected or not assented to, given their obscurity. And this is the point where his typology becomes Academic.

The Academics, let us recall, argued that, since there were no *kataleptic* impressions, knowledge accompanied by certainty was not attainable, and we could only have belief or opinion based on more or less persuasive or probable impressions. Once their degree of persuasiveness had been determined by skeptical examination, these beliefs could be approved for practical reasons, but never assented to, that is, never taken as true. To assess their degree of persuasiveness, they followed the complementary criteria of the "persuasive," "tested," and "irreversible" impressions. There are some parallels between these criteria and Hume's two remaining types of beliefs. Hume distinguishes between proofs and probabilities, depending on whether or not the judgment under examination obtains in all present and past experiences. On the one hand, proofs correspond to judgments based on impressions that, for the Academics, were not only "persuasive" but also "tested" and "irreversible," because a thorough examination, both in present and past circumstances, showed a very high degree of persuasiveness. On the other, Hume's probabilities accord with the Academics' impressions that were deemed persuasive enough, but not "tested" and "irreversible"—that is, sufficiently forceful impressions that are, however, connected to opposite ones in present or past experiences.

Let us delve more into this parallel by using some examples. First, the stability of proofs seems to be due to the fact that these claims meet both the Academic criteria of a "tested" impression and of the "irreversible" or "undiverted" ones. Upon skeptical examination, proofs such as "bread is nutritious," "the sun will rise tomorrow," and "men are mortal," would be "persuasive," "tested," and "irreversible," since they are the same in all observed cases and agree with all other present and past experiences on the matter. And second, Hume's probabilities, such as "parents love their children" and "friends help friends," broadly correspond to the Academics' merely "persuasive" impressions, since, although they are likely to be the case, experience has shown them not to obtain sometimes. These claims are still very likely but, unfortunately, not entirely sanctioned by experience. They should, then, be accepted with a more cautious attitude.

Some commentators, however, reject the parallel just sketched for reasons that I find plausible but insufficient. P. Smith, for example, finds that Hume's three kinds of judgments do not correspond to Carneades's levels of persuasiveness, since the latter did not distinguish between "relations of ideas" and "matters of fact," and did not accept the sort of certainty that Hume seems to attach to mathematical "knowledge."[39] Evidently, the Hellenistic schools and, among them, the New Academy, did not make the very modern distinction between abstract and empirical thinking. For them, "impressions" in general, as mental representations, constituted the tapestry of both our empirical and non-empirical – even metaphysical – judgments, as we have seen in Chaps. 1 and 2. However, this objection is easily met if we consider, as I suggested earlier, that the kind of belief Hume calls knowledge includes the sort of claims the Stoics judged to be founded on *kataleptic* impressions. Moreover, since Hume's typology is based on the sort of assuredness or certainty attached to the kind of belief in question, I think we do not need to make the distinction between abstract and empirical claims but are allowed to simply argue that knowledge, for Hume, accords with Stoic certainty.

Smith also claims that the parallel is unwarranted because Hume's concept of probability refers to mathematical frequency while Carneades's indicates persuasiveness:

> In Hume, probability has a mathematical-statistical meaning, which reflects the frequency of the relationship between kinds of ideas, as this relationship has been observed in experience. In his three levels, Carneades spoke of persuasion, not of frequency, and the translation of *pithanê* by *probabile*, although correct in Cicero's time, does not correspond to our notion of probability, and, for this reason, is liable to equivocation and misunderstandings. Thus, the interpreters who seek to relate Hume's probabilism to that of Carneades make a mistake (Smith 2007, p. 111, my translation).

Smith is right to point to the fact that our contemporary concept of probability indicates mathematical frequency and is very far removed from Cicero's notion. However, I disagree with him on the claim that Hume's use of the concept is restricted to mathematical or statistical probability. We should take a look at Hume's examples to see how, indeed, Hume's concept of probability rather accords with Cicero's. He speaks, for instance, about the probability of the testimonies of miracles, which we would never deem as a statistical issue. He also includes among probabilities those based on analogies, like the probability involved in the argument from the design of the universe—namely that there is a god, creator of the universe, who is a mind similar to the human mind. Given that these examples are hardly a matter of mathematical frequency, I believe he has in mind a concept of probability that is rather close to Carneades' persuasiveness or Ciceronian probability. Perhaps our own tendency to understand probability in terms of mathematical frequency makes us read this meaning into Hume's concept, while he really had in mind the ancient meaning of the term.

---

[39] Smith claims: "Hume's three levels (knowledge, proofs and probabilities) have no correspondence whatsoever with Carneades' three levels of probability, as Norton indicated. In Hume, there is always a relation of ideas, while for Carneades, that does not occur in the first level. Hume distinguishes between two kinds of knowledge (mathematical and empirical). Carneades never considers anything similar and for him the three levels, far from separating the two fields of knowledge, share increasing degrees of certainty" (Smith 2007, p. 111, my translation)

Still, someone could object that Hume's use of the Academic criteria to distinguish degrees of persuasiveness in his typology of belief does not necessarily entail that this typology is normatively intended. In other words, while it is very clear that the Academics' put forward these criteria to evaluate claims rationally, Hume's division of judgments into "knowledge," "proofs," and "probabilities" could be merely descriptive, providing only a classification of judgments according to the sort of assent we *naturally* give to them. I concede that in many sections of the *Treatise* it is difficult to determine to what extent Hume sees the adoption of belief as the result of voluntary reflection or as the product of sheer natural mechanisms. Indeed, the doubt that accompanies judgments contradicted by past irregularities does not always come across as a voluntary and reflective act. Doubting is, in principle, as natural as believing. If an idea lacks sufficient force and vivacity, we simply do not assent to it.

Nonetheless, I think this is not a completely valid objection. In the first place, we should not forget that, when presenting his definition of belief and describing the mechanisms of association of ideas at their basis, Hume referred mostly to unjustified beliefs, like those of devout Catholics who are more convinced of their doctrine when they are in contact with the relics of saints (*THN* I, 99). The choice of these examples and his emphasis on how education and the passions reinforce prejudices implies that these mechanisms sometimes are undeserving of trust and should be corrected. Second, we should have in mind that Hume's treatment of probabilities involves a very detailed examination of their internal functioning, and this suggests that, unlike knowledge and proofs, this type of judgment requires a more thorough assessment, which, again, cannot be left entirely to our natural inclination to believe. And, third, Hume devotes a section of his argument to "unphilosophical probabilities" and offers a set of rules by which we are to control the causal inferences on which most probabilities rests. This section, in my view, clearly shows that Hume's intention is not only descriptive but also prescriptive. Let us now see how these arguments hold, with an examination of Hume's account of the formation of probabilities and the type of assent we should accord them.

4.4.1 Probabilities and Their Internal Mechanisms of Association

Even though Hume claims that there are two kinds of probabilities, depending on whether they are founded on chances or causes and effects,[40] in this section, I will only refer to the latter, which is most relevant for our purposes. Probabilities concerning causes and effects are based on the association principle that bears the same name, namely, the principle of causality. For Hume, this principle confers the greatest force and vivacity to their connected ideas. For example, when we hear a knock on the door, we transfer the force and vivacity of its impression to the idea of a

[40]The same classification is found in the *Enquiry* (*EHU* VI, 56–59).

person causing the sound behind the door, and we assent to this idea or *believe* that someone is knocking on the door. In this way, the principle of cause and effect makes us assent to the most basic empirical claims and have a connected experience of the world around us. However, when judging from present causes to future effects, we exert or should exert caution. In these cases, our particular causal inferences may be countered by opposing impressions and ideas obtained in past experiences. Hume explains:

> A contrariety of events in the past may give us a kind of hesitating belief for the future after two several ways. First, by producing an imperfect habit and transition from the present impression to the related idea. When the conjunction of any two objects is frequent, without being entirely constant, the mind is determined to pass from one object to the other; but not with so entire a habit as when the union is uninterrupted, and all the instances we have ever met with are uniform and of a piece (*THN* I, 132).[41]

Cases in which past experiences are not entirely uniform are typical of probabilities so that, when predicting future effects, doubts proportional to these variations usually arise. In other words, specific predictions, such as "the temperature will drop significantly next winter," are never completely certain, particularly when past experiences have not been consistent, and, thus, the belief is a mere probability. Hume's account of how this happens is quite interesting:

> [...] when considering past experiments, we find them of a contrary nature, this determination, though full and perfect in itself, presents us with no steady object but offers a number of disagreeing images in a certain order and proportion. The first impulse, therefore, is here broke into pieces, and diffuses itself over all those images, of which each partakes an equal share of that force and vivacity, that is derived from the impulse (*THN* I, 134).

Hume's imagery and vocabulary in this paragraph certainly remind us of Cicero's and Sextus' terminology when describing the "tested" and "irreversible" impressions (*Acad. II.* 99; *Pyr. I*, 227–229). These impressions, they argued, are not contradicted, hindered, or pulled in another direction by related ones. In the same vein, Hume says that when we find contradicting past experiences, the initial impulse to believe the causal inference is shattered and distributed among all the contrary impressions, so that we cannot assent to it. Hume's re-creation of the Academic's view is undoubtedly imaginative and suggests an interest to make a psychological interpretation of the Academics' methodological criteria. Beyond this, it is clear that, in probabilities, both the Academics and Hume think that the impulse to assent to the

[41] The argument appears in the *Enquiry* thus: "There are some causes, which are entirely uniform and constant in producing a regular effect; and no instance has ever yet been found of any failure or irregularity in their operation. Fire has always burned, and water suffocated every human creature: The production of motion by impulse and gravity is a universal law, which has hitherto admitted of no exception. But there are other causes, which have been found more irregular and uncertain; nor has rhubarb always proved a purge, or opium a soporific to everyone, who has taken these medicines [...]. As a great number of views do here concur in one event, they fortify and confirm it to the imagination, beget that sentiment which we call *belief*, and give its object the preference above the contrary event, which is not supported by an equal number of experiments, and recurs not so frequently to the thought in transferring the past to the future" (*EHU*, 57–59).

impression is hindered or diverted by opposing impressions and this lessens or ought to lessen our inclination to believe in the particular inference.

Besides these probabilities, in the *Treatise,*[42] Hume also speaks, albeit tangentially, about probabilities derived from analogies. This type of probability combines two principles of association of ideas, namely cause and effect, and resemblance. He describes it as follows:

> [..] all kinds of reasoning from causes and effects are founded on two particulars, i.e., the constant conjunction of any two objects in all past experience and the resemblance of a present object to any one of them. The effect of these two particulars is that the present object invigorates and enlivens the imagination; and the resemblance, along with the constant union, conveys this force and vivacity to the related idea; which we are therefore said to believe or assent to it. If you weaken either the union or resemblance, you weaken the principle of transition, and of consequence that belief, which arises from it (*TNH* I, 142).

Probabilities based on analogies are complex inferences involving two sets of ideas related by the principle of cause and effect. What makes possible the relation is the similarity between the ideas involved so that we confer the force and vivacity of the first set, which is directly experienced, to the second one, in which one term is not experienced. As a result, we conclude that the causal relation in the first set also holds for the second. Unfortunately, Hume does not give us examples of this kind of probability in the *Treatise*. We know that, before publishing the book, Hume wanted to send it to the theologian Joseph Butler and, to that purpose, omitted in the final text some passages that could be considered frontal attacks on religious dogma. In his letter of 1737 to Henry Home, Hume said that he was "castrating" the *Treatise* of all his arguments on miracles, in his words, "cutting off its nobler parts" so that it would give "as little offence as possible" to Butler's convictions—an operation without which he considered himself unable to put the book "into the Doctor's hands" (*Letters*, Vol. I, p. 25; cf. Gaskin 1978, pp. 1–2; Fosl 2020, p. 160). In my view, the lack of significant development of this type of probability in the *Treatise* may also be due to Hume's having cut from the text the critical references to the design argument. Hume may have thought that his critique of this argument, just as his objections to the testimonies of miracles, would not be welcomed by Butler. As we know, Hume's critique of the design argument had to wait to be published posthumously in the *Dialogues Concerning Natural Religion;* even though we do have a taste of it in section 11 of the first *Enquiry* (EHU XI, 132–148). Since Hume had worked on it throughout his life, it could have followed the description of probabilities based on analogies in an earlier version of the *Treatise.* But, of course, this is only a speculation. In any case, Hume thought that the theistic belief of the argument from the universe's design was a probability based on analogy, and should be subjected to skeptical examination. This suggests that his reflection on probabilities had the normative purpose of indicating when we should assent to beliefs of this sort and when we should suspend them. Since the most explicit expression of this normative intention is his outline of a set of rules for controlling causal inferences, I will now turn to discuss these rules.

[42] There is no reference to this kind of probability in the corresponding arguments of the *Enquiry*.

4.4.2 Ciceronian Rules for Judging of Causes and Effects

In the section of the *Treatise* devoted to "unphilosophical probabilities," Hume discusses the causes of prejudice and how it can be overcome (Fogelin 2009, pp. 35–37). Hume argues:

> An [. . .] unphilosophical species of probability is that derived from general rules, which we rashly form to ourselves, and which are the source of what we properly call prejudice. [. . .] From this principle, I have accounted for that species of probability, derived from analogy, where we transfer our experience in past instances to objects which are resembling but are not exactly the same with those concerning which we have had experience (*THN* I, 147).

In brief, Hume thought that prejudices arise from the principle of resemblance and cause and effect, particularly in probabilities based on analogies. They form whenever we extend the characteristics of objects known to us to similar but unknown objects through the mentioned principles of association. For example, national or cultural stereotypes are based on the transfer of the characteristics of a particular individual known by experience to all the unknown members of the same social group (*THN* I, 146; Fogelin 2009, pp. 35–37). In other cases, the principle of resemblance is complemented by cause and effect, and we have a more complex probability, based on an analogy. For instance, people may argue that since Italians and Latin Americans tend to be undisciplined and disobedient to the law, the behavior of COVID-19 contagion in Italy during the first months of the pandemic was also to be expected in Latin American countries. But of course, not only the similarities between Italians and Latin Americans are not so prominent, but the cause of a high rate of mortality in Italy or Latin America was not necessarily due to disobedience to sanitary measures. Thus, the final inference is not valid, even though the association mechanisms do transfer the force of vivacity of the first set of ideas to the second.

In the passage mentioned above, it is important to draw our attention to Hume's allusion to "general rules." It is precisely in the context of "unphilosophical" probabilities that Hume expresses most clearly the normative aspect of his theory of belief. He argues that causal inferences should be regulated by a set of second-order rules, which help us make better causal conjectures. Generally speaking, he argues that "unphilosophical probabilities" are probabilities to which we assent spontaneously, based on specific "general rules," which seem to comprise the three principles of association of ideas already explained. But there is another set of rules, as it were, second-order rules, that help us control these inferences. He explains:

> Thus, our general rules are in a manner set in opposition to each other. When an object appears that resembles any cause in very considerable circumstances, the imagination naturally carries us to a lively conception of the usual effect, though the object be different in the most material and most efficacious circumstances from that cause. Here is the first influence of general rules. But when we take a review of this act of the mind and compare it with the more general and authentic operations of the understanding, we find it to be of an irregular nature and destructive of all the most established principles of reasoning; which is the cause of our rejecting it. This is a second influence of general rules and implies the condemnation of the former. Sometimes the one, sometimes the other prevails, according to the disposition and character of the person. The vulgar are commonly guided by the first, and wise men by the second (*THN* I, 150).

In this paragraph, we are told that our mechanisms of belief, based on the principles of association of ideas, can easily lead to prejudice and superstition, but that reflection – "when we take a review of this act of the mind and compare it with the more general and authentic operations of the understanding" – counteracts this tendency. Once again, Hume sees this conflict as one of the imagination versus the understanding. The former faculty, supported by the principles of association of ideas, makes us assent to insufficiently justified causal inferences, while the latter counteracts the first tendency on the grounds of its own set of principles. Commentators debate whether there are two sets of rules, the first identified with the imagination and the second with reason, or whether there is only one set of rules, the product of the imagination, on which the understanding reflects and exerts a corrective force (Calvo 2012, pp. 456–482; Falkenstein 1997, pp. 47–49; Fogelin 2009, pp. 37–38; Fosl 2020, p. 323; Hickerson 2013, pp. 1144–1149; Loeb 2002, pp. 25–29). Be it as it may, the normativity of the second set – whether an independent one or a derivation from the first – is not contested. Some second-order rules correct the imagination's original inclination to believe in "unphilosophical" probabilities.

Nonetheless, along with this normative declaration, we can also find naturalistic aspects in Hume's argument. For example, when Hume says that "sometimes the one, sometimes the other prevails, according to the disposition and character of the person," he seems to be claiming that, after the combat between the faculties, and because suspension of judgment cannot be maintained without the consequences we already mentioned, we *naturally* assent to one of the inferences made. In other words, we either believe our imagination or our understanding, "depending on our disposition and character." However, at the end of the paragraph, he also says that "the vulgar are commonly guided by the first, and wise men by the second," which implies that the second-order rules may actually be chosen as rational standards for adopting beliefs. Hume's argument seems to be that uneducated people are carried out by their imagination and habits – that is, by first-order general rules – while philosophers or experts consciously and reflectively adopt second-order rules to counterbalance the first ones. For this reason, he considers it necessary to offer a compilation of the latter rules, which control causal inferences by not letting our imagination lead us to assent to fanciful or not sufficiently proven causes:

> We shall afterwards take notice of some general rules, by which we ought to regulate our judgment concerning causes and effects; and these rules are form'd on the nature of our understanding, and on our experience of its operations in the judgments we form concerning objects. By them, we learn to distinguish the accidental circumstances from the efficacious causes; and when we find that an effect can be produced without the concurrence of any particular circumstance, we conclude that that circumstance makes not a part of the efficacious cause, however frequently conjoin'd with it (*THN* I, 149).

In the *Treatise*'s section entitled "Rules by Which to Judge of Causes and Effects," Hume lays out these rules.[43] Their purpose is to help the philosopher make more refined causal inferences, accurately determining which effects follow which causes,

[43] The rules do not appear in the *Enquiry*.

even when they seem to be the consequence of other causes. They also help distinguish between the essential and accidental aspects of a cause. For example, in the case of a judgment as "Illegal immigrants are driving the country to economic recession," the second-order rules should make us revise habitual motives that cause this judgment – racism, fear of losing economic opportunities, cultural stereotypes that immigrants are lazy, etc. – and take into account other possible causes of the situation, as the impact of global economic measures, reduced wages, inflation, etc., on the local economy. In this way, applying second-order rules allows the philosopher, social scientist, or, simply, the well-informed citizen to draw more accurate conclusions about the real causes of the country's economic recession.

Hume lists eight rules by which judgments about causes and effects must be guided:

1. The cause and effect must be contiguous in space and time.
2. The cause must be prior to the effect.
3. There must be a constant union between the cause and effect [...].
4. The same cause always produces the same effect, and the same effect never arises but from the same cause.
5. [...] Where several different objects produce the same effect, it must be by means of some quality which we discover to be common amongst them.
6. [...] The difference in the effects of two resembling objects must proceed from that particular in which they differ. For as like causes always produce like effects, when in any instance, we find our expectation to be disappointed, we must conclude that this irregularity proceeds from some difference in the causes.
7. When any object increases or diminishes with the increase or diminution of its cause, it is to be regarded as a compound effect, derived from the union of the several different effects, which arise from the several different parts of the cause.
8. [...] An object, which exists for any time in its full perfection without any effect, is not the sole cause of that effect but requires to be assisted by some other principle, which may forward its influence and operation (*THN* I, 173–174).

Many commentators (Hendel 1963; Passmore 1968, pp. 42–64; Chappell 1963; Flew 1961; Connon 1979, p. 142; Fosl 2020, pp. 150–152) have drawn parallels between this section of the *Treatise* and the rules that Newton sets out in his *Principia*. In general terms, the interpretation that Hume followed Newton in this and other respects of his science of human nature originates in Hume's indication, in the subtitle of the *Treatise,* that he intended to "introduce the experimental method of reasoning into moral subjects." Persuaded by this comment, scholars have stressed in different ways Hume's indebtedness to Newton. Fosl, for example, accepts that "Newtonian science [...] informs both the rhetorical framing of Hume's 'Attempt to Introduce the Experimental Method of Reasoning in Moral Subjects' (the subtitle of the *Treatise*) and the philosophical content of his naturalism" (2020, p. 150). While I acknowledge the rhetorical force of the *Treatise*'s subtitle, I concur instead with Noxon (1964), Jones (1982), and others who find that Newton's influence on Hume has been overemphasized. As I intend to show in what follows, it was Cicero, not Newton, who primarily influenced Hume's rules for causal inferences. But before I draw the parallels between Hume's rules and the Academics' system, let us examine for a moment the extent to which a close relation between Hume and Newton can be established.

At least three of Noxon's and Jones's clarifications on the relation between Hume and Newton are worth mentioning. First, they both claim that, in the historical context of the *Treatise*, extending the experimental method to human affairs was a generalized aspiration and did not necessarily allude to the particular adoption of Newtonian principles (Noxon 1964, pp. 27–33; Jones 1982, p. 7). Second, they argue that Hume rarely references Newton in either his philosophical works or his correspondence (he only mentions the latter seven times in all his works), and the few references that exist are irrelevant from a philosophical point of view (Noxon 1964, pp. 68–76; Jones 1982, p. 11). Based on this, they argue that it is probable that Hume's acquaintance with Newton's theories came from popular, second-hand sources, and not from a direct reading of his writings (Noxon 1964, pp. 68–76; Jones 1982, p. 11). Third, they both circumscribe any influence of Newton on Hume to some remarks in the *Treatise* and argue that in his mature writings, Hume showed himself increasingly concerned with the refutation of theological doctrines based on Newtonian ideas, particularly those of natural theology (Noxon 1964, p. 65; Jones 1982, p. 16) and, thus, he progressively escaped the shadow of the great scientist. Noxon concludes:

> Hume's relations to newtonianism were too complex and equivocal to be adequately expressed by calling him the Newton of moral sciences. That formula states a half-truth, which needs to be supplemented by stressing his extreme aversion to the theological developments of the Newtonian synthesis, which were so fashionable in his day (Noxon 1964, p. 77).

To these arguments, I should add that we do not have sufficient empirical evidence of Hume's reading of the *Principia*. Perinetti claims, for example, that the text was not available at *La Flèche*'s library during the time of Hume's composition of the *Treatise*.[44] Of course, he could have obtained the book before his visit to *La Flèche*, in another library, or from a seller in Great Britain, but, at least, this attests to the fact that Newton's *Principia* was not so widely read in Europe at the moment. We, then, should take with a grain of salt the view that Hume modeled his own rules for causal inferences on Newton's *Principia*.

Nonetheless, to be as cautious as Hume suggested, before accepting that the parallel is not plausible, it is reasonable to compare the two lists of rules and determine their similarities and differences. Newton's rules are as follows:

> 1. No more causes of natural things should be admitted than are both true and sufficient to explain their phenomena. [. . .] For nature is simple and does not indulge in the luxury of superfluous causes.
> 2. Therefore, the causes assigned to natural effects of the same kind must be, so far as possible, the same [. . .].

[44]"Newtonian philosophy – says Perinetti – had very little influence on the curriculum of French colleges before 1740. It is, then, not surprising if the only pre-1737 work by Newton in the catalogue of *La Flèche* library is a 1720 translation of the *Optiks*. The edition of the *Principia* held by the college is only of 1760." (2018, p. 51).

3. Those qualities of bodies that cannot be intended and remitted [i.e., qualities that cannot be increased and diminished] and that belong to all bodies on which experiments can be made should be taken as qualities of all bodies universally.
4. In experimental philosophy, propositions gathered from phenomena by induction should be considered either exactly or very nearly true, notwithstanding any contrary hypotheses, until yet other phenomena make such propositions either more exact or liable to exceptions (Newton 1999, pp. 794–796).

The most remarkable similarity between the two sets of rules lies in Hume's fourth rule – i.e., "the same cause always produces the same effect, and the same effect never arises but from the same cause" – and Newton's second rule – i.e., "the causes assigned to natural effects of the same kind must be, so far as possible, the same." Hume said of his fourth rule that "this principle we derive from experience and is the source of most of our philosophical reasonings" (*TNH* I, 173). Given this similarity, it is indeed possible that he recognized this Newtonian principle as fundamental both for empirical science and for the "science of human affairs." I grant that this could be one of the few points of Newtonian influence on Hume's rules. Apart from this, however, the remaining rules bear little similarity to the Newtonian principles. Most of Newton's rules are guidelines for generalization, while Hume's have a rather critical function. Newton's third rule, for example, recommends attributing universality to those qualities of objects which have been observed to be stable: "Those qualities of bodies that cannot be intended and remitted [i.e., qualities that cannot be increased and diminished] and that belong to all bodies on which experiments can be made should be taken as qualities of all bodies universally." Equally, his fourth rule advises considering as "exactly or very nearly true" inferences "gathered from phenomena, notwithstanding any contrary hypotheses, until yet other phenomena make such propositions either more exact or liable to exceptions." This is a rule clearly rejected by Hume since his view is instead that, in the case of causal inferences involving contradictory hypotheses, we should beware of falling into prejudices and inexact causal predictions and refrain from assent. Hume's rules emphasize skeptical caution when making causal inferences and indicate how to control inferences to avoid mistaken inductive generalizations by regulating or consciously redirecting the natural operations of our imagination. They do not provide ways to connect hypotheses into larger theoretical systems, as Newton's seem to do. This difference, I believe, is crucial to understand the normativity embedded in Hume's theory of belief as an expression of his Academic skepticism. Hume's rules advise a critical attitude, based on attention to irregularities and conscious control of our naturally or spontaneously made inferences. This, of course, does not mean that Newton's rules are naïve or suggest us to believe in all sorts of causal generalizations. As Fosl argues, Newton also had skeptical propensities, mainly expressed in his *non-fingo* principle, which seems to enact the ancient skeptical idea that we cannot know anything about what lies hidden beyond *phenomena*.[45] However, Newton's rules are more constructive than Hume's because the

[45] Since, according to Fosl, Hume "acknowledges the skeptical qualities of Newton's work" both in the *History of England* and the first *Enquiry*, Newton can be listed as one of Hume's possible skeptical influences (Fosl 2020, pp. 151–152).

former set indicates how to produce an organized empirical theory, while the latter aims to provide ways to control causal inferences.

Now, if Newton did not exert influence on Hume's rules, why should we believe that Cicero or the Academy did? First, because, as we have seen, his conception of belief and probability bears remarkable resemblance with the Academics'. And, second, because in this discussion about the rules for causal inferences, Hume constantly refers to the distinction between the wise person and the vulgar (*THN* I, 132–150; *EHU* VI, 58), which we have seen is an essential distinction in the dispute between Stoics and Academics.

Indeed, Hume's use of this distinction has caused much surprise among commentators. For example, Passmore noted that Hume does not answer the question about the difference between the wise and the ordinary person, but assumes, according to him without any basis, that the reader knows what the difference is (Passmore 1968, pp. 42–64). Noxon also accepted that Hume does not explain the grounds of the distinction and simply responded that the wise person is, for Hume, someone who follows the rules for causal inferences (Noxon 1964, pp. 84–85). More recently, Calvo argued that Hume's wise person should be equated to Rawls' "reasonable" individual—a person who has not only epistemic expertise but also social sympathy and thoughtfulness (Calvo 2012, pp. 454–455). I agree with Noxon that the wise person, or the philosopher, employs the rules for causal inferences when judging probabilities. Besides this, I think that Hume's characterization of the wise person also includes, as Calvo's describes:

> [...] the intellectual virtue of prudence involved in adjusting one's beliefs to evidence, the strength of mind necessary for calmed passions to rule over violent ones, [...] judging based on fine moral sentiments, and exerting the sort of perfect virtue typified by Cleanthes in the *Enquiry on the Principles of Morals* (Calvo 2012, p. 487, my translation).

However, I believe all these commentators fail to understand Hume's distinction between the sage and the ordinary person because they do not connect Hume's terms in these passages with Academic skepticism. Neither of them has explored the possibility that Hume's wise person is inspired by the figure of the Academic sage, who, as Cicero described, evaluates impressions, decides on their degree of persuasiveness, and approves only the most probable ones to act on their grounds. The reason for this, I believe, is that they were thinking of Hume's rules as having a prominent scientific function. And, in fact, in one passage, Hume distinguishes between uneducated people and natural philosophers or scientists saying that, when a particular causal prediction fails, the uneducated person tends to attribute the failure to a defect in the causal power itself, while the researcher of nature is disposed to ascribe it to a set of other possible and unperceived natural causes (*THN* I, 132). However, most of his examples are not scientific but technical or practical. For instance, Hume claims that when a clock functions improperly, an ordinary person might say the clock is broken and simply dispose of it, while the watchmaker would look for a different cause of the failure – a particle of dust obstructing the mechanism, for example – and repair it (*THN* I, 132). Despite the elitism of Hume's argument, the example is important to understand the philosophical context of Hume's discussion. If we keep in mind that the Hellenistic schools, particularly the Stoics and Academics, considered the sage an "expert" who carefully analyzes particular impressions before assenting to them and acting on their basis, then, the example becomes clear. Hume was placing himself in the context of the dispute between Academics and Stoics. His idea of a wise person seems to be that of a reflective individual who examines her usual causal inferences and cautiously decides to make more observations before passing judgment when facing something that seems unusual. She attempts to explain the exception by other possible causes or refrains from generalization when the probability is based on an imperfect analogy or similarity. More importantly, as the example of the watchmaker illustrates, this wise person uses the newly acquired piece of knowledge to act accordingly. Thus, the person Hume describes in these passages– and I grant that, by calling this person a "sage," he does it in a rather outmoded way – is a cautious or prudential agent, someone who does not leave her beliefs to natural inclinations or is easily satisfied with the most apparent explanation, but rather continues the inquiry until she obtains sufficiently probable judgments to act on their basis.

For this reason, stressing on the humanistic inheritance of Hume's thought, Jones claimed: "Hume's lack of interest in science finds an explanation in his deep commitment to Ciceronian humanism, with its distinctive attitude to man and to philosophy; such views separated him from Newton" (Jones 1982, p. 42). I would add to this assessment that Hume's rules for causal inferences are essential for evaluating all types of probabilities, but, coming from Cicero's Academic skepticism, they are especially suited to claims involved in practical rationality and decision making. As we have seen in Chaps. 1 and 2, the relationship between probabilities and practical life is a prominent feature of Academic skepticism. The skeptical sage is not primarily a natural scientist who has perfected the inductive method, but rather a prudent person who knows how to correct her natural leanings towards dogmatism, prejudice, and speculative obstinacy, and orients her action by probable, carefully examined, and non-dogmatically adopted beliefs. For this reason, I suggest that Hume's normative theory, particularly his rules for controlling causal inferences, has more of a practical goal in mind than a scientific one.[46] Indeed, *probabilities* are as much involved in practical deliberation as in empirical science. Pondering probabilities is precisely the core of our prudential reasoning, and, thus, learning how to evaluate probabilities using the rules of causal inferences is essential for decision-making and acting.

Let me finish this section with an example. A woman who has been taught by past experiences to believe that hard work leads to success may consistently think that her efforts will allow her to achieve her goals. But, as it is all too common, she may work very hard and adeptly in a given case, and yet be unable to reach her ends. Then, she may realize that no matter how capably, effectively, and hard she works for her company, she will not get a promotion. In this case, she should pause for a moment and try to find other causes that produced such an outcome – for example, gender biases that make it difficult for women to be taken as seriously in the workplace as men – and she should try to take these factors into account before acting in the future.

---

[46]For a discussion of Hume's normativity regarding morals see: Calvo 2012, pp. 468–472; Fosl 2020, pp. 311–314.

In doing this, she would be following Hume's sixth rule, according to which: "[...] as like causes always produce like effects, when in any instance, we find our expectation to be disappointed, we must conclude that this irregularity proceeds from some difference in the causes." (*THN* I, 173–174). We can imagine that this woman would have a strong tendency to maintain her first belief in the causal relation between hard work and success and perhaps continue with her high achievement standards. But if she is indeed a prudent, rational agent – or in Hume's words, a wise person – in her future deliberation, she should counter this belief with her newly acquired information that sexism may be a real cause of her struggles as well as of other hard-working women's. Thus, when trying to obtain a promotion, she would at least not be too harsh on herself and try to effect changes in her working environment through other means—promoting reflection on gender biases, for example. As we see from this example, practical deliberation involves the sort of causal inferences that Hume thinks we should control on the basis of his set of rules. These rules do apply to morals as much as to natural science.

Now, one more aspect of Hume's normative theory is important not to lose sight of. After applying the rules for causal inferences and obtaining a comprehensive picture of the causes involved in the issue at hand, the wise person, in the Carneadean fashion, does not *assent* to the resulting judgment but merely *approves* it. This is to say that the application of rules or criteria for the evaluation of probabilities does not provide a new dogmatic outcome—the wise person not only is wise to the extent that is disposed to examine her beliefs, but also to the extent that, once examined, she still does not take them as evident or certain, but merely accepts their probability, and use them to guide action. Stressing this aspect of Hume's Academic debt, Fosl has called the type of belief Hume recommends after examination a "non-dogmatic philosophical assent" (Fosl 2020, p. 321). With this denomination, he recognizes that Hume's Academic belief – the result of applying his normative theory – is neither the sort of belief held by dogmatists nor the kind of yielding to appearances of the Pyrrhonists (Fosl 2020, p. 322). Fosl argues that this undogmatic belief corresponds to the Clitomachean interpretation of Carneades's system, emphasizing the Academic's "approving" of impressions instead of "assenting" to them. As I mentioned in Chap. 2, on this interpretation – which I take to be Cicero's too – Carneades's sage does not dogmatically assent to the beliefs that have resisted skeptical examination, but only approves them for practical reasons, that is, in order to act, and continues to submit them to theoretical doubt. Hume, says Fols, inherits this understanding of non-dogmatically accepting sufficiently justified probabilities:

> When probability [...] [is] naturally and fatefully restored in the wake of those skeptical arguments, what sort of probability is it? If probability were to be understood in a Metrodorian sense, Hume would remain, even in the face of those unrefuted skeptical arguments, an epistemological dogmatist. He would be a qualified dogmatist, who holds that humans can achieve limited epistemic gains but not perfect knowledge. It would render Hume a dogmatist concerning knowledge of the natural world in the way that Locke is a dogmatist. This reading, however, would be mistaken (Fosl 2020, p. 325).

I agree with Fosl's assessment that Hume accords to probabilities a "non-dogmatic assent." However, this aspect of Hume's normative theory of belief is not so easy to grasp because it does not involve overtly stated precepts or rules. But, we can find it obliquely stated in Hume's praise of Academic skepticism, to which I have referred before. The "diffident" attitude of this skeptic does speak of an undogmatic acceptance of beliefs. Let us remember what Hume said in this respect: "A true sceptic will be diffident of his philosophical doubts, as well as of his philosophical conviction" (THN I, 273). We can read the phrase as stating that the true skeptic approves her "philosophical convictions," even though she does not assent to them and continues to doubt. In other words, they continue to be mere "probabilities," although, for the moment, sufficiently justified. We will go back to this epistemic attitude in the next chapter, where I attempt to show Hume's normative theory at work. For the moment, I would like to conclude this already too long chapter with a few remarks and a summary.

Before abandoning the discussion, however, we should notice the differences between Hume's treatment of belief and probability in the *Treatise* and the *Enquiry*. The latter work has a more concise account, in which some of the aspects mentioned before are entirely omitted or left out. This is the case, for example, with his comments on education, prejudices, and cultural stereotypes, but also with the reference to the passions as causes of belief, the discussion about probability by analogy, and the rules for causal inferences. Why are these reflections absent from his description of belief and probability? Are these problems not important for him anymore? Allow me to speculate, once again, on this matter. I believe that the younger Hume of the *Treatise* was already acquainted with the Academic system but was still hesitant about fully committing to it. For this reason, we find a mix of descriptive and normative indications in his account of belief and probability. Now, we would expect that, in the *Enquiry*, where his Academic orientation became explicit, he should have been more decisive about the normative aspects of his theory, but, in fact, he did not even mention them. The reason for this is, in my view, is that the mature Hume made yet another choice—he no longer saw it necessary to describe theoretical principles and was eager to move on to their application to metaphysical matters.[47] Thus, rather than explaining in detail the mechanisms by which belief is formed, explaining their causes, and outlining rules for evaluating judgments and making causal inferences, he preferred to establish the objects of his anti-metaphysical attacks – the testimonies on miracles, freedom, and the argument from the intelligent design of the universe – and apply the normative aspects sketched in the *Treatise* to examining their probability. Now, having said so, the place where this normative theory is most clearly at play is neither the *Treatise* nor the *Enquiry*, but the *Dialogues concerning Natural Religion*. As we will see in the next chapter, the *Dialogues* are a magnificent application of Hume's typology of belief and rules for causal inferences to the argument for God's existence from the

[47] Fogelin, on the contrary, claims that the main difference between the *Treatise* and the *Enquiry* lies in the treatment of the imagination, particularly of the functioning of the principles of association of ideas, which lead to internal contradictions. See: Fogelin 2009, pp. 147–148.

design of the universe. Through a series of skeptical arguments, Hume seeks to determine the degree of probability of natural theism's thesis and specify the type of belief that, because of withstanding the test, should be non-dogmatically adopted.

* * *

Before moving to the next chapter, let us review some of the core points of this chapter. First, I claimed that Humean skepticism is not, as Popkin and other scholars have affirmed, a form of modern Pyrrhonism but instead an adaptation of Academic probabilism, which can be identified in the normative features of his theory of belief. Second, Hume's most significant objection to Pyrrhonism was not, as is often presumed, that the skeptic cannot act because of suspending judgment on all matters—the famous *apraxia* objection. Instead, his primary concern was that adopting Pyrrhonism or a radical form of skepticism produces the undesirable consequences of melancholy and ostracism. In light of this, Hume identified some advantages of Academic skepticism: it is directed by a sincere love for truth, strengthens moderate passions, and helps control radical or Pyrrhonian doubt by recommending the philosopher to refrain from metaphysics and to focus her reflections on human affairs. Finally, to show more precisely how the New Academy influenced Hume, I provided an analysis of the *normative* aspects of Hume's theory of belief, in particular, his definition of belief, his typology of judgments into "knowledge," "proofs," and "probabilities," and the general and specific rules for judging on causal inferences. After this account, it became clearer how Hume's normative theory accords with the method and criteria of the New Academy to assess impressions and claims. Finally, I argued that despite the common opinion on the matter, Hume's rules for causal inferences are not primarily based on Newton's *Principia* but provide a set of Academic instructions to control prejudice and precipitated inferences, and are aimed at correcting not only scientific inferences but also practical ones. Finally, I argued that the attitude towards belief that Hume recommends after the skeptical examination is the non-dogmatic acceptance, a sort of Ciceronian "approval," of the probabilities shown to be sufficiently justified. This "non-dogmatic assent," as Fosl calls it, is well portrayed in Hume's praise of Academic skepticism, particularly in his description of this skeptic's diffident attitude towards both philosophical doubt and conviction. I turn, now, to examine how Hume applied this normative theory of belief to the metaphysical claims concerning the existence and nature of God.

Bibliography

Aguirre, Lisandro. 2010. David Hume y su adhesión inconsciente al escepticismo pirrónico. *Revista de Filosofía y Teoría Política* 41: 13–40.

Baier, Annette C. 1991. *A Progress of Sentiments: Reflections on Hume's Treatise*. Cambridge: Harvard University Press.

Battersby, Christine. 1979. The Dialogues as Original Imitation: Cicero and the Nature of Hume's Skepticism. In *McGill Hume Studies*, ed. D.F. Norton, N. Capaldi, and W.L. Robinson, 239–252. San Diego: Austin Hill Press.

Bayle, Pierre. 2005. *Dictionnaire historique et critique*. Vol. XII. Paris: Elibron Classics.

Burnyeat, Myles. 1997. Can the Sceptic Live His Scepticism? In *The Original Sceptics: A Controversy*, ed. Myles Burnyeat and Michael Frede, 25–57. Indianapolis: Hackett Publishing Company.

Calvo, Ángela. 2012. *El carácter de la "verdadera filosofía" en David Hume*. Bogotá: Ediciones Universidad Javeriana.

Chappell, Vere Claiborne. 1963. *The Philosophy of David Hume*. New York: The Modern Library.

Connon, R. W. 1979. The Naturalism of Hume Revisited. In McGill Hume Studies, eds. David Fate Norton, Nicholas Capaldi, and Wade L. Robinson, 121–145. San Diego: Austin Hill Press.

de Crousaz, Jean Pierre. 1733. *Examen du pyrrhonisme ancien et moderne*. La Haye: Hachette Livre.

de Olaso, Ezequiel. 2018. *Escepticismo e Ilustración. La crisis pirrónica de Hume y Rousseau*. Buenos Aires: Centro de Investigaciones Filosóficas.

de Pierris, Graciela. 2001. Hume's Pyrrhonian Skepticism and the Belief in Causal Laws. *Journal of the History of Philosophy* 39 (3): 351–383.

Falkenstein, Lorne. 1997. Naturalism, Normativity, and Scepticism in Hume's Account of Belief. *Hume Studies* 23 (1): 29–72.

Flew, Anthony. 1961. *Hume's Philosophy of Belief: A Study of His First Inquiry*. New York: Routledge.

Fogelin, Robert. 2009. *Hume's Skeptical Crisis. A Textual Study*. Oxford: Oxford University Press.

Fosl, Peter S. 1994. Doubt and Divinity: Cicero's Influence on Hume's Religious Skepticism. *Hume Studies* 20 (1): 103–120.

———. 1998. The Bibliographic Bases of Hume's Understanding of Sextus Empiricus and Pyrrhonism. *Journal of the History of Philosophy* 36: 261–78.

———. 2020. *Hume's Scepticism: Pyrrhonian and Academic*. Edinburgh University Press.

Frede, Michael. 1997. The Sceptic's Two Kinds of Assent and the Question of the Possibility of Knowledge. In *The Original Sceptics: A Controversy*, ed. Myles Burnyeat and Michale Frede, 127–151. Indianapolis: Hackett Publishing Company.

Gaskin, John C. A. 1978. *Hume's Philosophy of Religion*. London: Macmillan Press.

González, Catalina. 2020. Modern Skeptical Disturbances and Their Remedies. In *Sceptical Doubt and Disbelief in Modern European Philosophy: A New Pan-American Dialogue*, ed. Plínio Junqueira Smith and Vicente Raga Rosaleny, 78–100. Cham: Springer.

Hendel, Charles W. 1963. *Studies in the Philosophy of David Hume*. Indianapolis: Bobbs-Merrill.

Hickerson, Ryan. 2013. What the Wise Ought to Believe: A Voluntarist Interpretation of Hume's General Rules. *British Journal for the History of Philosophy* 21 (6): 1133–1153.

Hume, David. 1896. *A Tratise of Human Nature*, ed. L.A. Selby-Bigge. Oxford: Clarendon Press.

———. 1902. *Enquiries Concerning the Human Understanding and Concerning the Principles of Morals*, ed. L.A. Selby-Bigge. Oxford: Clarendon Press.

———. 1932. *The Letters of David Hume (Vol. I and II)*, ed. J.Y. Greig. New York: Oxford University Press.

———. 2007. *Dialogues concerning Natural Religion and Other Writings*, ed. Dorothy Coleman. Cambridge: Cambridge University Press.

Immerwhar, J. 1979. A Skeptic's Progress: Hume's Preference for the First *Enquiry*. In McGill Hume Studies, eds. David Fate Norton, Nicholas Capaldi y Wade L. Robinson, 227–238. San Diego: Austin Hill Press.

Jaffro, Laurent. 2011. Le sceptique humian est-it modéré? Le role du Pyrrhonismdas la genèse causale du scepticism mitigé. *Revista Internacional de Filosofía* 52: 53–69.

Jones, Peter. 1982. *Hume's Sentiments. Their Ciceronian and French Context*. Edinburgh: Edinburgh University Press.

Laird, John. 1932. *Hume's Philosophy of Human Nature*. London: Methuen & Co. Ltd.

Livingston, Donald. 1998. *Philosophical Melancholy and Delirium. Hume's Pathology of Philosophy*. Chicago: University of Chicago Press.

Loeb, Louis E. 2002. *Stability and Justification in Hume's Treatise*. New York: Oxford University Press.
Maia Neto, Jose R. 1997. Academic Skepticism in Early Modern Philosophy. *Journal of the History of Ideas* 58 (2):199–220.
Newton, I. (1999). *The Principia. Mathematical Principles of Natural Philosophy*. Trans. I. Bernard Cohen and Anne Whitman. Berkeley: University of California Press.
Norton, David Fate. 1973. *Hume's Philosophical Development*. Oxford/New York: Clarendon Press.
———. 1982. *David Hume: Common-Sense Moralist, Sceptical Metaphysician*. Princeton: Princeton University Press.
Noxon, James. 1964. Hume's Agnosticism. *The Philosophical Review* 73 (2): 248–261.
O'Connor, David. 2001. *Routledge Philosophy Guidebook to Hume on Religion*. London/ New York: Routledge.
Olshewsky, Thomas M. 1991. The Classical Roots of Hume's Skepticism. *Journal of the History of Ideas* 52 (2): 269–287.
Passmore, John A. 1968. *Hume's Intentions*. London: Duckworth.
Peirce, Charles Sanders. 1992. *The Essential Peirce. Vol. 1*, ed. Nathan Houser and Christian Kloesel. Bloomington: Indiana University Press.
Penelhum, Terence. 1979. Hume's Skepticism and the Dialogues. In *McGill Hume Studies*, ed. D.F. Norton, N. Capaldi, and W.L. Robison, 253–278. San Diego: Austin Hill Press.
———. 1983. *God and Skepticism: A Study in Skepticism and Fideism*. Dordrecht: Reidel Publishing Company.
Perin, Casey. 2010. *The Demands of Reason*. Oxford: Clarendon Press.
Pernietti, Dario. 2018. Hume at La Flèche: Skepticism and the French Connection. *Journal of the History of Philosophy* 56 (1): 45–72.
Plutarch. 1967. Adversus Colotem (Reply to Colotes). In *Moralia Vol. XIV*. Trans. B. Einarson and P.H. De Lacy, 153–315. Cambridge: Harvard University Press.
Popkin, Richard H. 1951. David Hume: His Pyrrhonism and His Critique of Pyrrhonism. *The Philosophical Quarterly* 1 (5): 385–407.
Price, John Vladimir. 1965. *The Ironic Hume*. Austin: University of Texas Press.
Sanfélix, Vicente. 1994. Del delirio melancólico a la serenidad reflexiva. In *Mirar con cuidado: Filosofía y escepticismo*, ed. J. Marrades and N. Sánchez, 55–70. Valencia: Pre-textos.
Schmitt, Charles B. 1972. *Cicero Scepticus: A Study of the Influence of the Academica in the Renaissance*. The Hague: Martinus Nijhoff.
Sextus, Empiricus. 2000. *Pyrrhoniae Hypotyposes (Outlines of Scepticism)*, ed. Julia Annas and Jonathan Barnes. Cambridge: Cambridge University Press.
Smith, Norman Kemp. 1905. The Naturalism of Hume (I). *Mind* 14 (54): 149–173.
———. 1941. *The Philosophy of David Hume: A Critical Study of Its Origins and Central Doctrines*. London: Macmillan and Co.
Smith, Plínio J. 2007. Hume y el escepticismo antiguo. *Signos Filosóficos* 19 (18): 105–126.
———. 2011. Cómo Hume se volvió escéptico. *Revista Internacional de Filosofía* 52: 71–84.
Stroud, Barry. 1991. Hume's Scepticism: Natural Instincts and Philosophical Reflection. *Philosophical Topics* 19 (1): 271–291.

Chapter 5
Hume and Cicero on Metaphysics: Philo's Non-Dogmatic Deism

Abstract In this chapter, I discuss how Hume applies his normative theory of belief – consisting of a typology of beliefs and a set of rules to guide causal inferences – to the theses of natural theism in the *Dialogues concerning Natural Religion*. Given the significant influence that Cicero's *De Natura Deorum* had on Hume's *Dialogues*, I begin the chapter by surveying the most important references to Cicero in Hume's correspondence and philosophical works. After this, I draw a parallel between the two dialogues' philosophical arguments and narrative strategies, to prepare the ground for the most detailed analysis of Hume's application of his normative theory of belief in the *Dialogues*. The third section of the chapter, centers on determining what kind of belief, in Hume's typology, is held by natural theism. I show that this belief is a probability based on an analogy by, first, examining the possibilities that it is a natural belief or a proof, and discarding both options based on Philo's use of some of the rules for causal inferences in his skeptical objections. I close the chapter with a discussion of the last part of the *Dialogues*, usually referred to as "Philo's reversal," to evaluate whether Philo's sudden profession of faith can be understood as an avowal of theism, fideism, or deism, or simply as a disguise of his atheistic inclinations. I claim that, after the skeptical examination of natural theism, Philo accepts a non-dogmatic adoption of a much less metaphysically robust belief—namely the belief that Gaskin called "attenuated deism" and that I would like to name "non-dogmatic, anti-religious deism."

In the previous chapter, I detailed how Hume adopted Academic skepticism in his *normative* theory of belief, consisting of a typology of beliefs and a set of rules to guide causal inferences. Now I turn the discussion toward how Hume used this normative theory to examine natural theism—the belief in the existence and providence of God based on the argument from the intelligent design of the universe. As we have seen in Chap. 3, Cicero applied the method of the New Academy to the Hellenistic theories on the existence and nature of gods in *De Natura Deorum (DND)*. Given the significant influence that this particular treatise had on Hume's philosophy of religion, we should now examine how this influence is apparent in his *Dialogues concerning Natural Religion* and how this work reflects Hume's normativity.

C. González Quintero, *Academic Skepticism in Hume and Kant*, Synthese Library 449, https://doi.org/10.1007/978-3-030-89750-5_5

Scholars have long acknowledged the influence of Cicero's *DND* on both the *Dialogues'* topics and stylistic features.[1] Some influential accounts have been N.K. Smith's introductory study of the *Dialogues* in his edition of 1947 (Smith 1947, pp. 1–75); C. Battersby's characterization of the work as an "original imitation" of Cicero's *DND* (Battersby 1979, pp. 239–252), P. Jones' study on Hume's adoption of Cicero's humanistic views (Jones 1982);[2] and R. Fosl's analysis of the parallels between the two works (Fosl 1994, pp. 103–120).[3]

Despite this broad recognition, however, I believe that scholars have not yet fully elucidated the depth of Hume's debt to Cicero in this work. More specifically, commentators have failed to see that Hume used *De Natura Deorum* as a *methodological* model and not only as a source of skeptical topics and rhetorical strategies to disguise his frontal attacks on religion. In other words, just as in *DND,* Cicero used the Academic method described in the *Academica* to assess the Hellenistic theological doctrines, in the *Dialogues,* I believe, Hume applied the normative aspects of his theory of belief described in the *Treatise* and the second *Enquiry* to evaluate Christian natural theism.

With this parallel in mind, in the following pages, I illuminate how Hume uses his Academic skepticism to examine the question of the existence and nature of God. But before turning to this point specifically, I will begin by surveying Hume's most important references to Cicero in his correspondence and philosophical works. While other scholars have examined this aspect of Hume's debt to Cicero already (Jones 1982, pp. 29–43; Fosl 2020, pp. 43–48), I hope to complement their analysis by tracing a general line of evolution from Hume's youthful interest in Cicero's ethical treatises to his more mature involvement with the latter's skeptical and metaphysical discussions. I also advance some hypotheses on whether or not Hume read the *Academica*. This review will show that Hume identified Cicero as a skeptic, even though the standard view in the eighteenth century was to consider him as either an eclectic or a Stoic. Hume accurately understood Cicero's skepticism as a moderate one, particularly helpful to examine metaphysical issues such as the existence and nature of God.

After this, based on Price's and Fosl's accounts (Price 1964, pp. 97–106; Fosl 1994, pp. 103–120), I draw a parallel between the two dialogues' philosophical arguments and narrative strategies. This comparison should prepare the ground for the most detailed analysis of Hume's application of his normative theory of belief in the *Dialogues*. The third section of the chapter centers on what kind of belief, in Hume's typology, is held by natural theism. I show that this belief is a probability based on an analogy, by first, examining the possibilities that it is a natural belief or a

---

[1]See: Battersby 1979, pp. 239–252; Fosl 1994, pp. 103–120; Jones 1982, p. 29; Penelhum 1979, pp. 253–278; Price 1964, pp. 97–106; Smith 1947, pp. 60–61.

[2]For Jones: "Every educated reader could discern, at the time of their posthumous publication, that Hume's *Dialogues concerning Natural Religion* was modeled on Cicero's *De Natura Deorum*" (Jones 1982, p. 3).

[3]Fosl argues: "Hume drew upon the *Academica* and *De Natura Deorum* in developing his own "Academical" philosophy of religion" (Fosl 1994, p. 104).

proof, and discarding both options based on Philo's use of some of the rules for causal inferences in his skeptical objections. I close the chapter with a discussion of the last part of the *Dialogues,* usually referred to as "Philo's reversal," to evaluate whether Philo's sudden profession of faith can be understood as an avowal of theism, fideism, or deism, or simply as a disguise of his atheistic inclinations.

By the chapter's end, I hope to have established how Hume's exam of natural theism – by applying his typology of belief and his rules for causal inferences – led him to a non-dogmatic adoption of a much less metaphysically robust belief—namely the belief that Gaskin called "attenuated deism" and that I would like to name "non-dogmatic, anti-religous deism." This belief was accepted by Philo, since it resisted the *Dialogues*' skeptical examination and it proved helpful to avoid the damaging practical consequences of theism, such as religious wars and factions, intolerance, enthusiasm, and superstition. I conclude that Hume's non-dogmatic deism corresponds to an Enlightened attitude, which supports moderate religious passions, promotes toleration, and recommends the state's political control over the different religious sects. That being said, let us begin our investigation into Hume's references to Cicero in his correspondence and works.

5.1 Hume's Cicero: Moralist, Orator, and Skeptic

Various sources confirm that Hume did not simply read Cicero and other Latin ancient authors during his youth, but these authors accompanied him throughout his life.[4] The most famous passage in which he speaks of them appears in his autobiography, *My Own Life* (*E.* xxxi–xli), where Hume describes how avidly he read Cicero and Virgil in the first stages of his education:

> My studious disposition, my sobriety, and my industry, gave my family a notion that the law was a proper profession for me; but I found an unsurmountable aversion to everything but the pursuits of philosophy and general learning; and while they fancied I was poring upon Voet and Vinnius, Cicero and Virgil were the authors which I was secretly devouring (E. xxxiii).

Hume also professed his preference for Latin authors in his correspondence. A clear instance is the 1727 letter to Michael Ramsay, written when he was barely sixteen years old (Fosl 2020, p. 44). There, Hume described how, while studying at the library, he used to read at the same time both literary and philosophical works, to compare different perspectives on a subject and find suitable examples for his reflection:

> [...] for I hate task-reading and I diversify them at my pleasure; sometimes a philosopher, sometimes a poet, which change is not unpleasant nor disserviceable neither; for what will more surely be engraved upon my mind a Tusculan Dispute of Cicero's *De Aegritudine*

[4] According to Jones, in the first year of his undergraduate studies at Edinburgh University, Hume learned, "Greek grammar, and the rules of rhetoric, with special reference to Cicero and Ramus." (Jones 1982, p. 10).

> *Lenienda* than an *Ecloge* or *Georgick* of Virgil's; the philosophers' wise man and the poet's husbandman agree in peace of mind, in a liberty and independency of fortune and contempt of riches, power and glory (*Letters*, Vol. I, p. 10).

This letter, among others, is filled with praise for Cicero and includes memorized citations of his main works and discourses. Indeed, in his correspondence alone, Hume discusses Cicero's most famous orations (*Letters*, Vol. I, 40–43), uses him as a historical source to describe past events (*Letters*, Vol. I, 150–153), and even mentions specific aspects of the Roman's private life.[5] All these references reveal Hume's deep knowledge and admiration for Cicero, both as a philosopher and as an orator. But perhaps his most important allusions appear in two letters: first, the letter to the French physician George Cheyne (1734), in which he related how the zeal of his youthful studies led to a weakening of his mental health; and, second, his 1739 note to Francis Hutcheson, where he commented on his mentor's moral theory in relation to Cicero's *De Officiis* and *De Finibus* (Fosl 2020, p. 44). Let us dive into these letters a little more.

In the first letter, Hume claims that one of the causes of his mental illness may be his active engagement with the works of classical moralists, among them Cicero:

> There was another particular, which contributed more than anything to waste my spirits and bring on me this distemper, which was, that having read many books of morality, such as Cicero, Seneca, and Plutarch, and being smit with their beautiful representations of virtue and philosophy, I undertook the improvement of my temper and will, along with my reason and understanding. [. . .] These [readings] no doubt are exceedingly useful, when joined with an active life [. . .] but, in solitude, they serve to little other purpose than to waste the spirits (*Letters,* Vol. I, 13–14).

As we can see, the melancholy that Hume attributed to his skeptical reflections in the *Treatise* may have begun in these early years while reading Cicero and other ancient moralists. However, in this piece of correspondence, Hume did not yet connect the debilitation of his spirits with his skeptical doubts, but only with an excess of reflection and self-improvement discipline. The reasons for this could be that Hume either did not want to refer specifically to his conflicted relation to skepticism or did not hold yet skeptical doubts, so that his spirits' weakening was due to the sheer effort of philosophical reflection. In any case, the link between his reading of Cicero and the general symptoms of melancholy is already present in this letter, at least five years before the publication of the *Treatise*.

In his 1739 letter to Hutcheson, we see a more mature, although still young, Hume – he was 28 years old and publishing the *Treatise* – claiming that it was a mistake on his correspondent's part to derive all virtues from benevolence (Fosl 2020, p. 44). Instead, Hume expressed his preference for the Ciceronian typology of virtues:

> Were benevolence the only virtue, no characters could be mixt, but would depend entirely on their degrees of benevolence. Upon the whole, I desire to take my catalogue of virtues from

[5] For example, he refers to Terentia, Cicero's first wife, perhaps to insinuate to his addressee, by analogy, that she is not behaving as well as he would wish (*Letters*, Vol. II, 8–11).

> Cicero's *De Officiis*, not from the *Whole Duty of Men*. I had, indeed, the former book in my eye in all my reasonings (*Letters*, Vol. I, p. 34).

By affirming that he was more inclined to derive his typology of virtues from *De Officiis* than from *The Whole Duty of Men,* Hume revealed his preference for a secular system of morality. What is more, in case his advice had not been sufficiently persuasive, in the postscript to the letter Hume insists that Hutcheson revise his position on the motives of virtue by drawing on yet another of the Roman' treatises, *De Finibus*:

> You are a great admirer of Cicero, as well as I am. Please to review the 4th book, *De Finibus bonorum et malorum*; where you find him prove against the Stoics that if there be no other goods but virtue, it is impossible there can be any virtue; because the mind would then want all motives to begin its actions upon: and it is on the goodness or badness of the motives that the virtue of the action depends (*Letters*, Vol. I, p. 35).

These two letters show that Hume was deeply engaged with Cicero's moral treatises during his twenties and had recognized already the secular, republican, and humanistic spirit of the Roman's practical philosophy. This influence would continue until the publication of *An Enquiry Concerning the Principles of Morals* (*EPM*) in 1751, where Hume mentions Cicero fifteen times (Fosl 2020, p. 44). The most important references in the second *Enquiry,* are a review of the social virtues of *De Officiis* (*EPM*, 140) and an oblique mention to *De Natura Deorum,* when discussing Cicero's virtue of social utility in relation to the Epicurean gods. Hume argues there that social utility is such an essential virtue that Cicero was right to criticize the Epicurean doctrine of the gods because, if they are not helpful to human beings, we should not emulate them (*EPM*, 143).[6]

Unfortunately, in the *Treatise,* his only explicit reference to Cicero is found in the Appendix (*THN*, 623–636), where Hume seeks to clarify his views on the mechanisms of association of ideas. Explaining the particular mechanism of spatial contiguity, he introduced a rather extensive quotation from *De Finibus*:

> I cannot say whether it is a natural instinct or a kind of illusion, but when we see the places where we are told that the notables of the past spent their time, it is far more moving than when we hear about their achievements or read their writings. This is how I am affected right now. I think of Plato, whom they say was the first philosopher to have regularly held discussions here. Those little gardens just nearby not only bring Plato to mind but actually seem to make him appear before my eyes. Here come Speusippus, Xenocrates, and his pupil Polemo, who sat on that very seat we can see over there. Even when I look at our own Senate-house [...], I often think of Scipio, Cato, Laelius, and above all, my grandfather. Such is the evocative power that locations possess. No wonder the training of memory is based on them (*De Fin.* V, 1–2).

The exact quotation appears in the footnote corresponding to the explanation of the mechanisms of association in the *Enquiry concerning Human Understanding* (EHU V, 42). It is interesting to note that this is the only citation of a Ciceronian

[6]Interestingly, this idea is quite the opposite of Hume's final remarks in the *Dialogues*, where he sides with the Epicureans, although, of course, he omits their idea that we should emulate the gods.

work in the *Treatise.* There are no mentions to the *Academica*, particularly in his treatment of probabilities. This suggests that the young Hume of the *Treatise* was not yet acquainted with Cicero's *Academica*, and his knowledge of Academic skepticism either came from other dialogues, such as *De Finibus* and *De Natura Deorum*, or from his acquaintance with other modern Academics, such as Mersenne, Huet, and Foucher (Fosl 2020, pp. 39–7; Maia Neto 1997, p. 2175). I will return to these hypotheses later. For the moment, we can accept Jones's view that "there is no evidence that Hume read Cicero's works as a unified and structured progression from *Academica* to *De Officiis* [. . .] although he does seem to have been fairly clear which views Cicero was merely reporting for discussion and which were his own" (Jones 1982, p. 30). It seems that Hume definitely began his reading of Cicero with the latter's moral treatises. Still, we cannot pinpoint the moment, if there is any, in which he read the *Academica.*

In the *Enquiry concerning Human Understanding*, Hume's most famous allusion to Cicero appears in the opening section while discussing the two kinds of philosophy—the "easy and obvious" and the "accurate and abstruse." Hume said that "[. . .] the fame of Cicero flourishes at present, but that of Aristotle is utterly decayed" (*EHU*, 7). The passage indicates how Hume's admiration for the Roman philosopher remained steady throughout the years and bears witness to the modern, anti-Aristotelean context in which Ciceronian philosophy reappeared—namely the Renaissance attacks on Scholastic philosophy (Schmitt 1972, pp. 43–77). Other than this, his repeated and very appreciative references to Academic philosophy (*EHU* V, 40–41; XII, 161) confirm that, at the time of writing the first *Enquiry* (it was published in 1748 when Hume was 37), he was fully aware of the school's mitigated skepticism.

Now, we find the most significant number of references to Cicero in the *Essays.* There, Hume mentions the Roman on at least 35 occasions, in which he primarily employs Cicero's discourses and treatises as historical sources to confirm specific information. For example, he uses Cicero's discourses to support the claim that the Roman state distributed corn equally between rich and poor citizens (*E.*, 441). Beyond this, his allusions to the Roman's rhetorical views are also noteworthy. Indeed, Hume shows a deep and detailed knowledge of both Cicero's speeches and the rudiments of classical rhetoric. In "Of Eloquence," he describes Cicero as the best orator of ancient Rome: "But the greatest of that age was inferior to CICERO, the most eloquent speaker that had ever appeared in ROME" (*E.*, 98). Yet this assessment is not devoid of criticism. In the same essay, he notes some of the defects of Cicero's grandiloquent style:

> Some objections, I own, notwithstanding his vast success, may lie against some passages of the ROMAN orator. He is too florid and rhetorical: His figures are too striking and palpable: The divisions of his discourse are drawn chiefly from the rules of the schools: And his wit disdains not always the artifice even of a pun, rhyme, or jingle of words (*E.*, 105).[7]

[7] He argues this in the context of an assessment of Great Britain's lack of grandiloquent oratory. For a good treatment of the ambiguities and the political implications of this essay, see: Potkay 1994, pp. 24–58.

Along with other similar passages, this fragment shows Hume's knowledge of and admiration for the art of discourse. He was well versed in matters of invention and style, efficiently identified the different parts of rhetorical speech, and could even connect some aspects of Ciceronian oratory with the rules of ancient treatises (*Letters*, Vol I, pp. 40–43). It is curious, however, that in "Of the Standard of Taste," Hume praises the same grandiloquent style that he criticizes in "On Eloquence," arguing that, unlike Cicero's oratory, his philosophy is no longer esteemed: "The abstract philosophy of Cicero has lost its credit: the vehemence of his oratory is still the object of our admiration" (*E*, 243). When we compare this fragment to the famous passage of the first *Enquiry*, where Hume opposes the abstruse philosophy of Aristotle to Cicero's easy one (*EHU*, 7), we cannot but wonder whether Hume is sincere in his observations. Contradictions aside, it is clear that Hume's admiration for Cicero's rhetoric and philosophy never diminished, even if he sometimes criticized his oratory and philosophical style.

Now, by in 1742, when "Of the Rise and Progress of the Arts and Sciences" was published, Hume already knew about Cicero's skepticism and was familiar with *De Natura Deorum*:

> [...] and it is remarkable that Cicero, being a great sceptic in matters of religion, and unwilling to determine anything on that head among the different sects of philosophy, introduces his friends disputing concerning the being and nature of the gods, while he is only a hearer; because forsooth, it would have been an impropriety for so great a genius as himself, had he spoke, not to have said something decisive on the subject, and have carried everything before him as he always does on other occasions (Hume 1826, p. 144).[8]

In this essay, Hume refers to Cicero's dramatic strategies, such as giving himself only a marginal role in *De Natura Deorum* to avoid influencing the reader's opinion with his authorial voice. Yet, we can sense an ironic tone in Hume's comment since, as we know, Cicero did give the final verdict in favor of the Stoic doctrine in his own voice. This point will be decisive for Hume's own dramatic decisions in the *Dialogues*, where he is neither a main character nor a hearer, and the character that gives the final verdict is a witnessing student. We can infer, from these indications, that the writing of the *Dialogues* was already taking place when Hume wrote this essay.

Finally, it is worth mentioning the essay "Of the Dignity or Meanness of Human Nature" because, in it, Hume refers to Cicero as a wise man: "Were the lowest of our species as wise as Tully, or Lord Bacon, we should still have reason to say that there are few wise men." (*E*, 83–84). This passage supports the view that Hume was acquainted with the Hellenistic schools' notion of the sage and even identified Cicero with it.

Besides the *Essays*, Hume depicts Cicero's skepticism as the highest form in the *Letter to a Gentleman* (*LG*, p. 155) and openly recognizes his religious skepticism in the *Natural History of Religion* (*NHR*, 171). The latter work is filled with references

[8]I cite here the 1826 edition of Hume's *Essays* since, unfortunately, the Liberty Fund edition overlooked this footnote.

to Cicero. In one of them, Hume openly discusses the Roman philosopher's public and private opinions on the matter:

> If ever there was a nation or a time, in which the public religion lost all authority over mankind, we might expect, that infidelity in Rome, during the Ciceronian age, would openly have erected its throne, and that Cicero himself, in every speech and action, would have been its most declared abettor. But it appears that, whatever skeptical liberties that great man might take in his writings or in philosophical conversation; yet he avoided, in the common conduct of life, the imputation of deism and profaneness (*NHR*, 171).

Based on this passage, Fosl argued that Hume may have decided to emulate Cicero in disguising his irreligious convictions:

> Could that characterization of Cicero present an oblique confession of Hume's own insincerity? Indeed, in the *Natural History of Religion*, Hume implicitly seems to align himself with the Academic sceptic Carneades, and therefore, symbolically, with the Academic movement in general, whose "chief business," he says, is nothing less than exposing the lack of "foundation" for religious belief. For this reason, Hume's personal declarations of religious conviction and true religion should be taken with a grain of salt (Fosl 2020, p. 47).

I agree with Fosl's view in this respect. I would only add that, when Hume affirms, in the quoted passage, that Cicero evaded "the imputation of deism," he might be attributing to Cicero his own, disguised and Academic, deism. I will come back to this claim at the end of the chapter. For the moment, I should conclude this section with a general balance of Hume's reception of Cicero.

Because of the higher number of allusions to Cicero's rhetoric and his unambiguous preference for the Roman's moral treatises, Jones has argued that Cicero's humanistic views most heavily influenced Hume:

> Hume's interest in Cicero, like that of his contemporaries, centered on the moral, political, and religious tenets, on the one hand, and on the rhetorical views, on the other. Cicero's epistemological reflections were generally subordinate to those tenets and conducted in the light of them, and Hume himself moved towards a similar policy (Jones 1982, p. 30).

More recently, Fosl has challenged this view, arguing that Hume admired not only Cicero's rhetoric and moral philosophy but also his skepticism (Fosl 2020, p. 45). He also claims that, since all these aspects of Cicero's thought are interrelated, "Hume appropriated the Ciceronian idea that philosophy and rhetoric are intimately connected" (Fosl 2020, p. 45). Fosl's balance is accurate, and I believe we can confidently assert that Cicero's epistemological views did play a crucial role in Hume's philosophy. Hume recognized Cicero's skepticism in a way that many eighteenth-century philosophers, among them Kant, did not.[9] He was aware of both Cicero's humanism and his skeptical, Academic stance. Whether Hume read the *Academica* is a question that remains difficult to answer, but we can imagine at least two scenarios in this regard. First, as I already suggested, we can suppose that he actually did not read it or did not thoroughly read it, and became acquainted with Academic skepticism through other Ciceronian treatises, Sextus's works or some

---

[9] As I show in this chapter, Kant considered Cicero as a Stoic in moral matters (*VL*-Jäsche, 31; *VL*-Vienna, 803).

modern Academic philosophers. And, second, we may consider the possibility that having read the text, Hume did not find it necessary to refer to it explicitly[10] or did not want to mention it directly to avoid censure. I think the answer lies somewhat between these two interpretations. The younger Hume may not have been directly acquainted with the *Academica,* and his knowledge of Academic skepticism may have proceeded from other Ciceronian treatises, such as *De Finibus* and *De Natura Deorum,* from Sextus's references to the Academic skeptics, and/or from his acquaintance with some modern Academic philosophers. If this is the case, then, the influence of Carneades's criteria to evaluate impressions on his normative theory of belief in the *Treatise*, most probably came from Cicero's comments, in the prefaces of his dialogues, about Academic philosophy and probability, from Sextus's Empiricus descriptions, or from Mersenne's, Foucher's, and Huet's appropriations, who, as Fosl and Maia Neto have claimed, are important sources of Hume's Academic allegiances (Fosl 2020, pp. 39–75; Maia Neto 1997, p. 217). However, I believe that, by the time Hume was writing the first *Enquiry*, he probably had already read the *Academica* and found in Carneades' moderate skepticism the best way to overcome Pyrrhonism. His allusions to "Academic philosophy" as a sort of mitigated skepticism support the hypothesis that Hume knew more about the details of Academic skepticism by the end of the 1740s, but did not consider it necessary to describe the school's views in his *Enquiry*. As for the possibility that Hume had read the *Academica* but did not want to refer to it in his works to avoid censure, I grant that it also has some degree of plausibility. Perhaps Hume chose only to mention Cicero's ethical treatises, which were more congenial with the public at large, as a disguising strategy. However, this explanation would require much more historical elaboration than I can provide in this book. Therefore, I prefer to rely on the first option, namely, that his particular knowledge of Academic probabilism and Carneades's criteria for evaluating impressions came either from the introduction of other Ciceronian treatises, from Sextus Empiricus's texts, or from other modern Academics.

To summarize this brief survey, I have argued that Hume was deeply impressed by Cicero's rhetorical discourses, moral treatises, such as *De Officiis* and *De Finibus,* and metaphysical works, particularly *De Natural Deorum*.[11] Hume was not only familiar with Cicero's humanistic and rhetorical views but also with his skepticism concerning religious matters (Fosl 2020, p. 45). Having suggested this, since Cicero's skeptical influence is most prominently displayed in Hume's *Dialogues concerning Natural Religion*, I turn to this work for the remainder of the chapter.

[10] Jones argues that "[. . .] explicit allusion [to Cicero] was often unnecessary since Cicero was the one classical writer familiar to and admired by almost every educated person in France and England in the late seventeenth and early eighteenth centuries." (Jones 1982, p. 30).

[11] I will refer to the influence of his metaphysical treatise *De Fato*, in this chapter, where I discuss Hume's treatment of freedom in relation to Kant's Antinomies.

5.2 *De Natura Deorum* and the *Dialogues Concerning Natural Religion*

Among his philosophical writings, the *Dialogues concerning Natural Religion* is undoubtedly the work where Hume's debt to Cicero can be most immediately appreciated. While some commentators argue that the *Dialogues* are built on Cicero's treatise (Battersby 1979, pp. 239–252), others suggest that *De Natura Deorum* is the direct source of many of its arguments against natural theism.[12] I agree with both interpretations and add to them that, in the *Dialogues,* we see Hume's Academic skepticism, that is, his normative theory of belief, at work.

Hume's masterful use of the dialogue as a philosophical genre is the first sign of his commitment to Cicero's Academic skepticism. Another sign is his choice of names for the dialogue's leading characters—Cleanthes, Philo, Demea, and Pamphilus. Cleanthes, the representative of natural theism, is named after Cleanthes of Assos, the second leader of the Stoa and author of the first formulation of the argument from the universe's design. Philo, Hume's skeptical character, takes his name from Philo of Larissa, the last head of the New Academy and Cicero's mentor. The third main character, Demea, is, admittedly, more difficult to identify. However, as Laird argued, his name may allude to a character of Terence's comedy *Adelphos* (Laird 1932, pp. 294–295), who represents a traditional patriarch that advocates for an orthodox and severe education for his children. Finally, the name of the student who witnesses the dialogue, Pamphilus, seems to be taken from one of Plato's pupils at the Academy in Samos, mentioned by Cicero in *De Natura Deorum* (*DND* I, 72).

Beyond this explicit reference to the Hellenistic schools through some of their head's names, these characters also appear to be a disguise of eighteenth-century Great Britain's important theological figures. Commentators have usually argued that Cleanthes represented Joseph Butler's Newtonian theism – also termed "experimental theism" or "empirical theism" – which based theistic doctrine on the argument from intelligent design of the universe.[13] They also think that Demea could represent the orthodox theism of Samuel Clarke, who championed the *a priori* argument for the existence of God (Smith 1947, p. 26, n. 2; Mossner 1936, pp. 334–349). I believe that Demea also speaks for the fideists, since he argues that it is impossible to determine God's attributes rationally and, therefore, faith in revelation should be the sole basis of the belief in God's providence. Finally, Philo represents what Cleanthes calls "speculative skepticism," or skepticism concerning metaphysics, which, very likely, is Hume's own stance on the issue.

Allow me now to give a brief overview of the work. The *Dialogues* constitute a detailed evaluation of the argument from intelligent design of the universe and are divided into twelve sections. In the Introduction, Pamphilus, the student, offers a brief reflection on the suitability of the dialogical genre to discuss matters of natural

[12] See: Smith 1947, pp. 1–75; Penelhum 1979, pp. 253–278; Jones 1982, pp. 44–92; Price 1964, pp. 97–106; Gaskin 1978, pp. 209–231; Mossner 1936, pp. 334–349; Noxon 1964, pp. 248–261.

[13] According to Pyle (2006, p. 4), this position is also shared by Colin Maclaurin, Newtonian mathematician and defender of natural theology.

religion and presents the characters. The conversation opens with Cleanthes outlining the argument from intelligent design and wielding the familiar Humean adaptation of the *apraxia* charge against Philo. Philo responds by raising the banners of mitigated skepticism, namely the rejection of speculative philosophy and the limitation of inquiry to human affairs. In sections two and three, Philo puts forward his first objection to the argument from design—that the likeness between human artifacts and the universe is imperfect; hence, the analogy lacks cogency. In the fourth and fifth sections, this objection is followed by Philo's criticism of Cleanthes's anthropomorphism, and the latter's rather desperate efforts to maintain his posture. In the sixth section, the characters debate the Stoic thesis according to which God is the world. After this, in sections seven and eight, Philo defends the possibility that the order of the universe originates in matter itself—the Epicurean or materialistic thesis. The ninth part is a sort of digression, advanced by Demea, about the *a priori* argument for the existence of God. Given their empiricist allegiances, both Philo and Cleanthes refute it in a cursory way. In sections ten and eleven, Philo outlines what is perhaps his most important objection to the argument—that the existence of evil and its pervasiveness in human life contradicts divine providence. Finally, in the dialogue's concluding section, once Demea has left the conversation, Philo makes his sudden confession of faith. He says that he accepts the argument from the intelligent design, and both characters express their opinion that their recent disagreements were only due to their "spirit of controversy." They also debate the social utility or harmfulness of religion, the passions it arouses, and the dangers of superstition. The dialogue concludes with Pamphilus, the student, giving a summary of the discussion and declaring himself in favor of Cleanthes's theism.

There is a long-standing scholarly controversy about who represents Hume's position in the *Dialogues*. Following N. K. Smith's Introduction to his 1947 edition of the *Dialogues*, most contemporary commentators identify Hume with Philo.[14] Smith argued:

> I shall contend that Philo, from start to finish, represents Hume; and that Cleanthes can be regarded as Hume's mouthpiece only in those passages in which he is explicitly agreeing with Philo, or in those other passages in which, while refuting Demea, he is also being used to prepare the way for one or other of Philo's independent conclusions (Smith 1947, p. 59).

Smith's assessment is, in my view, absolutely accurate. However, before Smith's influential account, some scholars, particularly in the nineteenth- and early twentieth- centuries, saw Cleanthes or Pamphilus as Hume's representatives (Smith 1947, pp. 58–59). At the beginning of the twentieth century, other scholars held that, since the dialogical genre serves to show different perspectives on one matter, Hume's point of view was not represented by any particular character. J. Laird seems to have initiated this line of interpretation: "[. . .] the writer of a dialogue has the privilege of mixing and confounding his persons, and, unless he wants to, does not need to commit himself in any way whatsoever" (Laird 1932, p. 297; Laird et al. 1939, pp. 206–207). More recently, A. Pyle argued for an intermediate position: "My own

[14]See: Gaskin 1976, pp. 301–311, 1978, pp. 209–231; O'Connor 2001, p. 24; Penelhum 1979, p. 270; Price 1964, pp. 97–106.

view is that both these readings – that Philo speaks for Hume and that we should read the *Dialogues* as dialogues, i.e., without looking for a single authorial voice – are legitimate and reasonable interpretations of the text" (Pyle 2006, p. 26). As will become apparent throughout my analysis, I agree with Smith's view that, even if Hume may have shared some of Cleanthes's opinions, Philo is the character who most closely represents his intimate views on religion.

Now that we have been reacquainted with the work, we are ready to return to the main point of this chapter—detailing the *Dialogues*' connection to *De Natura Deorum*. Battersby claims that the *Dialogues* pertain to the literary genre called "original imitation," used by eighteenth-century authors to discuss a contemporary issue while re-visiting the structure and language of a classical literary piece (Battersby 1979, p. 240). The *Dialogues,* indeed, are modeled on the philosophical and dramatic pattern of *De Natura Deorum* and share with the latter some essential features. Some of these features were already identified by J.V. Price (1964, pp. 97–106), on whose account I build in what follows. In his view, although the similarities between Cotta's and Philo's skeptical arguments are prominent, Hume further developed Cicero's views to incorporate his empiricist allegiances. For this reason, he says:

> We should not conclude yet that the exclusive source of Philo's doctrine was Cotta. Whatever similarities there may be between them, the arguments derived from Cotta are wholly transformed by Hume and carefully stamped with his own brand of skepticism (Price 1964, p. 106).

Notwithstanding this crucial qualification, we can enumerate the similarities between both dialogues as follows: (i). the difference explicitly made between the questions on the existence and nature of God; (ii). the centrality of the argument from intelligent design of the universe; (iii). the critique of anthropomorphism; (iv). the refutation of God's providence through the discussion on evil, and (v). the dialogues' closing with a verdict in favor of one of the views under examination. I will explain each aspect in detail now.

5.2.1 *The Questions of the Existence and Nature of God*

While God's existence is presumed to be beyond doubt in both dialogues, God's nature is questioned and given careful examination. Despite this, the arguments introduced to refute the divine attributes necessarily challenge the evidence of God's existence. For example, in *DND*, the problem becomes explicit at the opening of Book III, when Cotta begins his refutation of Stoic doctrine. He asks Balbus why, having said at the beginning of his exposition that God's existence is evident, he nevertheless uses the Stoic arguments to prove it:

> [...] why was it that after saying that this part of your subject did not even need discussion, because the fact of the divine existence was manifest and universally admitted [...] you did not really feel confident that the doctrine of the divine existence was as self-evident as you could wish, and for that reason, you attempted to prove it with a number of arguments [?] (*DND*, III, 8–9).

An attentive reader of Cicero, Hume picked up and developed this suggestion (Price 1964, pp. 99–100). In the Introduction of his *Dialogues*, He makes Pamphilus claim that the dialogical genre is particularly adequate to discuss matters such as the *existence* of God, which is "so *obvious*, that it scarcely admits of dispute, but at the same time so *important*, that it cannot be too often inculcated," as well as the *nature* of God, which is "so *obscure* and *uncertain*, that human reason can reach no fixed determination with regard to it" (*DNR* I, 3–4; 127–128). After this presentation, Demea zealously holds on to the distinction throughout the dialogue. He insists that questioning the existence of God is the business of atheists and impious people, while educated and devoted Christians investigate, instead, God's nature or attributes:

> No man, no man, at least, of common sense, I am persuaded, ever entertained a serious doubt with regard to a truth so certain and self-evident. The question is not concerning the being but the nature of God. This, I affirm, from the infirmities of human understanding, to be incomprehensible and unknown to us (*DNR* II, 1; 141).

However, even if Cleanthes and Philo initially agreed on this point, they engaged in arguments that clearly put into question the existence of God. Philo's criticism of the design argument not only cast doubts on some divine attributes, such as his benevolence, wisdom, and power, but also on the very idea that a divine mind created the universe. Claiming, for instance, that the universe's order and regularity may have been caused by an internal, material principle – as Philo argues in sections seven and eight – necessarily leads to doubting the existence of a creator of the universe. In addition to this, Cleanthes's acceptance of Philo's arguments' plausibility, even if only to refute them, tacitly concedes that the answer is far from evident. For this reason, at some point in the conversation, Demea realizes that the discussion goes in the direction of accepting the possibility that God does not exist and accuses Cleanthes of impiety:

> I must own, Cleanthes [. . .] that nothing can more surprise me, than the light in which you have, all along, put this argument. By the whole tenor of your discourse, one would imagine that you were maintaining the being of God, against the cavils of atheists and infidels; and were necessitated to become a champion for that fundamental principle of all religion. But this, I hope, is not by any means a question among us (*DRN*, II, 1; 141).

After the reprimand, Philo restates the difference and, in an apparently pious remark, claims that philosophers who dare question God's existence should be justly punished:

> [. . .] But surely, where reasonable men treat these subjects, the question can never be concerning the being, but only the nature of the Deity. The former truth, as you well observe, is unquestionable and self-evident. [. . .] Whoever scruples this fundamental truth deserves every punishment, which can be inflicted among philosophers, to wit, the greatest ridicule, contempt, and disapprobation (*DNR* II, 3; 142).

As expected, Cleanthes notices Philo's insincerity and redirects the conversation to the argument from design: "Not to lose any time in circumlocutions," said Cleanthes, addressing Demea, "much less in replying to the pious declamations of Philo, I shall briefly explain how I perceive this matter [. . .]" (*DNR* XI, 5; 143). However, toward the end of the discussion, Demea acknowledges that the existence of God has been at stake all along and decides to leave the conversation expressing his irritation with Philo's skepticism:

> Hold! Hold! Cried Demea: Whither does your imagination hurry you? I joined in alliance with you, in order to prove the incomprehensible nature of the divine Being, and refute the principles of Cleanthes, who would measure everything by a human rule and standard. But now I find you running into all the topics of greatest libertines and infidels; betraying that holy cause, which you seemingly espoused (*DNR* XI, 18; 213).

These fragments should be sufficient to show that both Cicero and Hume used the distinction between the existence and nature of God to hide their true intentions. Both acknowledged the danger of openly accepting that the topic under debate was the existence of God, and, even though they put a pious acceptance of it in the mouths of their characters, the context reveals that the claim is called into question throughout the whole conversation (Price 1964, p. 99). While Cicero's reference to the issue was much briefer – indeed, just an ironic remark – Hume dramatically elaborated on it. The reason for this difference in both length and approach probably is that debating God's existence in Republican Rome, while not wholly welcomed, was not a matter of persecution and censure, as it was in Hume's cultural environment. Demea's dramatic role, it seems to me, served the purpose of showing the risk involved in daring to discuss this matter in the context of eighteenth-century religious censure. As Jones argues:

> [. . .] Hume follows the practice established by Cicero in making Philo apparently agree with Demea that their discussion concerns not "the being, but only the nature of the Deity," the former being "unquestionable." This transparent device, which Hume had used before [. . .], and which, indeed, would deceive only "the most careless, the most stupid thinker" (*DNR*, 214), was needed to avoid jeopardizing the whole inquiry at the outset (Jones 1982, p. 84).

That the existence of God is questioned throughout the dialogue does not mean, however, that once the skeptical examination was concluded, the characters could not accept that it was either a true or a probable assertion. Nevertheless, to reach this point, it was necessary to examine the claim rigorously, which was the ultimate aim of Philo's critique of the argument from design.

5.2.2 The Argument from the Intelligent Design of the Universe

In both dialogues, the argument from intelligent design of the universe is a central piece of the discussion on God's existence and nature. In Cicero's work, Balbus, the Stoic, expresses the argument in the following way:

> [. . .] when therefore we behold the compass of the heaven moving with revolutions of marvelous velocity and executing with perfect regularity the annual changes of the seasons with absolute safety and security for all things, how can we doubt that all this is effected not merely by reason, but by a reason that is transcendent and divine? (*DND* II, 97).

Less poetically, Hume has Cleanthes state the same idea:

> The curious adapting of means to ends, throughout all nature, resembles exactly, though it much exceeds, the productions of human contrivance [...] Since therefore the effects resemble each other, we are led to infer, by all the rules of analogy, that the causes also resemble; and that the Author of nature is somewhat similar to the mind of man (*DNR* II, 5; 143).

The argument's weight in both dialogues is notable, even though in *DND,* the characters also discuss the Epicurean's universality of the preconception of the gods. Evidently, Hume chose to focus on this argument given its centrality for empirical theism and the disputes among orthodox and unorthodox believers in seventeenth and eighteenth-century Great Britain. Hume attempted to show how extending scientific empirical presuppositions to theology led to the sort of skeptical contradictions that Cicero had already illustrated:

> It seems to me that Philo and Cotta are making essentially the same argument: that is, whether a necessary causal connection can be found between the phenomena of the external world and the perfect mind. Both are asking what justifiable, logical process can possibly be used to infer a god or supremely creating mind from our impressions of sense data in the external world (Price 1964, p. 103).

Having delineated this similarity, we may want to pause for a moment to observe the differences between the two formulations of the argument. In contrast to Cicero's poetic language, Hume presents the argument in a logical form, consistent with his empiricist epistemology. The resemblance of perceived effects – the universe and human-made artifacts – leads by analogy to infer a universe's cause similar to the human mind. As I explained in the previous chapter, in an analogy, two principles of association operate to enliven the ideas involved and produce assent or belief—namely resemblance and cause and effect. Through these mechanisms, the impressions and corresponding ideas of purposeful artifacts and the ordered universe transfer their force and vivacity to the causally associated ideas of a human and a divine mind. This explains how the imagination assents to God's existence and nature as a mind that has purposively created the universe. Hume attacked empirical theism on the grounds of these epistemological principles. As Fosl claims, this is "an important instance of Hume making used of his own philosophical accomplishment (his theory of causation) to deepen on one of Cicero's skeptical arguments (his attack upon natural religion)" (Fosl 1994, p. 109). Hume's adaptation of Cicero's view in this regard is, without doubt, the novelty and philosophical import of the *Dialogues*.

5.2.3 The Critique of Anthropomorphism

One of the most critical objections used by both Cicero and Hume against God's providence is the anthropomorphic charge. Price claims that "the rejection by Philo and Cotta of the anthropomorphic argument constitutes one of the most cohesive similarities between them" (Price 1964, p. 100; cf. Fosl 1994, p. 110). Nonetheless, the argument is employed in different contexts and with different argumentative aims in the two dialogues. On the one hand, in *De Natura Deorum*, Epicureanism is singled out as most egregiously committing this mistake by arguing that the preconception of the gods, acquired through visions or dreams, involves a similar shape to humans (*DND* I, 46). Cotta railed on, at length, against the Epicureans, objecting that this similarity would imply that the gods have certain facial and bodily features, live in specific places, pursue certain activities, feel passions, are mortal, etc., all of which are, evidently, absurd (*DND* I, 57–116). In conclusion, he claimed that the anthropomorphic preconception of the gods is just an arbitrary product of the imagination: "We have an idea of god implanted in our minds, you say. Yes, an idea of Jupiter with a beard, and Minerva in a helmet; but do you, therefore, believe that those deities are really like that?" (*DND*, I, 100).

On the other hand, in Hume's *Dialogues*, Cleanthes defends anthropomorphism against the fideistic view, presented by Demea, that the nature of God is beyond the grasp of reason, and, in consequence, we can only affirm that God exists and is a simple, infinite, and incomprehensible being. Against this view, Cleanthes claims that this is an irrational concession to skepticism:

> [...] For though it be allowed, that the Deity possesses attributes of which we have no comprehension; yet ought we never to ascribe to him any attributes, which are absolutely incompatible with that intelligent nature, essential to him? A mind whose acts and sentiments and ideas are not distinct and successive; one that is wholly simple and totally immutable; is a mind which has no thought, no reason, no will, no sentiment, no love, no hatred; or in a word, is no mind at all (*DNR* 4. 3; 159).

Of course, Cleanthes's acceptance of anthropomorphism gave rise to a large number of objections from Philo. In essence, the skeptic argued that anthropomorphism requires, first, renouncing the infinity and perfection of God; second, admitting the possibility that there is more than one god (polytheism); and, third, accepting that God (or the gods) may be mortal (*DNR* V, 5–10; 166–168). Then, Philo asked Cleanthes to completely embrace anthropomorphism and attribute to the deity a body, as the Epicureans did:

> And why not become a perfect anthropomorphite? Why not assert the Deity or Deities to be corporeal and to have eyes, a nose, mouth, ears, etc.? Epicurus maintained that no man had ever seen reason but in a human figure; therefore, the gods must have a human figure. And this argument, which is deservedly so much ridiculed by Cicero, becomes, according to you, solid and philosophical (*DNR* V, 11; 168).

Philo's argument is a short version of Cotta's and even includes a direct reference to Cicero. Furthermore, Hume uses here a skeptical *sorites*—the ancient logical form of argument through which once a similarity between two terms is accepted (in this case, God and the human mind), other similarities are also required to be accepted (that God has human drives and desires, a human body, human smells, etc.). In this way, the Academic skeptics sought to show the absurdity of the first accepted similarity. In my view, the allusion to Cicero in this paragraph does not require more commenting. As Cotta did in *De Natura Deorum*, Philo argues that anthropomorphism is utterly absurd and assuming such a crass and superstitious doctrine should embarrass any eighteenth-century natural theist.

5.2.4 The Argument on Evil

Cicero's *De Natura Deorum* and Sextus's *Outlines of Pyrrhonism* are the primary sources of Hume's extended treatment of evil, his most important argument against a purposively made universe and God's moral attributes (Price 1964, pp. 104–105). In *DND*, Cotta claimed that the existence of evil, particularly of moral evil, is sufficient reason to conclude that the universe is not ordered according to a design. Let us recall Cotta's words:

> [. . .] Do we then really think that this enormous crop of evil was sown by the immortal gods? For if the gods gave man reason, they gave him malice, for malice is the crafty and covert planning of harm; and likewise, also the gods gave him trickery and crime and all the other wickedness [. . .] (*DND* III, 75).

Hume's use of the argument is longer and more complex than Cicero's relatively brief mention. Section X of the *Dialogues* is devoted to the Epicurean idea, reported by Sextus, that given the existence of evil, God is either not wise, not benevolent, or not powerful (*Pyr.* III, 9–11). Philo's attack begins with a long circumlocution about how poets and historians of all times have ceaselessly depicted the overwhelming misery of human life. Given the pervasiveness of human suffering, he argues, either God is not wise, not all-powerful, or not benevolent:

> And is it possible [. . .] that after all these reflections [. . .] you can still persevere in your anthropomorphism, and assert the moral attributes of the Deity, his justice, benevolence, mercy, and rectitude, to be of the same nature with these virtues in human creatures? His power we allow infinite: Whatever he wills is executed. But neither man nor any other animal are happy: Therefore, he does not will their happiness. His wisdom is infinite: He is never mistaken in choosing the means to any end: But the course of nature tends not to human or animal felicity: Therefore, it is not established for that purpose (*DNR*. X, 24; 198).

Cleanthes responds by claiming that the argument on evil is flawed because, in the final account, the amount of good in the universe is always greater than the amount of evil (*DNR* X, 31; 200). But both Philo and Demea rapidly dismiss Cleanthes's Leibnizian optimism, and the section concludes with Philo's triumphal exclamation that the argument on evil is the most evident of all: "Nothing can shake the solidity of this reasoning, so short, so clear, so decisive." Therefore, we should adopt metaphysical skepticism and accept that reason is too limited to solve theological questions: "[. . .] these subjects exceed all human capacity, and [. . .] our common measures of truth and falsehood are not applicable to them" (*DNR* 10. 34; 201). Given the evidence of the existence of evil, we should limit our research to matters that stay within the reach of our capacities and refrain from speculative efforts. And this final step of the argument can be taken as Hume's up-front avowal of Academic skepticism (Fosl 1994, p. 104).

5.2.5 The Dialogues' Conclusion

As we saw in Chap. 2, Cicero ended *De Natura Deorum* by claiming, in his voice, that Stoic theism is the most verisimilar theological doctrine:

> Here the conversation ended, and we parted, Velleius thinking Cotta's discourse to be the truer [*verior*], while I felt that that of Balbus approximated more nearly to a semblance of the truth [*ad veritatis similitudinem videretur esse propensior*] (*DND* III, 95).

Hume also closed his *Dialogues* by advancing a final verdict in favor of Cleanthes's position, but this time the verdict is delivered by the student Pamphilus:

> Cleanthes and Philo pursued not this conversation much farther; and as nothing ever made greater impression on me, than all the reasonings of that day; so I confess that, upon a serious review of the whole, I cannot but think that Philo's principles are more probable than Demea's; but that those of Cleanthes approach still nearer to the truth (*DNR*, XII, 34; 228).

The passage speaks for itself about how Hume modeled his *Dialogues* after Cicero's *De Natura Deorum* (Price 1964, pp. 105–106). In both cases, the conclusion has produced a great deal of perplexity among interpreters, who wonder why, if both Cicero and Hume were skeptics, they decided to end their dialogues by accepting the superiority of theism over agnosticism. I already explained Cicero's Academic grounds for his final adherence to Stoicism in Chap. 2. Concerning Hume, some scholars have argued that Pamphilus's declaration was ironic, while others believe that he really accepted Cleanthes's theism. I will discuss these hypotheses in the last section of this chapter. For the moment, I suggest paying attention to the difference between Cicero's and Hume's dramatic choice for the character who gives the final verdict. In *DND* it is Cicero himself, while in the *Dialogues* it is the student Pamphilus (Battersby 1979, p. 245). As I have already mentioned, Hume's choice may result from his disapproving Cicero's dramatic strategy of pronouncing the verdict in his voice, because his authority unbalanced the conversation in favor of Stoicism (Hume 1826, p. 144). Probably, not to make the same mistake, Hume chose Pamphilus, a student with no authority, to adhere to Cleanthes's theism at the end of the dialogue. But he may have had other reasons as well, among them his desire to avoid censure (Smith 1947, pp. 60–62).

In conclusion, the commonalities between the dialogues confirm that Hume took Cicero's *DND* as a source for his arguments against theism and a model for his skeptical strategies. Hume further elaborates on some of Cicero's most important arguments, like the charge of anthropomorphism and the discussion on evil. But he also follows some of Cicero's disguising resources, such as his insistence that only the nature of the gods and not their existence is at stake and his dramatic choice of giving a final verdict in favor of theism. I can now turn to a more detailed analysis of the *Dialogues*, focusing on how Hume applied his Academic typology of beliefs and rules for causal inferences to the belief in God as a creator of the universe. I begin by determining whether we should place this analysis in the context of Hume's naturalism, that is, whether this belief should be considered a "natural belief" or not. After this, I examine the passages in the dialogues where Hume discusses whether it is a "proof," and finally, I support my interpretation that, for Hume, the belief held by natural theism is a "probability."

5.3 The Status of Empirical Theism: Natural Belief, Proof, or Probability?

Given the significance that the naturalistic interpretation of Hume's theory of belief had in both the twentieth and early decades of the twenty-first century, it is a commonplace for Hume scholars to debate whether the belief in a God as a creator of the universe is a "natural belief." Some commentators have argued that it is (Butler 1960, pp. 73–100; Tweyman 1986; Penelhum 1979, pp. 253–278; Fosl 1994, pp. 103–120), while others that, despite its pervasiveness among human beings, it is not (Smith 1947, pp. 1–75; Gaskin 1978, pp. 126–140; Jones 1982, p. 80; Foley 2006, pp. 83–112). I place myself in the latter camp, along with scholars who contend that only the beliefs in the existence of the external world, causality, the identity of the self, and the reliability of the senses can be considered "natural beliefs."[15] In particular, I follow Gaskin's line of argument, according to which, although the belief in God is widespread among humans, it is not a natural belief. His reasons are, first, that this belief is not indispensable to act and have an ordinary human life; second, that natural beliefs are adopted mechanically or instinctively, and are not the result of a more-or-less valid rational inference, while this particular belief in God is precisely the product of analogical inference, namely the argument from design; and, third, that natural beliefs are universal while Hume acknowledges in the *Natural History of Religion* (*NHR*) that there are atheistic societies: "Some nations have been discovered, who entertained no sentiments of religion, if travelers and historians may be credited; and no two nations, and scarce any two men, have ever agreed precisely in the same sentiments" (*NHR*, 124).[16] Thus, Gaskin concludes, the belief in God examined in the *Dialogues* is instead a *rational* belief:

---

[15] According to Gaskin: "Hume distinguishes at least three species of "natural instincts" or essential "natural beliefs:" (1). Belief in the continuous existence of an external world independent of our perception. (2) Belief that the regularities, which have occurred in our experience, form a reliable guide to those which will occur. (3) Belief in the reliability of our senses [. . .]. (Gaskin 1978, p. 116).

[16] There is an interesting debate on this point. Gaskin claims that, given Hume's views in the *Natural History of Religion* that there are atheistic communities, this belief cannot be universal. For M. Badía, however, Hume's position in this text does not allow for such a clear-cut interpretation. He says: "While Hume appears to suggest that belief in an invisible and intelligent power is not strictly universal, he also recognizes in the same place that the "propensity to believe in an invisible, intelligent power" (*NHR*, 135) is a general tendency of human nature. In essence, Hume is conceding to this belief the universality that he seemed to deny it at the beginning of the work" (Badía 1996, p. 100, my translation). In fact, Badía's quotation appears at the end of *NHR*, where Hume says that belief in God is not instinctive, although it does accompany human nature: "The universal propensity to believe in invisible, intelligent power, *if not an original instinct,* being at least a general attendant of human nature, may be considered as a kind of mark or stamp, which the divine workman has set upon his work." (*NHR*, 135, my italics). On this basis, Badía claims that this belief is "natural" but not "instinctive." (1996, pp. 96–109). However, at the beginning of the text, Hume had argued that, "this preconception springs not from an original instinct or primary impression of nature" (*NHR*, p. 125). Thus, despite the plausibility of Badía's argument, I believe Gaskin's view is more accurate.

> I would argue that Philo does not concede that belief in design must be a natural rather than a rational belief. Instead, he concedes that belief in a design is after all a *rational* belief: one to which the mind has a strong propensity but which is sometimes ambiguous, probable rather than certain, incapable of rational development or extension, dependent upon an imperfect or very limited analogy, and which commits the believer neither to action nor to forbearance from action (Gaskin 1978, pp. 125–126).

The instinctive force of the belief in causality or in the external world is simply not comparable to that of the belief in God. Those who suspend judgment on the existence of God – the agnostics – can act spontaneously. In contrast, those who do the same about the reality of the external world are forced to inactivity (Gaskin 1978, p. 129). Let us recall that the ancient *apraxia* objection was precisely based on this presupposition—that if the skeptics suspended belief in the existence of the external world, they would not be able to act. By contrast, the skeptics' life would not be confined to quietude if they suspended judgment on the existence of God. Probably they could not engage in religious practices (however, as we have seen, in the case of the Pyrrhonists, this could be avoided by following the observance of abiding by the customs of society) but, other than that, they would continue to behave like a normal human being. Thus, I concur with Gaskin that Hume's naturalism does not apply to this belief. While nature does force us to surrender our doubts about the existence of the objects of experience, causal relations, and the unity of the self, it does not make us give up our doubts about the existence of God.

Nonetheless, to assert this thesis conclusively, we need to examine the textual evidence more closely since, at some point of the discussion, Cleanthes argues that the theistic belief has the character of a "natural belief." In the passage, he criticizes Philo for exhibiting a fanciful and extreme skepticism that calls into question even the most commonsensical beliefs:

> The declared profession of every reasonable sceptic is only to reject abstruse, remote and refined arguments; to adhere to common sense and the plain instincts of nature; and to assent, wherever any reasons strike him with so full a force, that he cannot, without the greatest violence, prevent it. Now the arguments for natural religion are plainly of this kind; and nothing but the most perverse, obstinate metaphysics can reject them. [...] What degree, therefore, of blind dogmatism must one have attained, to reject such natural and such convincing arguments? (*DNR* III, 7; 154).

In this passage, Cleanthes uses an *ad hominem* argument to reject Philo's view that the belief of theism is not natural. This kind of argument, as we know, does not provide positive grounds for an opinion but simply aims at refuting the opponent's views by pointing to her character flaws. Later in the dialogue, Demea criticizes Cleanthes's tendency to use this type of argument. Understanding it is a fallacious accusation, Philo does not respond to Cleanthes but remains silent, and, immediately afterward, Pamphilus says that "Philo was a little embarrassed and confounded," and Demea diverts the course of the conversation (*DNR* III, 10; 155). The fact that it is Pamphilus who says this of Philo, I believe, should call our attention. Only an inexpert student would take an *ad hominem* argument for a good one and believe that Philo was refuted. The dramatic strategy, however, has been interpreted in several ways. For some commentators, Pamphilus's comment indicates Hume's intention to distance himself from Philo and his acceptance of Cleanthes's view that the belief of theism is a natural one (Penelhum 1979, pp. 253–278). Others think that this is yet another of Hume's strategies to balance the dialogue, which, up to that point, was somewhat weighted toward Philo's skepticism (Smith 1947, p. 65). I think that Hume may be sending a message to educated readers who can recognize Cleanthes's *ad hominem* strategy. It can also be a way to avoid censure since non-educated readers would probably mistake Pamphilus's observation for Hume's sincere endorsement of theism.

Gaskin's interpretation, once again, may shed light on this problem. He argues that, in his letter to Gilbert Elliot of 1751, Hume clearly expresses that he does not consider the theistic belief a natural one:

> In the *Dialogues*, Cleanthes [. . .] does indeed express himself in a manner that suggests that belief in God, via the argument from design, could be a natural belief.[. . .] But this part of Cleanthes' case is not endorsed by Philo, and in a letter, unequivocally in his own person, Hume admits that he cannot establish via the design argument that belief in God is unavoidably forced upon us like belief in our senses: "The propensity of the mind towards it [the design argument], unless that propensity were as strong and universal as that to believe in our senses and experience, will still, I am afraid, be esteemed a suspicious foundation." (Gaskin 1978, pp. 123–124; *cf. Letters*, Vol. I, 153–157).

The letter provides essential clues about Hume's position. There Hume declares that the "hero of the *Dialogues*" is Cleanthes and asks his friend Elliot to help him make this character's arguments stronger. This endorsement could be easily taken as an indication that Cleanthes expresses Hume's views, but immediately after this, Hume identifies himself with Philo. He says that he entertained Philo's doubts concerning religion since very young, and even wrote a book compiling them that he recently destroyed:

> Any propensity you imagine I have to the other side [Philo's side], crept in upon me against my will: and it is not long ago that I burned an old manuscript book, wrote before I was twenty; which contained page after page, the gradual progress of my thoughts on that head. It began with an anxious search after arguments to confirm the common opinion. Doubts stole in, dissipated, returned, were again dissipated, returned again, and it was a perpetual struggle of a restless imagination against inclination, perhaps against reason (*Letters*, Vol. I, p. 154).

Hume's description bears great similarity with his portrayal of the toils of skepticism in the Conclusion of the first book of the *Treatise*. He repeats the view that the combat takes place between imagination and reason and adds now inclination, seemingly on the side of reason. Perhaps here, by 'inclination,' he means the tendency caused by his calm passions, so often confused with reason (*THN* II, 417), to assent to the belief in the existence of a creator. If this is the case, then we can read the passage as saying that his inclination wanted to believe, while his imagination kept producing arguments against theism. Or, perhaps we should interpret that reason made him doubt, while his imagination, enlivened by the analogy of the argument, tried in vain to assent to the "common opinion" of God's existence. In any case, the combat is depicted, again, in the darkest way.

In the next paragraph of the letter, Hume goes on to say that, had he lived closer to his friend, he would have enjoyed co-writing the book, with him setting out Philo's case and Elliot contriving Cleanthes's:

> Had it been my good fortune to live near you, I should have taken on me the character of Philo in the dialogue, which you'll own I could have supported naturally enough: and you would not have been averse to that of Cleanthes (*Letters*, Vol. II, p. 154).

This testimony suggests that Hume was not a defender of natural theism. So, we should instead conclude that Hume did not share Cleanthes's view that the belief in God is a natural belief (Gaskin 1978, pp. 123–124).

Having established this, we can now turn to Hume's normative theory of belief and attempt to determine whether the belief of theism should be classified as "knowledge," "proof," or "probability," and what kind of assent should, accordingly, be given to it. To remind the reader of what I detailed in the previous chapter, in both the *Treatise* and the *Enquiry*, Hume classified judgments according to these three kinds. Judgments involving logical necessity or "relations of ideas" were labeled "knowledge." "Matters of fact" or empirical judgments sanctioned by an absolute regularity of the observed cases in past and present experiences, were called "proofs." And, other "matters of fact," which obtain in some cases but are contradicted by opposite past or present experiences, were termed "probabilities" (*THN* I, 124). Among probabilities, Hume named a subcategory based on analogies (*THN* I, 142), in which the resemblance and causal connection between the terms of the analogy transferred the force and vivacity of the known objects to the unknown ones (*THN* I, 142).

Now, we can quickly rule out the possibility that the belief under examination is "knowledge" on the grounds that it is empirical and does not involve logical necessity. Next, then, we should examine the possibility that it is a "proof." Philo begins this examination in the second part of the *Dialogues*:

> That a Stone will fall, that fire will burn, that the earth has solidity, we have observed a thousand and a thousand times; and when any new instance of this nature is presented, we draw without hesitation the accustomed inference. The exact similarity of the cases gives us a perfect assurance of a similar event [...]. But wherever you depart, in the least, from the similarity of the cases, you diminish proportionally the evidence; and may, at last, bring it to a very weak *analogy*, which is confessedly liable to error and uncertainty (*DNR* II, 7; 144).

Again, "proofs" are based on an absolute similarity between the cases observed in the past and the present. For example, I can be sure that if I put my hand into the fire, it will burn because whenever I have done so in the past, it has happened, and the present case is similar in every respect to the past. In addition, most people's experiences confirm this view, expressed in the common habit of not putting one's bare hand into the fire. Thus, I can say that the judgment "a bare hand will burn when in contact with fire" is a "proof." However, if circumstances are different – if, for example, my hand is covered by a thick glove – I might hope to have a different result. Then, the judgment "a hand covered by a glove will burn when in contact with fire" is not a "proof" but a "probability." Hume's argument here is just the same—when the cases under examination are not similar in all respects, we do not have a proof, but only a probability.

Analogously, in the *Dialogues*, Hume offers some examples of beliefs based on defective resemblances or, in his own words, "imperfect analogies":

> After having experienced the circulation of blood in human creatures, we make no doubt that it takes place in Titius and Maevius: But from its circulation in frogs and fishes, it is only a presumption, though a strong one, from analogy, that it takes place in men and other animals. The analogical reasoning is much weaker when we infer the circulation of the sap in vegetables from our experience that the blood circulates in animals; and those, who hastily followed that imperfect analogy, are found, by more accurate experiments, to have been mistaken (*DNR* II, 7; 144).

The same holds when the mechanism of cause and effect complements that of resemblance. If there is no complete resemblance of the effects, the inference of similar causes is imperfect. In these cases, the sixth rule should be taken into account: "The difference in the effects of two resembling objects must proceed from that particular in which they differ [. . .] we must conclude that this irregularity proceeds from some difference in the causes" (*THN* I, 174). In other words, if, upon examination, the resemblance between the known objects in an analogy is deemed weak or their causal connection feeble, assent should be suspended.

In the same vein, the effects of the analogy are not perfectly alike. We know from experience that a house is the product of an architect, but the likeness between a house and the universe is not forceful enough to grant the inference that the latter must also have an architect as its cause. Philo says:

> If we see a house, Cleanthes, we conclude, with the greatest certainty, that it had an architect or builder, because this is precisely that species of effect, which we have experienced to proceed from that species of cause. But surely you will not affirm that the universe bears such a resemblance to a house, that we can with the same certainty infer a similar cause, or that the analogy is here entire and perfect (*DNR* II, 8; 144).

Given Cleanthes's insistence that the similarity between the two effects is quite compelling, Philo advances a methodological constraint, namely to apply the same cautious attitude of natural scientists when deciding on causes and effects:

> But observe, I entreat you, with what extreme caution all just reasoners proceed in the transferring of experiments to similar cases. Unless the cases be exactly similar, they repose no perfect confidence in applying their past observation to any particular phenomenon. Every alteration of circumstances occasions a doubt concerning the event; and it requires new experiments to prove certainly, that the new circumstances are of no moment or importance (*DNR* II, 17; 147).

Again, Philo is seeking to convince Cleanthes that the belief in God is not a "proof," given the lack of similarity of the effects involved in the analogy. If we recall that Hume's fourth rule for judging causes and effects is that "the same cause always produces the same effect, and the same effect never arises but from the same cause" (*THN* I, 173), we can see why Philo is so adamant about this point. The researcher of nature is very cautious in making inferences of this kind and only determines that a causal judgment should be assented to when the observed cases have been similar in all their aspects. Hume is using an ancient skeptical strategy—to diminish the plausibility of the adversaries' doctrine by showing how, abiding by their own premises and methodological indications, they cannot make their alleged inferences without contradicting themselves. In this case, Philo seems to be tacitly arguing that if empirical theism is modeled after Newton's method, it should exhibit the same cautiousness of natural science before jumping to such definitive conclusions (*DNR* II, 27; 151). Thus, Philo urges Cleanthes to take the attitude of cautious scientists and accept that the belief in God of the design argument is not a "proof" but a "probability."

Having cleared this point, Philo resumes his skeptical attack, making Cleanthes explicitly accept that the belief of natural theism is a "probability" based on an analogy:

> But to show you still more inconveniences [. . .] in your anthropomorphism; please to take a new survey of your principles. *Like effects prove like causes.* This is the experimental argument; and this, you say too, is the sole theological argument. Now it is certain that the liker the effects are, which are seen, and the liker the causes, which are inferred, the stronger is the argument. Every departure on either side diminishes the probability and renders the experiment less conclusive. You cannot doubt of this principle: Neither ought you to reject its consequences (*DNR* V, 1; 165).

This strategy allowed Philo to situate the problem in the general framework of Hume's normative theory of belief. The theistic belief in God is a probability based on an analogy, and, as we know, probabilities can be rationally examined to decide whether or not to approve to them and to what degree. According to Hume, if the resemblance between the known objects in the analogy is weak or their causal connection feeble, the transference of the impression's vivacity should be reduced, and the inclination to assent refrained. Of course, this can occur naturally, for example, when we are not convinced by an analogy used in everyday conversation because we sense that its terms are too dissimilar or causally distant. But, given the influence of the passions, education, and prejudice, we can assent to the conclusions of some analogical reasoning even when the similarity and causal connection of the terms are not strong enough. In these circumstances, critical scrutiny is required to help us observe the argument's undetected flaws and refrain from assent. This is what the skeptic Philo is so masterfully doing in the *Dialogues.*

Having already shown that the resemblance between objects of human contrivance and the universe is not as evident as Cleanthes thinks, Philo moves to argue that the similarity between a human and a divine mind is also too weak, and for this reason, we should not, as theism does, give anthropomorphic attributes to the creator. In Sect. 5, then, Philo begins his argument against anthropomorphism by quoting Velleius's – Cicero's Epicurean spokesperson – criticism of the argument from design:

> If Tully esteemed this reasoning so natural as to put it into the mouth of his Epicurean: "What sort of mental vision enabled your teacher Plato to envisage the construction of so massive a work, the assembling and building of the universe by the god in the way which he describes? What was his technique of building? What were his tools and levers and scaffolding? Who were his helpers in so vast an enterprise? How could the elements of air and fire, water and earth knuckle under and obey the will of the architect?" (*DNR* V, 2; 166).

Immediately after this reference to Cicero – sympathetically called Tully – Philo asks Cleanthes to reaffirm his anthropomorphism. Cleanthes says that the discoveries of natural sciences show "the image of mind reflected on us from innumerable objects," and Philo corners him: "'Add, a mind *like the human*,' said Philo. 'I know of no other,' replied Cleanthes. 'And the liker the better,' insisted Philo. 'To be sure,' said Cleanthes" (*DNR* V, 4; 166). Cleanthes's unrestricted acceptance allows Philo to refute his anthropomorphism relatively easily by arguing that it entails renouncing God's infinity, perfection, unity, and immortality; attributes that belong to his very essence. Having accepted the absolute resemblance between the human and the divine mind and given the previously established principle that "like effects prove like causes," Philo says that an intelligent creator of this imperfect universe could be an "infant," "inferior," or "superannuated" deity (*DNR* V, 12, 169). The point, then, is that the analogy supports a kind of deity that is in no way similar to the god of theism. In other words, Philo confirms that the argument from intelligent design does not justify the theistic characterization of God.

Nonetheless, like with any other "probability," the belief in an intelligent creator does have some degree of persuasiveness, and Hume acknowledges it. For this reason, in the following sections of the *Dialogues*, Philo begins to proceed more constructively. I will focus my attention now on this side of Philo's argumentation.

5.3.1 Philo's Academic Hypotheses

In Chap. 2, I argued that the skepticism of the Academy has a constructive side, which can be best appreciated in Cicero's discussion of the immortality of the soul in the *Tusculan Disputations,* where he claimed that, to alleviate our fear of death, we could resort either to the Platonic thesis or the Epicurean one. Now we can see Hume rehearsing this aspect of Academic argumentation in a similar way. In sections six and seven, Philo advances two parallel hypotheses: first, that since the world is more akin to an animal or a vegetable than to an object of human contrivance, God may be the soul of this animal or vegetable; and, second, that the world may not have a creator but possess its own material principle of order and preservation. The latter argument is further elaborated in section eight of the *Dialogues*, where Philo claims that he will "revive the old Epicurean hypothesis," which, with some alterations, could "[...] bear a faint appearance of probability" (*DNR* VIII, 2, 182). This hypothesis affirms that the universe is a finite entity whose movement, transformation, and regularity originated from an initial, blind impulse:

> Suppose [...] that matter was thrown into a position, by a blind, unguided force; it is evident that this first position must in all probability be the most confused and most disorderly imaginable, without any resemblance to those works of human contrivance [...] But is it not possible that [the universe] may settle at last, so as not to lose its motion and active force (for that we have supposed inherent in it), yet so as to preserve an uniformity of appearance, amidst the continual motion and fluctuation of its parts? This we find to be the case with the universe at present (*DNR* VIII, 7–8; 184).

Philo suggests, in a cautious but sufficiently open way, that the order of the universe we commonly attribute to an intelligent cause may be the result of an original blind impulse—an initial material drive that produced the observed regularity and lawfulness of the universe before plunging it, once again, into absolute chaos. Even though Philo announces that this hypothesis is weak at the end of the section, he thinks it has some degree of probability and deems it a positive alternative to theism. However, he does not dare say that it is a more probable hypothesis than natural theism but cautiously chooses to put it at the same level of probability as Cleanthes's thesis. Doing this, he is forced to conclude that, given the equal weight of both claims, one should suspend judgment as the Pyrrhonists suggested:

> But all [religious systems], on the whole, prepare a complete triumph for the sceptic; who tells, that no system ought ever be embraced with regards to such subjects: For these plain reasons, that no absurdity ought ever to be assented to with regards to any subject. A total suspense of judgment is here our only reasonable resource. And if every attack, as is commonly observed, and no defense, among theologians, is successful; how complete must be his victory, who remains always, with all mankind, on the offensive, and has himself no fixed station or abiding city, which he is ever, on any occasions, obliged to defend? (*DNR* VIII, 12; 186–187)

From this passage, some commentators have concluded that Philo committed himself to Pyrrhonism in theological matters (Fosl 1994, p. 110). I doubt it, because his tone is not triumphal; rather, he suggests that a victory obtained without defending one's position is insignificant. In other words, Philo's lamentation may be a sort of sign-post, indicating that, while a Pyrrhonist may find these theological disputes a suitable terrain for advancing religious faith on irrational grounds, a moderate skeptic should defend "on any occasions," some sort of "abiding city,"[17]—that is, some claim that provides a rational basis for an active and good life. In my view, then, it is more plausible that this part of the dialogue serves the purpose of showing the consequences of behaving as a radical or Pyrrhonian skeptic, and as a way to show Philo's distance from this kind of skepticism.

Furthermore, after this apparent surrender to Pyrrhonian irrationality, Philo continues his attack on Cleanthes's theism in ways that lead to the final affirmation of the hypotheses mentioned above. First, Philo presents the most decisive attack on Cleanthes, that is, the skeptical argument on the existence of evil. The discussion begins with Philo and Demea acknowledging a human consensus on the existence of evil and arguing that whoever denies it holds a "bold and paradoxical" opinion (*DNR* X, 6–7; 194). After enumerating the various natural and moral evils that accompany human life, Philo asks Cleanthes:

> And is it possible [. . .] that after all these reflections, and infinitely more, which might be suggested, you can still persevere in your anthropomorphism, and assert the moral attributes of the Deity, his justice, benevolence, mercy, and rectitude, to be of the same nature with these virtues in human creatures? (*DNR* X, 24; 198).

Given the evidence of evil, Philo argues, God cannot possess the moral attributes that Cleanthes's anthropomorphism suggests, that is, power, wisdom, and benevolence. In particular, if God is all-powerful and wise, and has created the world

[17] The same reference to the skeptic as a nomad will appear in Kant (A IX).

rationally, he cannot also be benevolent, for the existence of evil brings no happiness to humanity. Thus, Philo concludes: "Epicurus's old questions are yet unanswered. Is he willing to prevent evil but not able? Then is he impotent. Is he able but not willing? Then he is malevolent. Is he both able and willing? Whence then is evil?" (*DNR* X, 25; 198).

As powerful as it is, the argument on evil does not deny the existence of a creator of the universe but only refutes the deity's providence, that is, his creating the universe for the benefit of humans, which is the Epicurean view on the gods. To this attack, Cleanthes responds that the argument is easy to overcome if we deny the existence of evil. However, in his following exchange, he does not deny its existence but only its prevalence:

> The only method of supporting divine benevolence (and it is what I willingly embrace) is to deny absolutely the misery and wickedness of man. Your representations are exaggerated: Your melancholy views mostly fictitious: Your inferences contrary to fact and experience. Health is more common than sickness: Pleasure than pain: Happiness than misery. And for one vexation which we meet with, we attain, upon computation, a hundred enjoyments (*DNR* X, 31: 200).

That Cleanthes does not deny the existence of evil is essential for Philo's subsequent refutation. He says that even accepting a balance between good and evil or the prominence of good over evil, the existence of the most insignificant evil in the world would contradict the infinitude of God's attributes (*DNR* X, 34: 201). And surprisingly, Cleanthes accepts this critique and proceeds to show his willingness to renounce the infinitude of God insofar as it preserves the similarity between the divine and human minds. He claims that, were we to accord the deity a higher, albeit merely finite, degree of perfection, we would be able to account for both the rational order of the universe and the existence of evil:

> But supposing the Author of nature to be finitely perfect, though far exceeding mankind; a satisfactory account may then be given of natural and moral evil, and every untoward phenomenon be explained and adjusted. A less evil may then be chosen, in order to avoid a greater: Inconveniences be submitted to, in order to reach a desirable end: And, in a word, benevolence, regulated by wisdom, and limited by necessity, may produce just such a world as the present (*DRN* XI, 1; 203).

Despite this argument being an enormous concession to the skeptic, Philo dismisses Cleanthes's acceptance of an imperfect God, arguing that the evils of the universe are still unwarranted, even if the creator merely has a finite mind, only superior in degree to the human. After this, having analyzed and discredited the plausibility of Manichaeism, Philo seems to side with a form of deism:

> There may *four* hypotheses be framed concerning the first causes of the universe: *that* they are endowed with perfect goodness, *that* they have perfect malice, *that* they are opposite and have goodness and malice, *that* they have neither goodness nor malice. Mixed phenomena can never prove the two former unmixed principles. And the uniformity and steadiness of general laws seem to oppose the third. The fourth, therefore, seems by far the most probable (*DNR* XI, 15; 212).

We find here a constructive way of reasoning, which begins to show Philo's allegiances. The fourth hypothesis – that the first cause of the universe has no moral attributes – is one of the main principles of "deism" (Force 1993, pp. 282–305, 1996, pp. 53–74). In other words, if the divine mind is not provident – that is, if it is utterly unconcerned with human affairs – we can attribute to it neither good nor evil intentions. As Philo argued in section eight, an initial blind force or impulse may have created the universe. Unsurprisingly, at his point in the dialogue, Demea, visibly irritated, accuses Philo of holding an openly impious position. He realizes Philo did not use his skeptical arguments, as he had supposed, to affirm the incomprehensibility of divine attributes and let theism be grounded on revelation – that is, to support fideism – but rather to remove the moral attributes from the deity and, in this way, give plausibility to deism. As a result, Demea asks Philo if he is "a more dangerous enemy than Cleanthes himself" (*DNR* XI, 18; 213) and leaves the meeting.

I believe that this dramatic exchange is essential to interpret the dialogue *vis-à-vis* modern fideism. Let us remember that Philo's allusions to a Pyrrhonian suspension of judgment in section III did not produce the same reaction in Demea. He was willing to accept skepticism if it led to fideism and its irrational acceptance of Christian revelation. However, once he saw that Philo's skepticism could result in deism, he no longer wanted to partake in the dialogue. Philo was, in fact, "a more dangerous enemy than Cleanthes himself" because, no matter how critical of Christian orthodoxy and superstition, Cleanthes accepted natural theism and revelation. But Philo's skepticism led to a claim that in eighteenth-century Britain was conflated with atheism, namely deism, since it opposed both natural theism and revealed religion. Paradoxically, then, such a "moderate" skeptic as Philo posed a much more serious threat to religion than the radical Pyrrhonian fideist.

Notwithstanding these passages, some commentators have identified Hume's views on religion with modern fideism. D. Livingston, for example, argues:

> A fideistic revealed religion that abandoned philosophical pretenses [...] could enter Hume's world of custom alongside philosophical theism. [...] In this regard, Hume's criticism of religion is radically different from a great variety of latter-day Enlightenment criticism which, refusing to acknowledge the humility of creatures embedded in custom, seek in the name of radical philosophical autonomy to purge human consciousness of the sentiment of the sacred (Livingston 1998, p. 152).

I take the opposite position. Hume's skepticism concerning religion does not involve the fideistic acceptance of religious custom and authority. Instead, it seeks "philosophical autonomy," a position in agreement with philosophical Enlightenment. For this reason, I believe it is necessary to broaden the scope of the ancient skeptical schools' influence to gain a more balanced view of the effect of skepticism on modern religious debates. In this way, while accepting that the incorporation of Pyrrhonism gave rise to modern fideism, we can appreciate how the adoption of Academic skepticism led to other religious attitudes, such as Hume's form of deism.

Now, to observe more justly Philo's constructive final hypothesis, namely his adherence to a sort of deism, we should move to the last section of the *Dialogues,* in which Philo gives his sudden "confession of faith." This confession, usually called "Philo's reversal," has posed many interpretative challenges. Since it involves a very concise and intricate reasoning, I will now give a more detailed account of its general line of argument.

5.3.2 Philo's *Reversal*: Non-Dogmatic, Anti-Religious Deism

Section twelve of the dialogue begins with Philo and Cleanthes conceding that, given Demea's religious orthodoxy, it was only natural that he found their tendency towards theological controversy irritating. Philo gives his sudden confession of faith in this general context, which I will describe here step by step. First, Philo says:

> You, in particular, Cleanthes, with whom I live in unreserved intimacy; you are sensible, that, notwithstanding the freedom of my conversation, and my love of singular arguments, no one has a deeper sense of religion impressed on his mind, or pays more profound adoration to the divine Being, as he discovers himself to reason, in the inexplicable contrivance and artifice of nature. A purpose, an intention, a design strikes everywhere the most careless, the most stupid thinker; and no man can be so hardened in absurd systems, as at all times to reject it (*DNR* XII, 2; 214).

After this initial claim, Philo continues praising the order and beauty of nature. He mentions the economy of natural laws and final causes in the universe:

> That nature does nothing in vain, is a maxim established in all the schools, merely from the contemplation of the works of nature, without any religious purpose; and, from a firm conviction of its truth, an anatomist, who has observed a new organ or canal, would never be satisfied till he had also discovered its use and intention. [...] The same thing is observable in other parts of philosophy: And, thus, all the sciences almost lead us insensibly to acknowledge a first intelligent Author; and their authority is often so much the greater, as they do not directly profess that intention (*DNR* XII, 2; 214–215).

To press the point, Philo offers the example of Galen, whose highly complex description of human beings stimulated the faith of others, despite his own religious disbelief, unlike the atheist philosophers of his own time who are excessively stubborn and even question that the order of nature provide sufficient proof of God's existence:

> Could I meet with one of this species [of philosophers] (who, I thank God, are very rare) I would ask him: Supposing there were a God, who did not discover himself immediately to our senses; were it possible for him to give stronger proofs of his existence, than what appears on the whole face of nature? (*DNR* XII, 4; 215).

Finally, he observes that this belief should be wholeheartedly accepted on account of the number and diversity of arguments that support it:

> Now according to the rules of just reasoning, every fact must pass for undisputed, when it is supported by all the arguments which its nature admits of, even though these arguments be not, in themselves, very numerous or forcible: How much more, in the present case, where no human imagination can compute their number, and no understanding estimate their agency? (*DNR* XII, 4; 216)

These paragraphs have been the object of many scholarly discussions. Commentators are divided on whether Philo's claims are ironic (Battersby 1979, p. 251; Mossner 1936, p. 344; Price 1965, pp. 134–152; Priest 1985, p. 349) or constitute Hume's recognition that the belief of theism is either a natural belief or at least a very pervasive one (Penelhum 1979, pp. 253–278; Fosl 1994, p. 108).[18] In my opinion, Hume is ironic in some of the section's passages but not in all of them. More importantly, I do not think that his "reversal" is an acceptance of theism. I believe that Philo accept a belief that is very close to deism's—the belief in a god creator of the universe, who is not provident and does not have human attributes, except that it may be something "similar" to the human mind. As we shall see, while this is a sufficiently plausible belief for Philo, it is a very insubstantial one from a religious point of view. To support this claim, I shall refer first to the passages in Philo's argumentation I believe to be ironic, and then to those where I find he expresses his positive and definitive views.

First, when Philo affirms that "a purpose, an intention, a design strikes everywhere the most careless, the most stupid thinker; and no man can be so hardened in absurd systems, as at all times to reject it" (*DNR* XII, 2; 214), he must be ironic since, of course, throughout the dialogue, Philo showed himself very skeptical of this purported evidence. He was, indeed, so "hardened in absurd systems" – those of skepticism! – that could reject several aspects of the analogy. Perhaps we should emphasize that the passage says that no one can deny the evidence of the argument "at all times," and hence, he could do it before, but now accepts theism as a natural or common belief. However, the reference to "absurd systems" can also be a veiled criticism of theological doctrines, such as empirical theism. Those who hold this theism could also be "so hardened in an absurd systems" that they cannot see the ease with which others doubt its purported evidence. Moreover, as Foley has argued (2006, pp. 98–99), Philo is not the only person in the dialogue who does not believe that the argument from the design of the universe supports the existence and attributes of God. Demea also rejects the argument on the grounds that it is insufficient to prove God's existence and leads to anthropomorphism. For Demea, in contrast, the *a priori* argument should be used to prove the existence of God, and revelation should supply the divine attributes.

Second, when Philo says that some philosophers only dare to deny this "supreme intelligence" because of their "pertinacious obstinacy," I believe he may be referring to himself ironically. The comment, "Could I meet with one of this species [of philosophers] (who, I thank God, are very rare) I would ask him ..." (*DNR* XII, 4; 215), seems to confirm this point. Indeed, Philo's judgment about the fortunate lack of disbelieving philosophers cannot be taken seriously by the author of the *Dialogues,* who enjoyed so much the company of the French atheistic *philosophes.*

Third, the line immediately following this passage – "Supposing there were a God, who did not discover himself immediately to our senses, were it possible for him to give stronger proofs of his existence, than what appears on the whole face of nature?" – must be in jest. Is it necessary to point out that God does not manifest to the senses and, therefore, "supposing" a God who does not do that is preposterous?

[18] There is yet a third thesis, advanced by Foley, according to which Philo is simply inconsistent and Hume wishes to portray him as such (2006, pp. 83–112).

Finally, Philo's claim that "according to all rules of just reasoning, every fact must pass for undisputed, when it is supported by all the arguments which its nature admits of, even though these arguments be not, in themselves, very numerous or forcible" (*DNR* XII, 4; 216) seems to me to be either a mistake or an ironic assertion. The rules of just reasoning do not allow us to accept without discussion a fact supported by few and weak arguments, even if they are the only ones admitted by its nature. Our reason may accept these arguments, but only when our thinking is clouded by other influences, such as prejudices or violent passions, not when it is sound and rigorous. Someone could argue that Hume is here mistaking the so-called general rules of the imagination to which he referred in the *Treatise* (*THN* I, 147) for the "rules of just reasoning." However, this hypothesis does not make sense because he explicitly speaks about "*just* reasoning" here, and does not mention the imagination or the understanding.

Moreover, Philo continued by stating that, in the case of the arguments supporting the theistic hypothesis, "no human imagination can compute their number and no understanding estimate their cogency" (*DNR* XII, 4; 216). The conclusion seems to be that, given our acceptance of much weaker claims, we have even more reasons to accept the theistic thesis since the arguments in its favor are innumerable and escape our understanding. But, is this a genuine compliment to an argument? I do not think so. As I see it, if neither the imagination can survey nor the understanding assess all the arguments in support of a thesis, we should not accept it. Particularly if we are following the "rules of just reasoning."

Now, these ironic references may not be enough to claim that Philo's reversal is not a sincere avowal that the belief in a creator of the universe is a natural one. At the beginning of the analysis, I outlined the main reasons the theistic belief cannot be classified as a natural belief. However, a closer analysis of these particular paragraphs may shed more light on that thesis; therefore, it is worth coming back to the problem for a moment.

On the one hand, we should note that whenever Hume talks about natural beliefs, he speaks of *instinct, custom,* or *mechanism.* In this paragraph, however, Philo says that his religious conviction is based on how the deity "discovers himself *to reason* in the inexplicable contrivance and artifice of nature" (*DNR*, XII, 2; 214; my emphasis). Thus, Hume acknowledges that the theistic belief is rational, not instinctive or mechanical. On the other hand, the passage that most inclines commentators towards accepting the hypothesis that theism is a natural belief is Philo's avowal that "all the sciences *almost lead us insensibly to* acknowledge a first intelligent Author" (*DRN* XII, 2; 215). Indeed, the expression "lead us insensibly" may be interpreted as saying that the belief in question is somehow mechanical or instinctive. However, I do not think the textual evidence supports this conclusion. We must bear in mind that Philo did not affirm that the sciences, in fact, lead us insensibly to this conclusion, but only that they "almost" do. It seems to me that Hume, a master in the art of concealing his views, would know how to make use of all kinds of rhetorical devices, such as this one, to hide his intentions.

Still, another possible objection is that, when Philo says that "a purpose, an intention, a design strikes everywhere the most careless, the most stupid thinker" (*DNR* XII, 2; 214), he accepts that this belief is natural. However, a careful reading of the passage reveals that he only asserts that no one can deny a purpose, a design, and an intention *in nature*, not that such purpose, design, or intention has a divine origin. As he claims right after this: "That nature does nothing in vain is a maxim established in all the schools, merely from the contemplation of the works of nature, *without any religious purpose*" (*DNR* XII, 2; 214, my emphasis). In conclusion, most of the initial passages of Philo's confession of faith seem to be ironic and none of them actually justifies the view that for Hume belief in God is a natural belief. Let us move now to identify Philo's earnest views on the matter.

Immediately after Philo's unexpected profession of faith, Cleanthes reaffirms his position, claiming that the design argument is, indeed, the most evident and natural one:

> The comparison of the universe to a machine of human contrivance is so obvious and natural, and is justified by so many instances of order and design in nature, that it must immediately strike all unprejudiced apprehensions, and procure universal approbation (*DNR*, XII, 5; 216).

Cleanthes affirms that no one could suspend judgment on God's existence and providence for a very long time without once again assenting to a belief that is "supported by strong and obvious reason, by natural propensity, and by early education" (*DNR* XII, 5; 216). In other words, he insists that this belief is a natural one. As a response, Philo also declares himself opposed to suspension of judgment and offers what I consider to be his actual position:

> So little, replied Philo, do I esteem this suspense of judgment in the present case to be possible, that I am apt to suspect there enters somewhat of a dispute of words into this controversy, more than is usually imagined. That the works of nature bear a great analogy to the productions of art is evident; and according to all the rules of good reasoning, we ought to infer, if we argue at all concerning them, that their causes have a proportional analogy. But as there are also considerable differences, we have reason to suppose a proportional difference in the causes (*DNR* XII, 5; 216).

In this passage, again, Philo's claim that suspending belief about God's existence and provident nature is not possible may lead us to think that he is accepting theism as a natural belief. But, this is not the only possibility. Philo may be saying – in the fashion of the Academics his name represents – that, after careful examination, it is not possible for a just reasoner to suspend belief on the issue because the argument actually gives enough rational justification to adopt *another* belief. This is clear from the last part of the paragraph, where Philo recommends accepting the analogy, but with qualifications. The similarity between the objects of human contrivance and the universe is, for him, reasonable ground for believing in the existence of a creator of the universe somewhat analogous to the human mind. However, when determining this creator's attributes, differences between the observable effects of the analogy should be taken into account, and, as a result, some restrictions should be in place. First, says Philo, we "ought to attribute a much higher degree of power and energy to the supreme cause than any we have ever observed in mankind" (*DNR* XII, 6; 217). Second, we must accept that the difference of degree in the quality of the two minds cannot be determined. On this point, Philo reminds us of how "verbal disputes" on philosophical matters, especially those relating to the degrees of quality, usually center on ambiguous language and offers the very controversial example of the equal opinion theists and atheists would hold, were they asked about the nature of the divine mind:

> That the dispute concerning theism is of this nature, and consequently is merely verbal, or, perhaps, if possible, still more incurably ambiguous, will appear upon the slightest enquiry. I ask the theist, if he does not allow, that there is a great and immeasurable, because incomprehensible difference between the human and the divine mind: The more pious he is, the more readily will he assent to the affirmative. [...] I next turn to the atheist, who, I assert, is only nominally so, and can never possibly be in earnest; and I ask him, whether, from the coherence and apparent sympathy in all the parts of this world, there be not a certain degree of analogy among all the operations of nature [...] and the structure of human thought, be not energies that probably bear some remote analogy to each other: It is impossible he can deny it: He will readily acknowledge it (*DNR* XII, 7; 218).

As Jones argues, at this point, Hume is following Cicero's "tendency to dismiss apparently substantive disagreements as merely verbal disputes" (Jones 1982, p. 39). And, indeed, this is a Ciceronian, rhetorical device that Hume also employed in his discussion of freedom and necessity in the *Enquiry* (*EHU*, VII, 62–63). We can count, both here and there, on this argument to be a disguising strategy, used to win the favor of those who think that philosophers make distinctions that are too thin and practically useless. The controversial example goes as follows: if one asked a theist about the difference between the divine and the human mind, she would surely respond that it is incommensurable; and if one asked the atheist, having first persuaded her of the existence of an ordering principle of the universe, if she believed that this principle might be similar to the human mind, she would respond that the likeness would be probably minimal, but possible (*DRN* XII, 7; 218). The parallelism between the theist and atheist tends to confuse contemporary readers because we think that an atheist, on principle, cannot be convinced of the existence of a first principle resembling a human mind without ceasing to be an atheist. But, Hume probably intends to refer to other opponents of Christianity, namely the deists, who were considered atheists in seventeenth and eighteenth-century Britain. We know now that there was a blurred difference between the epithets "deist" and "atheist" during Hume's time. J. Force notes: "As with 'priestcraft,' 'atheism,' or 'freethinking,' 'deism' is one of the dirty words of the age" (Force 1993, p. 282). Thus, to openly advocate deism in this historical period, was like accepting atheism, and accepting atheism was, first and foremost, rejecting Christian theism. William Blake's harsh appraisal of deism also illustrates the common opinion on the movement:

> Deism is the worship of the God of this world by the means of what you call Natural Religion and Natural Philosophy, and of Natural Morality or Self-Righteousness, the Selfish Virtues of the Natural Heart. This was the Religion of the Pharisees who murder'd Jesus. Deism is the same and ends in the same (Blake 1957, p. 243, quoted by Force 1993, p. 67, n. 2).

Despite this characterization, the deists accepted a rational ordering principle and a creator of the universe. For this reason, Hume's description of the dispute between the theist and the atheist is actually meaningful— if pushed sufficiently enough on the similarity of this rational principle with the human mind, the deists would probably accept it (Force 1993, pp. 253–278, 1996, pp. 53–74). Referring to the deist as an atheist, then, Philo concludes:

> The theist allows that the original intelligence is very different from human reason: The atheist allows that the original principle of order *bears some remote analogy to it*. Will you quarrel, Gentlemen, about the degrees, and enter into a controversy, which admits not of any precise meaning, nor consequently of any determination? (*DNR* XII, 7; 218, my italics).

Philo's view that the human and divine mind are only *remotely* similar is also stated at the end of the dialogue (*DNR* XII, 33; 227). To strengthen his position at this point of the dialogue, Philo continues arguing that the difference between the effects of the analogy – again, the objects of human contrivance and the universe – entails accepting a greater resemblance between the *natural* attributes of God and human minds than between their corresponding *moral* traits:

> And here I must also acknowledge, Cleanthes, that, as the works of nature have a much greater analogy to the effects of our art and contrivance, than to those of our benevolence and justice; we have reason to infer that the natural attributes of the deity have a greater resemblance to those of man, than his moral have to human virtues. But what is the consequence? Nothing but this, that the moral qualities of man are more defective in their kind than his natural abilities. For as the supreme being is allowed to be absolutely and entirely perfect, whatever differs most from him departs the farthest from the supreme standard of rectitude and perfection (*DNR* XII, 8; 219).

This argument is a very cunning one. Hume seems to be implying that we cannot, based on the argument from design, infer God's moral attributes—a view also advanced by deism. But immediately after suggesting this, he contradicts himself by arguing that this only means that human features, particularly moral traits, are inferior in degree to the deity's. The piety of the second part of the paragraph seems, in my opinion, to be intentionally placed to divert the reader's attention from the first one. Openly asserting that the analogy does not allow us to attribute moral qualities to God would amount to expressing a deistic – or, again, atheistic view – too directly.

Philo reaffirms, then, that what he previously said constitutes his "unfeigned sentiments on this subject" (*DRN* XII, 9; 219). This time, I find that Philo's assertion is actually not ironic, for, up to now, his confession only includes the following statements:

1. That, based on the argument from design, it is *reasonable* to believe in the existence of God as a creator of the universe.
2. That when it comes to determining the attributes of this God, we should attend to the differences between the terms of the analogy.
3. That we can assign to God greater power and energy than humans have.
4. That we can only attribute to God a mind *remotely similar* to that of humans, a point upon which both theists and "atheists" (deists) ultimately agree.

I prefer to exclude from this list anything definitive about the moral qualities of God, since Hume's argument on this point seems inconsistent. However, I think that he sides with the deists in denying that God has moral attributes.

Given this analysis, I agree with the interpretations that consider Philo's reversal to be consistent with the skeptical position he maintained throughout the dialogue. His apparent confession does not commit him to natural theism, but only to some sort of deism (Laird et al. 1939, pp. 207–217; Gaskin 1976, pp. 310–311, 1983, pp. 219–229).[19] Given the characteristics mentioned above, it seems that Philo's belief shares the general tenets of British deism. Granting that the movement was not as uniform as some historians portray it (Hudson 2009, pp. 1–27; Hudson et al. 2014, pp. 13–23), I think we can accept Force's (1996, pp. 53–74) and Gay's (1968, pp. 9–26) characterization, based on the testimonies of eighteenth-century Christian apologists, such as Samuel Clarke and William Whiston. According to them, the deists' held:

1. A *rational* belief in God as the creator of the universe, based on the argument from intelligent design of the universe.
2. The belief that God does not directly rule the world but only does so through the natural laws established from creation—that is, God does not intervene in human affairs and does not carry out miracles.
3. The rejection of revealed religion, especially of the Scriptures.
4. The rejection of the belief in the immortality of the soul.

Based on this characterization, we can argue that Philo has accepted the first two aspects of deism up to this point in the dialogue. The first one is implied in his assessment of the belief resulting from the argument from design. For him, it is not a natural belief but rather one based on a *rational* inference, namely a probability based on an analogy, which has resisted *to some extent* the skeptical examination. Philo's denial that the divine mind is provident and his acceptance that both minds bear only a very *remote* similarity correspond to the second characteristic of deism. And, as we will discuss next, the rest of the section supports the third point of the list, because the dialogue turns then to the question of the practical utility of revealed religion.

Just as Cicero did at the end of *De Natura Deorum*, Hume thought it was necessary to finish his dialogue by examining the practical utility of Christian theism. Is this revealed religion practically advisable? Is it an ally for morality? For cautionary reasons, the debate is situated within the broader issue of the difference between "true religion" and "superstition." Philo claims that he is a fervent enthusiast of true religion and that his previous skepticism was only motivated by his "abhorrence of vulgar superstitions," which led him to "indulge a peculiar pleasure, in pushing such principles, sometimes into absurdity, sometimes into impiety" (*DRN* XII, 9; 219). Cleanthes responds by insisting on the social utility of religion, including the superstitious one: "Religion, however corrupted, is still better than no religion at all" (*DRN* XII, 10; 219). He asserts that the belief in an afterlife, where we are assigned rewards and punishments for our behavior on earth, is always helpful for restraining humans' sinful inclinations. Indignant, Philo responds with the long list of evils that religion has brought upon humanity:

> How happens it, then, said Philo, if vulgar superstition be so salutary to society, that all history abounds so much with accounts of its pernicious consequences on public affairs? Factions, civil wars, persecutions, sub-versions of government, oppression, slavery; these

[19]Kant himself offers this interpretation in his *Prolegomena:* "For let us assume at the outset (as Hume in his *Dialogues* makes Philo grant Cleanthes), as a necessary hypothesis, the deistic concept of the First Being [. . .]" (Ak. IV, 359).

> are the dismal consequences which always attend its prevalency over the minds of men. If the religious spirit be ever mentioned in any historical narration, we are sure to meet afterwards with a detail of the miseries, which attend it (*DNR* XII, 11; 220).

Despite Cleanthes's efforts to defend revealed religion, Philo continues his tirade, consistently arguing that his target is not true religion but superstition. To begin with, he claims, the rewards and punishments of the next life, which Cleanthes deems so effective for checking the passions, do not influence incontinent souls who do not refrain from immediate pleasures because these rewards and punishments are too uncertain or too distant in time (*DNR* XII, 13; 220). Moreover, religion does not promote moral education but only encourages a superficial observance of its precepts, which, complemented by fanaticism and ignorance, ultimately corrupts the natural sources of morality:

> But even though superstition or enthusiasm should not put itself in direct opposition to morality; the very diverting of the attention, the raising up a new and frivolous species of merit, the preposterous distribution which it makes of praise and blame, must have the most pernicious consequences, and weaken extremely men's attachment to the natural motives of justice and humanity. [. . .] where the interests of religion are concerned, no morality can be forcible enough to bind the enthusiastic zealot (*DNR* XII, 16–18; 222).

For Philo, religious enthusiasm is detrimental to morality because it corrupts the moral tendencies and sentiments of human nature. It inflames the violent passions of fanaticism and enthusiasm, and obscures the calm passions of benevolence, justice, and sympathy, on which morality is based.

Finally, neither in the political sphere is revealed religion a promoter of social harmony. Religion does not support the State in fostering social progress, as is attested by how, in all nations and times, states have enforced laws to put boundaries to ecclesiastic power:

> Is there any maxim in politics more certain and infallible, than that both the number and authority of priests should be confined within very narrow limits, and that the civil magistrate ought, forever, to keep his *fasces* and *axes* from such dangerous hands? But if the spirit of popular religion were so salutary to society, a contrary maxim ought to prevail (*DNR* XII, 21; 223).

Unlike Cicero, for whom the State religion was like the walls that protected a city from external threats, Philo thought that State politics should completely dissociate from religion. The "wise magistrate" ought not to accept a prevailing religion among its citizens, he said, for then he would "sacrifice, to an uncertain prospect of tranquility, every consideration of public liberty, science, reason, industry, and even his own independency" (*DNR* XII, 21; 223). The laws should, at most, allow the inevitable existence of diverse sects, hoping that none of them prevails in a way that challenges or disrupts the public powers. This review of the practical consequences of religion suggests that Hume shares the deists' apprehension for revelation or, at least, that he advocates for the anti-religious underpinnings of deism.[20]

[20] I will not refer to the last point, namely the rejection of belief in the immortality of the soul, but an analysis of Hume's essay "Of the Immortality of Soul" (E. 590–597) would easily confirm Hume's deistic inclination.

In conclusion, having examined the arguments for and against natural theism, what follows, for Philo – and I dare say, for Hume – is not suspension of judgment, but the adoption of a much less robust belief than theism: a rational belief, based on the analogy of the argument from design, in a God creator of the universe, whose similarity with the human mind is however so feeble that we cannot attribute anthropomorphic features to it, but only a power superior to human capacity and proportional to its effect. Besides this, given the harmful practical consequences of revealed religion, this belief should not encourage any institutional creed. Philo's conclusion is clear:

> If the whole of natural theology, as some people seem to maintain, resolves itself into one simple, though somewhat ambiguous, at least undefined proposition, that the cause or causes of order in the universe probably bear some remote analogy to human intelligence: If this proposition be not capable of extension, variation, or more particular explication: If it affords no inference that affects human life, or can be the source of any action or forbearance: And if the analogy, imperfect as it is, can be carried no farther than to the human intelligence; and cannot be transferred, with any appearance of probability, to the other qualities of the mind: If this really be the case, what can the most inquisitive, contemplative, and religious man do more than give a plain, philosophical assent to the proposition, as often as it occurs; and believe, that the arguments, on which it is established, exceed the objections, which lie against it? (*DNR* XII, 33; 227).

He insists that the only justifiable theological belief is that the cause or causes of the universe bear some remote similarity to the human mind. It should also be noted that Philo is not saying here that it is necessary to embrace only *one* cause of the universe. Neither does this remote similarity should compel us to think that the deity is an intelligent *being*, given that it may be a mere rational principle, a blind material force (or, again, a number of them). From a theoretical point of view, this is indeed a very general belief. And, from a practical one, not only does it not support any kind of religious practice but also suggests refraining from them, at least in their traditional and popular forms. In this way, as I had argued before, we arrive at an anti-religious deism (Gaskin 1983, pp. 160–173). Indeed, Philo's arguments about the uselessness of promises of rewards and punishments in another world and the morally corrupting force of established religious institutions seem to be sufficient to indicate that the theological belief accepted in Philo's reversal has significant anti-religious overtones. And, this is precisely the sort of belief he refers to with the term "true religion" or "philosophical religion," which, he says, does not beget the dangers mentioned above: "And so will all religion, said Philo, *except the philosophical and rational kind*" (*DNR* XII, 13; 220; my emphasis). N.K. Smith offers a good summary of Hume's stance:

> Hume's attitude to true religion can therefore be summed up in the threefold thesis: (1) that it consists exclusively in intellectual assent to the 'somewhat ambiguous, at least, undefined' proposition, 'God exists'; (2) that the 'God' here affirmed is not God as ordinarily understood; and (3) as a corollary from (1) and (2), that religion ought not to have, any influence on human conduct beyond, that is to say, its intellectual effects, as rendering the mind immune to superstition and fanaticism (Smith 1947, p. 24).

The most crucial aspect of Hume's "true religion" is that it is compatible with a civil foundation of morality, which derives moral precepts from the natural constitution of the human mind: "so pure a religion [. . .] represents the deity to be pleased with nothing but virtue in human behavior" (*DNR* XII, 15; 221). Hume's views are, then, sufficiently stated by Philo's so-called "reversal" but actual confession of anti-religious deism.

However, before Philo could delve into this (most profane!) view, Cleanthes, who had been displaying unlimited patience, decides to point out to Philo the precipice off which he is slipping:

> Take care, Philo, replied Cleanthes, take care: Push not matters too far: Allow not your zeal against false religion to undermine your veneration for the true. Forfeit not this principle, the chief, the only great comfort in life; and our principal support amidst all the attacks of adverse fortune (*DNR* XII, 24; 224).

Cleanthes's warning is important, given the deistic overtones of the position Philo seems to be espousing. That Philo's confession of faith was not embracing natural theism and that Christian orthodoxy would see it as dangerously irreligious becomes evident from Cleanthes's urging Philo not to continue with this argument. But Philo ignores his warnings, and the final two pages of the *Dialogues* are reserved for a re-enactment of the Epicurean arguments against religion. I will say a word about this before closing the chapter, but let me now clarify my claim that, in his reversal, Philo embraces a form of deism.

Gaskin's interpretation of Philo's reversal as involving an "attenuated" form of deism (Gaskin 1978, pp. 126–131, 1983, pp. 160–173) is an essential antecedent for my position. Gaskin claims:

> [...] I have concluded that Hume regards belief in a god as reasonable rather than unreasonable and he thinks that reasonableness is recognized by most men when they survey the order in nature (something of the design argument remains). But this reasonable belief amounts to so very little that the theist and the atheist can agree about it. The belief is without religious significance and leaves Hume free (a) to adopt an attitude of mitigated skepticism to all theological arguments because they are beyond our understanding; (b) to attribute belief in a particular revelation to an irrational faith which he sarcastically refers to as a miracle; and (c) to criticize and condemn religion 'as it has commonly been found in the world' (Gaskin 1978, p. 130).

Gaskin refers here not only to Philo's position in the *Dialogues* but also to Hume's views in the *Natural History of Religion* and the first *Enquiry*. The belief in God Hume concedes throughout his works to be reasonable is devoid of any religious significance. It is neither the belief of the natural theist nor of the fideist.[21] It is important to highlight, however, Gaskin's allusion to Hume's deism as an

[21] As Gaskin claims, Hume's apparently fideistic avowal at the end of the *Enquiry*, where he says that "whoever is moved by Faith to assent to it [Christian religion], is conscious of a continued miracle in his own person, which subverts all the principles of his understanding and gives him a determination to believe what is most contrary to custom and experience" (EHU X, 131) is, in reality, ironic, and cannot be seen as an endorsement of fideism. Hume's views, expressed in Philo's apparently pious "reversal," are consistent with this passage's irony.

"attenuated" one (Gaskin 1983, pp. 160–173). He chose this adjective to avoid a more robust identification of Hume with the English deists, who held their tenets dogmatically, and second, to stress his non-dogmatic attitude toward this belief.

Regarding the first motivation, Gaskin claims that Hume did not want to identify himself as a deist publicly, on the one hand, because he wanted to avoid censure[22] and, on the other, because he neither accepted the argument from design without reservations, as some deists did, nor supported other arguments used by them, like the *a priori* and cosmological proofs (Gaskin 1983, pp. 167–168). Thus, strictly speaking, Hume was not a deist, although he was sympathetic to deism. Concerning the second motivation, Gaskin claims that Hume's adherence to the belief in God from the argument from design is "diffident," a mere acknowledgment that it is a probability and not a proof:

> What remains [of the argument from design] is something like the conclusion of deism shorn of reliance upon *proofs* and hence diffidently affirmed: order in the natural world suggests an origin or source, which might warrant cautious use of the word intelligence (Gaskin 1983, p. 171).

I believe it is essential to highlight this diffident attitude to avoid misunderstandings. Indeed, surveying the influence of Hume on the American deists, Fosl argues that, given his skepticism, we should refrain from identifying Hume's religious ideas with deism (Fosl 1999, pp. 171–192). Fosl rightly contends that Hume was not a deist and could not be one of them, because the deists accepted without reservations the argument from design, claimed that knowledge of God's attributes is possible, and understood natural laws as God's commands:

> In the first place, it is important to see that Hume himself was not a deist. Indeed, many of the most basic philosophical themes and positions he develops directly militate against deism. Hume's skepticism with regard to reason and the senses, for example, undermines the crucial deist contention that reason possesses sufficient and irreproachable power both to disclose and to establish truths about God, humanity, and nature (Fosl 1999, p. 174).

Fosl's view has a firm basis since the deists seem to have held their theological convictions in a more dogmatic way than Hume could have endorsed. But he also argues that "Hume and the deists share, among other things, a commitment to what might be called, in their view, a reasonable secularism as well as a deep suspicion and critical posture toward orthodox religious belief" (Fosl 1999, p. 174). And, this reasonable secularism, I believe, is part of his "attenuated" or non-dogmatic way of accepting deism as a mere probability, more verisimilar than theism, but still only highly probable.

Other commentators, such as Norton, see Hume's skepticism as a "speculative atheism." For him, this sort of atheism holds that either there is no God at all or there is a God who does not engage in human affairs. Hume's religious views accord with the latter, since he does not deny the existence of God but only rejects divine providence, miracles, and the immortality of the soul. Norton concludes:

[22] However, he seems to have accepted "a tincture of deism" in his thought (Gaskin 1983, p. 168).

> If atheist is used in only the strict present-day sense, to refer to a person who explicitly denies the existence of a god or gods, one could conclude that Hume was not an atheist, for we have, so far as I know, no record of such an outright denial. I suggest, however, that such a conclusion is anachronistic and seriously misleading; by standards of his contemporaries, Hume was indeed an atheist (Norton 1991, p. 39).

Given the loose and derogatory use of the term "atheist" during the eighteenth century, we could accept both Gaskin's and Norton's views and say that Hume was a "non-dogmatic deist" or a "speculative atheist." However, I consider it helpful to keep the distinction between deism and atheism, even if it seems anachronistic, because, although the deists were often seen as atheists, they did not consider themselves as such, and I believe that, to some extent, Hume shared this assessment of his sentiments. As Gaskin argues, Hume was seen in Diderot's circle – renown for open atheistic avowals – as holding a rather conservative religious stance. Hence, James Macdonald's famous remark: "[. . .] poor Hume, who on [his] side of the water was thought to have too little religion, is here thought to have too much." (Cited by Gaskin 1983, p. 165). It seems to me, then, that Hume did not consider himself an atheist, and he had good reasons for this, since he did not deny that the design argument reasonably justified the belief in a god creator of the universe. He was, then, not an atheist who is, in the strict sense of the term, a negative dogmatist. But neither was he an agnostic, who would entirely suspend judgment on the matter, nor a fideist, who, suspending rational judgment on God's existence and attributes, would irrationally embrace the doctrines of revelation. Hume's position was very subtle, which is why we have such a difficult time trying to understand him. Now, let me finish this chapter by further explaining how Hume's position cannot be seen as a mild dogmatism but only as a non-dogmatic approval of an Epicurean kind of deism.

5.3.3 Hume's Non-dogmatic Deism: A Carneadean Approval of Epicureanism

I mentioned before that the last part of the *Dialogues* is devoted to examining the practical utility of religion. Philo's final position is greatly influenced by his assessment of the harmful social consequences of revealed religion. Let me add now that Philo not only expounds religion's social repercussions but also surveys its psychological effects. He argues that religion is less a form of consolation in the face of life's misfortunes than an inextinguishable source of fear and melancholy reflections. Since it is born out of the human fear for natural adversities and the hope that invisible powers spare us from them, fear is religion's natural and unavoidable attendant: "But still it must be acknowledged, that, as terror is the primary principle of religion, it is the passion which always predominates in it, and admits but of short intervals of pleasure" (*DNR* XII, 29; 225). Philo says that even the brief periods of hope and joy that religion allows for are detrimental to human mental health because, colored as they are by ecstatic enthusiasm, they quickly plunge us into sorrow and dejection:

> Not to mention, that these fits of excessive, enthusiastic joy, by exhausting the spirits, always prepare the way for equal fits of superstitious terror and dejection; nor is there any state of mind so happy as the calm and equable. But this state it is impossible to support, where a man thinks, that he lies, in such profound darkness and uncertainty, between an eternity of happiness and an eternity of misery (*DNR* XII, 30; 226).

Since the best philosophy, for Hume, promotes a balanced temper devoid of violent passions, Philo cannot adhere wholeheartedly to revealed theism. This final criticism closely resembles Lucretius's famous condemnation of religion as a source of superstition, mental distress, and fear of the afterlife.[23] And, this assessment of the emotional consequences of religion is, in my opinion, essential for understanding Hume's position in the *Dialogues* as thoroughly Academic. While in Cicero's dialogue, the Academic skeptic approved the Stoic theses concerning the existence and nature of the Gods as the most plausible or persuasive ones, in Hume's, Philo accepted the deistic hypothesis as the most probable one, in great measure because of its Epicurean anti-religious overtones. Even if Hume did not rely on a *preconception* of the gods to accept their existence but instead argued – in a way that befits an eighteenth-century philosopher – that the design argument provided sufficient empirical basis for this acceptance, he did conform to the Garden's practical orientation in religious matters. But, once again, Philo's acceptance of Epicurean-inspired deism is not dogmatic. In other words, he is diffident of this newly acquired conviction, recognizes it as merely probable and is therefore willing to put it to test whenever necessary, while temporarily adopting it for practical purposes.

Now, someone could object that the *Dialogues*' final verdict is expressed in Pamphilus's words at the end of the book, not in Philo's confession of faith, and, thus, Cleanthes's theism did win the dispute. Pamphilus says: "so I confess that, upon a serious review of the whole, I cannot but think that Philo's principles are more probable than Demea's, but that those of Cleanthes approach still nearer to the truth" (*DNR* XII, 34; 228). Should not this avowal be seen as Hume's actual views on the matter? Why should we, after all, think that Hume sides with Philo and not with Cleanthes, if we have seen that Cicero used this dramatic strategy to show his own allegiances at the end of *De Natura Deorum*? As I already argued, his avowal cannot be taken as Hume's sincere position on the issue, but should instead be regarded as an indication of Hume's indebtedness to Cicero's *DND*, as well as a way to avoid censure. I think that given Hume's misgivings with Cicero's use of his authoritative voice in *DND*, his decision to put this final verdict in the mouth of a student dramatically indicates that it should not be taken to heart. Instead, Philo's confession of Epicurean, anti-religious deism should be considered as his final position. With Pamphilus's naïve last words, Hume was only disguising his allegiance to Philo. Of course, he must have been conscious of the difference between his own strategy and Cicero's. And, perhaps, while ending the *Dialogues,* in a kind

[23]This very complaint can be found in other essays by Hume, such as "Of Suicide" (*E.* 493–502), "Of Superstition and Enthusiasm" (*E.* 98–103), and "Of the Immortality of the Soul" (*E.* 503–510).

of private dialogue with his favorite Roman philosopher, Hume thought something like this: "You are mistaken, my dear Tully. In matters of religion, it is the Epicurean argument, not the Stoic, that triumphs." Indeed, I believe that this mental conversation may have accompanied Hume all his life.

* * *

To close this chapter, I would like to highlight certain aspects of Hume's relationship with Cicero. Cicero was undoubtedly one of Hume's most admired philosophers, and *De Natura Deorum* served him as a model for creating one of the most critical anti-metaphysical works of the Enlightenment, the *Dialogues concerning Natural Religion*. In adopting this model, Hume heavily drew on Hellenistic arguments and skeptical strategies to disguise his intentions while following the New Academy's methodological indications to apply the normative aspects of his theory of belief to examining the theistic hypothesis. Hume's position in the *Dialogues* is primarily represented by the skeptic Philo, who attempts to refute the argument for the intelligent design of the universe and shows how the resulting belief in God is neither a natural belief nor a proof, but a mere probability based on a relatively weak analogy. However, Hume also offers a more constructive view in Philo's final avowal of faith—that the analogy sufficiently grounds the belief that there exists a God, creator, or originator of the universe, who bears a *remote likeness* to the human mind. In accepting this view, I suggest, Hume distances himself from Pyrrhonian fideism, with its irrational acceptance of revealed religion. More importantly, Philo's negative evaluation of traditional religion's practical and emotional effects, inspired by the Epicurean anti-religious arguments, leads him to argue that the resulting belief in God should not give rise to any religious practice. Given that this belief coincides with the basic tenets of British deism, and building on Gaskin's hypothesis that Hume held an "attenuated" form of deism, I argue that, in keeping with the Academy's recommendations, Hume proposes to approve or non-dogmatically accept a form of Epicurean anti-religious deism. In sum, by considering that this belief is the most probable and practically advisable, Hume followed his normative theory of belief and Cicero's Academic guidelines.

Bibliography

Badía, Miguel A. 1996. *La reflexión de David Hume en torno a la religión*. San Juan: Editorial de la Universidad de Puerto Rico.

Battersby, Christine. 1979. The Dialogues as Original Imitation: Cicero and the Nature of Hume's Skepticism. In *McGill Hume Studies*, ed. D.F. Norton, N. Capaldi, and W.L. Robinson, 239–252. San Diego: Austin Hill Press.

Blake, William. 1957. *Complete Writings with Variant Readings*, ed. G. Keynes. Oxford: Oxford University Press.

Butler, Ronald J. 1960. Natural Belief and the Enigma of Hume. *Archiv für Geschichte der Philosophie* 42 (1): 73–100.

Cicero, Marcus Tullius. 1997. *De Officiis* (*On Duties*). Trans. Walter Miller. Cambridge/London: Harvard University Press.

———. 2000. *De Natura Deorum (On the Nature of the Gods)*. Trans. H. Rackham. Cambridge/London: Harvard University Press.
———. 2001a. *De Fato* (*On Fate*). Trans. H. Rackham. Cambridge/London: Harvard University.
———. 2001b. *De Finibus (On Moral Ends*). Trans. Julia Annas. Cambridge: Cambridge University Press.
———. 2006a. *Academica (On Academic Scepticism)*. Trans. Charles Brittain. Indianapolis/Cambridge: Hackett Publishing Company.
———. 2006b. *De Divinatione (On Divination)*. Trans. D. Wardle. Oxford/New York: Oxford University Press.
Foley, Rich. 2006. Unnatural Religion: Indoctrination and Philo's Reversal in Hume's *Dialogues concerning Natural Religion. Hume Studies* 32 (1): 83–112.
Force, James E. 1993. Biblical Interpretation, Newton and English Deism. In *Scepticism and Irreligion in the Seventeenth and Eighteenth Centuries*, ed. Richard H. Popkin and Arjo J. Vanderjagt, 282–305. Leiden: E.J. Brill.
———. 1996. Samuel Clarke's Four Categories of Deism, Isaac Newton, and the Bible. In *Scepticism in the History of Philosophy: A Pan-American Dialogue*, ed. Richard H. Popkin, 53–74. Dordrecht: Kluwer Academic Publishers.
Fosl, Peter S. 1994. Doubt and Divinity: Cicero's Influence on Hume's Religious Skepticism. *Hume Studies* 20 (1): 103–120.
———. 1999. Hume, Skepticism, and Early American Deism. *Hume Studies* 25 (1–2): 171–192.
———. 2020. *Hume's Scepticism: Pyrrhonian and Academic*. Edinburgh: Edinburgh University Press.
Gaskin, John C.A. 1976. Hume's Critique of Religion. *Journal of the History of Philosophy* 14 (3): 301–311.
———. 1978. *Hume's Philosophy of Religion*. London: Macmillan Press.
———. 1983. Hume's Attenuated Deism. *Archiv für Geschichte der Philosophie* 65: 160–173.
Gay, Peter. 1968. *Deism: An Anthology*. Princeton: Van Nostrand Reinhold Inc.
Hudson, W. 2009. *The English Deists. Studies in Early Enlightenment*. London: Pickering & Chatto.
Hudson, Wayne, Diego Lucci, and Jeffrey R. Wigelsworth, eds. 2014. *Atheism and Deism Revalued. Heterodox Religious Identities in Britain, 1650–1800*. London: Ashgate Publishing.
Hume, David. 1826. Of the Rise and Progress of the Arts and Sciences. In *The Philosophical Works of David Hume, Vol. 3*. Edinburgh: Adam Black and William Tait.
———. 1896. *A Tratise of Human Nature*, ed. L.A. Selby-Bigge. Oxford: Clarendon Press.
———. 1902. *Enquiries Concerning the Human Understanding and Concerning the Principles of Morals*, ed. L.A. Selby-Bigge. Oxford: Clarendon Press.
———. 1932. *The Letters of David Hume (Vol. I and II)*, ed. J.Y. Greig. New York: Oxford University Press.
———. 1985. *Essays. Moral, Political and Literary*. Indianapolis: Liberty Fund.
———. 2007a. *Dialogues concerning Natural Religion and Other Writings*, ed. Dorothy Coleman. Cambridge: Cambridge University Press.
———. 2007b. Natural History of Religion. In *Dialogues concerning Natural Religion and Other Writings*, ed. Dorothy Coleman, 124–136. Cambridge: Cambridge University Press.
Jones, Peter. 1982. *Hume's Sentiments. Their Ciceronian and French Context*. Edinburgh: Edinburgh University Press.
Kant, Immanuel. 1992. *Lectures on Logic* (*VL*). Trans. J. Michael Young. Cambridge: Cambridge University Press.
———. 2000. *Critique of the Power of Judgment (KU)*. Trans. Paul Guyer. Cambridge: Cambridge University Press.
———. 2004. *Prolegomena to Any Future Metaphysics (P)*. Trans. G. Hartfield. Cambridge/New York: Cambridge University Press.
Laird, John. 1932. *Hume's Philosophy of Human Nature*. London: Methuen & Co.

Laird, John, Alfred Edward Taylor, and Thomas Edmund Jessop. 1939. The Present-Day Relevance of Hume's *Dialogues Concerning Natural Religion*. *Proceedings of the Aristotelian Society, Supplementary Volumes* 18: 179–228.
Livingston, Donald. 1998. *Philosophical Melancholy and Delirium. Hume's Pathology of Philosophy*. Chicago: University of Chicago Press.
Locke, John. 1975. *An Essay Concerning Human Understanding*. Oxford: Clarendon Press.
Maia Neto, Jose R. 1997. Academic Skepticism in Early Modern Philosophy. *Journal of the History of Ideas* 58 (2): 199–220.
Mossner, Ernest Campbell. 1936. The Enigma of Hume. *Mind* 45 (179): 334–349.
Norton, David Fate. 1991. Hume, Atheism, and the Autonomy of Morals. *Taula* 16: 29–58.
Noxon, James. 1964. Hume's Agnosticism. *The Philosophical Review* 73 (2): 248–261.
O'Connor, David. 2001. *Routledge Philosophy Guidebook to Hume on Religion*. London/New York: Routledge.
Penelhum, Terence. 1979. Hume's Skepticism and the Dialogues. In *McGill Hume Studies*, ed. D.F. Norton, N. Capaldi, and W.L. Robison, 253–278. San Diego: Austin Hill Press.
Plutarch. 1967. Adversus Colotem (Reply to Colotes). In *Moralia Vol. XIV*. Trans. B. Einarson and P.H. De Lacy, 153–315. Cambridge: Harvard University Press.
Potkay, Adam. 1994. *The Fate of Eloquence in the Age of Hume*. Ithaca/London: Cornell University Press.
Price, John Vladimir. 1964. Sceptics in Cicero and Hume. *Journal of the History of Ideas* 25 (1): 97–106.
———. 1965. *The Ironic Hume*. Austin: University of Texas Press.
Priest, Graham. 1985. Hume's Final Argument. *History of Philosophy Quarterly* 2 (3): 349–352.
Pyle, Andrew. 2006. *Hume's Dialogues Concerning Natural Religion: Reader's Guide*. London/New York: Continuum.
Schmitt, Charles B. 1972. *Cicero Scepticus: A Study of the Influence of the Academica in the Renaissance*. The Hague: Martinus Nijhoff.
Sextus, Empiricus. 2000. *Pyrrhoniae Hypotyposes (Outlines of Scepticism)*, ed. Julia Annas and Jonathan Barnes. Cambridge: Cambridge University Press.
Smith, Norman Kemp. 1947. Introduction. In *Dialogues Concerning Natural Religion*, ed. David Hume, 1–75. Indianapolis: Bobbs-Merrill Company.
Tweyman, Stanley. 1986. *Scepticism and Belief in Hume's Dialogues Concerning Natural Religion*. Dordrecht: Martinus Nijhoff Publishers.

Chapter 6
Skepticism and the Skeptical Method: Sources of Ancient and Modern Skepticism in Kant

Abstract Kant's relationship to the skeptical tradition has often been identified with his refutation of the main strands of skepticism in the Transcendental Analytic of the *Critique of Pure Reason* (*KrV*) – Humean skepticism on causality, Cartesian or "problematic" idealism, and Pyrrhonian or Baylean skepticism. In more recent decades, however, commentators have begun to acknowledged that Kant not only sought to refute the skeptics but also made use of what he called a "skeptical method" to undermine the illegitimate pretensions of traditional metaphysics. In this chapter, I seek to offer a general picture of the ancient and modern sources of Kant's skeptical method. I begin by describing the main features of this method, in order to identify them in possible ancient and modern sources. In discussing these sources, I observe, first, Kant's attitude towards Pyrrhonism in the *Lectures on Logic* (*VL*), and survey the modern sources that may have informed Kant's knowledge of this ancient tradition. Among them, I include the seventeenth- and eighteenth-century debates on Zeno of Elea's paradoxes regarding the infinite divisibility of space, time, and matter, particularly in Bayle's formulation (Smith, 2008, pp. 467–477). After this, I turn to Kant's contact with Cicero's works and examine his comments about Academic skepticism in the *Lectures on Logic*. In doing so, I note his negative assessment of Academic skepticism and his observations about suspension of judgment, which, paradoxically, are more aligned with Carneades's teachings than his comments on Academic skepticism would have us expect. Finally, I explore the role that Hume's skepticism played in the evolution of Kant's critical thought, especially in his conception of the Antinomies of Pure Reason. To make this connection clearer, I examine the scholarly debate on whether Hume or the antinomical problems produced Kant's "awakening from his dogmatic slumber"—i.e., his turn from traditional metaphysics to critical philosophy. I conclude that Kant's skeptical method involves important features of Academic skepticism, which may come from his reading of Hume's discussion of the cosmological problems in the *Enquiry on Human Understanding*. Among these discussions, Hume's treatment of the question of freedom – influenced, in turn, by Cicero's *De Fato* – is, in my view, a notorious link between Kant and Academic skepticism.

C. González Quintero, *Academic Skepticism in Hume and Kant*, Synthese Library 449, https://doi.org/10.1007/978-3-030-89750-5_6

Kant's relationship to the skeptical tradition has often been identified with his refutation of the main strands of skepticism in the Transcendental Analytic of the *Critique of Pure Reason* (*KrV*)—Humean skepticism on causality, Cartesian or "problematic" idealism, and Pyrrhonian or Baylean skepticism.[1] In general, Kant defines skepticism as "a principle of artful and scientific ignorance that undermines the foundations of all cognition, in order, if possible, to leave no reliability or certainty anywhere" (A 424). Kant also claims that the skeptics are "a kind of nomads who abhor all permanent cultivation of the soil" (A ix), for not only do they seek to destroy the edifice of science, but they also endeavor "to give short shrift to all metaphysics" (B xxxvi). These assessments seem to leave no doubt about his hostility towards the skeptics. Hence, it appears to be clear that Kant's critical efforts are directed primarily against skepticism.[2]

However, this traditional view has been challenged lately. In more recent decades, commentators have begun to acknowledge that Kant not only sought to refute the skeptics but also made use of what he call a "skeptical method" to undermine the illegitimate pretensions of traditional metaphysics (Makkreel 1998, pp. 101–109; Santozki 2004, pp. 102–103; Smith 2008, pp. 474–476, 2013, pp. 247–263). Kant explains:

> As harmful as this skepticism is, though, the skeptical method is just as useful and purposeful, provided one understands nothing more by this than the way of treating something as uncertain and of bringing it to the highest uncertainty, in the hope of getting on the trail of truth in this way. It is quite useful to the critical procedure, by which is to be understood that method of philosophizing in accordance with which one investigates the sources of his claims or objections and the grounds on which these rest, a method which gives hope of attaining certainty (*VL*-Jäsche, 84).

This characterization of the skeptical method[3] as a valuable device to examine the grounds of reason's metaphysical claims is essential to understand how, as P. Smith

---

[1]See: Forster 2008, pp. 4–5; Guyer 2003, pp. 1–37; Lazos 2014, pp. 61–205; Smith 2013, pp. 247–263, 2008, p. 465; Stepanenko 2005, pp. 35–46. P. Guyer sees Kant answering these three kinds of skepticism: Humean, Cartesian, and Pyrrhonian (Guyer 2003, pp. 1–37, 2008, pp. 23–70, 2014, pp. 10–15). Forster calls them "Humean skepticism," "Veil of perception skepticism," and "Pyrrhonian kepticism," respectively (Forster 2008, pp. 4–5). On whether this classification comes originally from P. Guyer or Forster, see: Guyer 2009, pp. 384–389. C. Jauregui focuses on Kant's arguments against Cartesian idealism (Jáuregui 2008, pp. 89–121), and E. Lazos on Berkeleyan idealism and Cartesian skepticism (Lazos 2014, pp. 61–205). P. Stepanenko finds three sorts of skeptical profiles that Kant would answer to – problematic or Cartesian idealism, skepticism concerning the objective validity of the categories (particularly, the category of cause), and Humean/Schulzean skepticism concerning the reality of external objects (Stepanenko 2005, pp. 35–46). For P. Smith, the three kinds of skepticism Kant is concerned with are Humean, Baylean, and Cartesian skepticism (Smith 2008, pp. 463–500, 2013, p. 247). I agree with Smith that Kant's antinomies owe more to Bayle than to Pyrrhonism.

[2]Guyer says: "Thus, from a methodological point of view, Kant's project in philosophy became that of undermining both Humean and Pyrrhonian skepticism in both theoretical and practical philosophy, and, much more incidentally, along the way refuting Cartesian skepticism about external objects as a nagging but by no means central problem in theoretical philosophy" (Guyer 2014, p. 15).

[3]For other references to the skeptical method see: *KrV* A424/B452; A486/B514; A507/B535; *VL*-Blomberg 208; *VL*-Vienna 885; and *VL*-Dohna-Wundlaken 744–745.

has claimed, "reformulating the diverse kinds of skepticism, Kant has deepened and given them a particular form" (Smith 2008, p. 464; my translation). Of course, Kant's use of skepticism does not imply that he was a skeptic or became a skeptic at some point in the evolution of his thought. However, since his critical project aimed at defending scientific knowledge from some strands of skepticism and attacking metaphysical dogmatism with the help of traditional skeptical strategies, it is worth analyzing his project within the history of modern skepticism.

With this in mind, in this chapter, I seek to offer a general picture of the ancient and modern sources of Kant's skeptical method. I begin by describing the main features of this method in order to identify them in possible ancient and modern sources. In discussing these sources, I observe, first, Kant's attitude towards Pyrrhonism in the *Lectures on Logic* (*VL*), and, following the lead of some scholars (Forster 2008, pp. 16–20; Smith 2008, pp. 463–500, 2013, pp. 247–263; Tonelli 1997, pp. 69–98), survey the modern sources that may have informed Kant's knowledge of this ancient tradition. Among them, I include the seventeenth- and eighteenth-century debates on Zeno of Elea's paradoxes regarding the infinite divisibility of space, time, and matter, particularly in Bayle's formulation (Smith 2008, pp. 467–477). After this, I turn to Kant's contact with Cicero's works and examine his comments about Academic skepticism in the *Lectures on Logic*. In doing so, I note his negative assessment of Academic skepticism and his observations about suspension of judgment, which, paradoxically, are more aligned with Carneades's teachings than his comments on Academic skepticism would have us expect. Finally, I explore the role that Hume's skepticism played in the evolution of Kant's critical thought, especially in his conception of the Antinomies of Pure Reason. To make this connection clearer, I examine the scholarly debate on whether Hume or the antinomical problems produced Kant's "awakening from his dogmatic slumber"—i.e., his turn from traditional metaphysics to critical philosophy. I conclude that Kant's skeptical method involves important features of Academic skepticism, which may come from his reading of Hume's discussion of the cosmological problems in the *Enquiry on Human Understanding*. Among these discussions, Hume's treatment of the question of freedom – influenced, in turn, by Cicero's *De Fato* – is, in my view, a notorious link between Kant and Academic skepticism.

6.1 Kant's Skeptical Method

In the first *Critique* and other writings, Kant states his renowned metaphor of the ages of reason.[4] He characterizes mature reason as the stage in which the capacity becomes mature enough to exercise self-criticism (A xi) and argues that, before

[4]The wording in *Dreams of a Spirit-Seer* is curious but indicates that, even in his pre-critical period, Kant had already conceived of the metaphor: "For that about which one knows a great deal early on in life as a child – of that, one can be sure, one will certainly know nothing later on in life when one

reaching this state, it must have gone through the previous phases of infantile dogmatism and juvenile skepticism:

> The first step in matters of pure reason, which characterizes its childhood, is dogmatic. The [...] second step is skeptical and gives evidence of the caution of the power of judgment sharpened by experience. Now, however, a third step is still necessary, which pertains only to the mature and adult power of judgment, which has at its basis firm maxims of proven universality, that, namely, which subjects to evaluation not the *facta* of reason but reason itself, as concerns its entire capacity and suitability for pure a priori cognitions; this is not the censorship but the critique of pure reason (A761/B789).

Only by overcoming the candid dogmatism of childhood and the arrogant skepticism of youth can the coming of age of reason be guaranteed. Hence, for Kant, dogmatism and skepticism are accurately conceived as immature rather than misguided exercises of reason, whose vices and virtues must be respectively overcome and incorporated by the mature faculty. For this reason, a philosophical critique should not only refute skepticism and dogmatism but profit from their valuable methods. As R. Makkreel remarks: "Kant sketches a history of pure reason in which his critical philosophy surpasses both dogmatism and skepticism. But by his own admission, critical philosophy still makes use of dogmatic and skeptical procedures" (Makkreel 1998, p. 101).[5]

There are, for Kant, negative skeptical outlooks, whose procedures are not suitable for the self-examination of reason, and positive ones that, although immature, are worthy efforts to trace the faculty's limits. Kant identifies Hume's skepticism about causality as the prototype of a destructive form of skepticism. He calls it a "censorship of reason" which "leads to doubt about all transcendent use of principles" (A760/B788-A761/B789) and compares it to a kind of heteronomous device, that is, a maxim of thought that restrains reason's dogmatic impulses by applying to it an empirical, and therefore external, doubt. For Kant, given that in Hume's analysis, the judgment "every event has a cause" is an empirical hypothesis, and all empirical hypotheses are in principle contingent, Humean doubt about causality, in fact, cancels itself out:

> [...] the same thing happens to [Hume] that always brings down skepticism, namely, he is himself doubted, for his objections rest only on *facta*, which are contingent, but not on principles that could effect a necessary renunciation of the right to dogmatic assertions (A767/B796-A778/B795).

Kant famously avoided this skeptical vicious circle by placing causality amid the *a priori* principles of our understanding (A176/B218-A210/B256). I will not discuss this aspect of Kant's critique further since my purpose here is rather to understand the features of his skeptical method. I only want to conclude from this contrast that,

has reached maturity. And the man of thoroughness will in the end at best be the sophist of his youthful delusions" (*Ak*. II, 320).

[5] I will not delve into Kant's use of dogmatic procedures, as this would involve a discussion of the Transcendental Analytic, and divert us from the purpose of this chapter. For a description of this method, see: *VL*-Blomberg, 206.

unlike Hume's censorship, Kant's skeptical method ought to be a product of reason itself. For him, if the skeptical method is to be helpful, it must involve an internal and autonomous mechanism through which reason can set its own limits. In other words, by applying the skeptical method, reason should be able to restrict its speculative impulses from within or based on its own *a priori* principles.

More specific characterizations of this method describe it in similar terms to the ancient procedure of opposing theoretical claims of equal validity. In the Paralogisms, for example, Kant says:

> The skeptical objection puts the proposition and its opposite over against one another, as objections of equal weight, each alternatively a dogma with the other as an objection to it; thus, on both sides it is dogmatic in appearance, in order to annihilate entirely every judgment about the object (A388).

Moreover, in the Vienna *Lectures*, Kant refers to the skeptical method as the "critical method," and argues that it serves the purpose of revealing the causes of the conflict of reason with itself:

> This method, then, where we do not merely doubt everything, but also investigate the cause of the conflict of the understanding with the understanding itself, in order thereby to illuminate the truth, is the critical method (*VL-Vienna*, 885).

Based on these and other descriptions,[6] we can see how Kant's skeptical method is similar to the one used by the ancient skeptical schools—that is the method of contrasting two opposing judgments of equal force or validity (Forster 2008, p. 4; Smith 2008, p. 474, 2013, pp. 253–254). However, in Kant's hands, this method of opposition is restricted to examining the claims of pure reason in its speculative flights. The skeptical method allows us to observe the equal logical validity of opposed metaphysical claims and, in doing so, provides an internal control of the faculty's speculative impulses. Reason can and ought to examine these internal conflicts, as a good skeptic who, observing contrary arguments of equal weight, suspects the rational legitimacy of their grounds. P. Smith describes Kant's skeptical method as "[. . .] an activity that aims at describing or provoking a conflict between contrary claims. [. . .] the function of the skeptical method is precisely to help us discover the presupposition that grounds dogmatic conflicts" (Smith 2008, pp. 474–475).

Up to this point, then, we can say that Kant's skeptical method is an internal rational device to examine the legitimacy of metaphysical views by examining the grounds of their opposite claims. However, there is a final characteristic of this method that commentators tend to overlook (one exception is Moller 2013, pp. 301–320). This characteristic is expressed in the metaphor of the judge that Kant frequently employs in the *Critique of Pure Reason* (A xii; B xiii; A 752/B779). At the beginning of the 1770s, in the *Blomberg Lectures*, Kant used the metaphor in

---

[6] See note 2.

relation to skepticism and the skeptic. Just as judges give verdicts in court, so too the genuine skeptic[7] should decide *practically* what cannot be proven *theoretically*:

> Thus, the *scepticus* constantly inquires, he examines and investigates, he distrusts everything, but never without a ground. In this, he resembles a judge, who weighs the grounds for something as well as against it, and who listens to the plaintiff as well as the defendant, prior to and before deciding the matter and passing judgment (*VL*- Blomberg 209–10).

In this passage, the *genuine skeptic* passes judgment on the case examined—she does not entirely suspend judgment but, at some point of the discussion, abandons *epochê*. However, this skeptic's verdict, just as the judge's, is more a practical imperative – that x or y should be done – than a theoretical claim deciding the issue. Although the skeptic/judge does affirm or deny, for example, that the defendant did commit the crime, the verdict aims at orienting action more than at offering theoretical certainty. As S. Moller has argued, Kant's critical examination involves a final verdict on reason's conflict that is modeled on legal maxims or rules for action:

> Kant also clarifies this method with the imagery from the legal sphere. [...] The metaphor here shows that the critique can draw new conclusions from the arguments put forward by disputing parties. [...] In the verdict over the ideas of reason, Kant refers to legal maxims as a model for the conclusions of the critique; Kant distinguishes between what reason can claim in speculative use and what it can presume in practical use. [...] Here, the analogy is used for capturing the necessity for finding a practicable solution to the disputes of reason (Moller 2013, pp. 315–316).

A final characteristic of Kant's skeptical method, then, is that it does not lead to suspension of judgment but instead to a sort of "practical verdict" on the issue at hand, which is modeled after legal maxims and, thus, is more practically than theoretically oriented. In the next chapter, I will further examine this feature when discussing his solution to the Antinomies of pure reason. For the moment, I want to summarize what I consider to be the main characteristics of Kant's skeptical method:

1. It provides an internal examination of reason's natural inclination to oppose metaphysical theses and antitheses of equal weight.
2. It is not oriented towards the suspension of judgment but seeks to examine the rational legitimacy of the conflicting claims' grounds.
3. Once the method is applied, the critical philosopher – like a judge – draws a conclusion that resolves the conflict from a practical point of view.

Having drawn the main characteristics of this method, I now turn to consider its possible sources, beginning with ancient Pyrrhonism. Keeping in mind this description of Kant's skeptical method throughout the chapter should help us obtain a clearer understanding of his assessment of the ancient skeptical schools and their most prominent representatives.

---

[7] In the *VL*-Blomberg, Kant's assessment of skepticism is still positive. Only later, in his critical phase, does he establish the contrast between skepticism and the skeptical method.

6.2 Kant and Pyrrhonism: Ancient and Modern Sources[8]

From the *Lectures on Logic,* we infer that Kant was familiar with the leading schools of ancient skepticism. It is common to find such statements as: "The skeptics can be divided into Pyrrhonists and Academics, the latter of whom pursued it furthest" (*VL*-Dohna-Wundlacken, 719).[9] In this and other similar statements, Kant appears to think that Pyrrhonian skepticism is closer to the spirit of scientific inquiry, while Academic skepticism sustains a sort of "negative dogmatism." He summarizes the positions of the two schools thus:

> I. The Academic doubter proved and demonstrated *per definitionem* that nothing could properly be proved, nothing defined, which was, however, a *contradictio in adjecto*, and this was dogmatic, seeming doubt.
>
> II. The Pyrrhonic, i.e., skeptical doubter, says, on the other hand, that to each and every one of our judgments, or at least to most, another judgment may always be opposed which maintains exactly the opposite of what is contained in the former judgment; this was actually to be regarded more as a kind of very fine and outstanding observation than as a reprehensible doubt (*VL-Blomberg* 209).

As we have already seen, this explanatory framework – whereby Academic skepticism is dogmatic while Pyrrhonian skepticism is non-dogmatic and leads to continuing inquiry – corresponds to how Sextus Empiricus described the different skeptical sects in the first paragraph of the *Outlines*.[10] According to Sextus, because of holding the principle of inapprehensibility of impressions, Academic skepticism precludes the very possibility of inquiry, while, refusing to accept any dogma, Pyrrhonism continues the search (*Pyr.* I, 1–4).[11] In accordance with this description, Kant shows in the *Lectures* a deep appreciation of Pyrrhonism. He considers the Pyrrhonist a meticulous inquirer who entertains healthy doubts about the hypotheses examined before approving them. This school's doubt is, in his account, a merely provisional, scientific device, which allows the researcher to approach certainty gradually:

> This sect did not really doubt all truths, but instead, it lengthened investigation and postponed decision. Thus, they proved through doubt that the grounds for decision were not yet complete, and thus they led the dogmatists, who wanted to decide everything, to the path of caution (*VL-Blomberg*, 213).

For Kant, Pyrrhonism's objective is to lead the dogmatist to self-reflection through examining various hypotheses before accepting one of them as true. As discussed in previous chapters, the Pyrrhonists did indeed attack the dogmatism of the Hellenistic

---

[8] A preliminary analysis of the issues I examine in this section has already appeared in González 2011, pp. 225–230.

[9] Other fragments like this are found in: *VL-Vienna*, 803, 885–886; *VL-Blomberg*, 209, 216; *VL-Dohna-Wundlacken*, 744.

[10] This description, in turn, seem to have been drawn from Aenesidemus's *Pyrrhonian Discourses.*

[11] "[. . .] with regard to the objects investigated by philosophy, this is probably why some have claimed to have discovered the truth, others have asserted that it cannot be apprehended, while others again go on enquiring. Those who believe they have discovered it are the "Dogmatists" [. . .] Cleitomachus and Carneades and other Academics treat it as inapprehensible: the Sceptics [Pyrrhonians] keep on searching" (*Pyr.* I, 1–4).

schools and, in this sense, as Kant affirms, they "led them to caution." However, here Kant does not seem to be aware of the Pyrrhonian conception of inquiry as positing arguments for and against a thesis to suspend judgment on the matter and attain *ataraxia* or peace of mind. Rather, he considers that the goal of Pyrrhonian inquiry is finding certainty:

> Skeptical [Pyrrhonian] doubt, on the other hand, consists in being conscious of the uncertainty with a cognition and thus in being compelled to inquire into it more and more, so that finally *one may nonetheless attain certainty* with the help of careful investigations (*VL-Blomberg* 209, my italics).

This idea that Pyrrhonism seeks to attain certainty seems to be a rather Cartesian approximation to skepticism and, in any case, an inaccurate portrayal of the ancient school. Now, Kant's admiration for Pyrrhonism appears principally in the *Blomberg Lectures*—the set of lectures that correspond to the earliest period of his teaching in the first half of the 1770s. Encouraged by these dates, M. Forster contends that Kant went through a kind of "Pyrrhonian crisis" during the mid-60s (Forster 2008, p. 19), in which this skeptical tradition heavily influenced him. Besides the *Lectures,* Foster argues, other sources that attest to this Pyrrhonian crisis include the text entitled *Announcement of the Programme of Lectures for the Winter Semester* of 1765–1766 (*Ak.* II, 303–314) and the pre-critical treatise, *Dreams of a Spirit-Seer,* from 1766 (A*k*. II 317–373). In the *Announcement*, Kant speaks of the "zetetic" method of setting against each other two opposite claims of equal weight, which, for Forster, is a direct indication that Kant had Pyrrhonism in mind during this time (Forster 2008, p. 18). The fragment of the *Announcement* reads:

> The method of instruction, peculiar to philosophy, is zetetic, as some of the philosophers of antiquity expressed it (from *zêtein*). In other words, the method of philosophy is the method of enquiry. It is only when reason has already grown more practiced and only in certain areas, that this method becomes dogmatic, that is to say, decisive (*Ak.* II, 307).

Similarly, Forster claims that the *Dreams of a Spirit-Seer* offers an example of Kant's employment of the Pyrrhonian method since this treatise presents two opposite perspectives on the existence of disembodied spiritual beings. Having observed the commonality between both sources, he concludes:

> This crisis in the mid-1760s was Pyrrhonian in two senses. First, it was Pyrrhonian in character. Specifically, it was so both in virtue of the general structure of the problems raised against metaphysical claims, namely, a setting of each claim against a contrary claim and demonstration that equally strong arguments could be given on both sides, and in virtue of their result, namely, a suspension of judgment about the issues involved. [...] Second, this crisis was also largely Pyrrhonian in inspiration (Forster 2008, p. 18).

While granting plausibility to Forster's argument, I find that it lacks sufficient textual support. In my opinion, it is far from clear that Kant's adoption of the ancient skeptical method of opposing claims indicates a specifically-Pyrrhonian crisis. In the first place, although the *Dreams of a Spirit-Seer* is an excellent example of how Kant uses this skeptical method, it is also plausible that Kant's skeptical crisis between 1760 and 1770 had Hume, instead of Sextus Empiricus, as a source. In this respect, L. Robinson's hypothesis that during this decade Kant went through an "empirical-skeptical" phase strongly influenced by Hume seems to be more accurate (Robinson

1924). Very generally speaking, the *Dreams* describes two opposite views on the hypothesis regarding the existence of disembodied spirits (*Ak.* II, 317–373). The first, which Kant calls "dogmatic," corresponds to traditional metaphysics and accepts the possibility that these entities exist. The second appears under the polemic title "Anti-Kabbalah, a Fragment of Ordinary Philosophy" and expresses the empiricists' denial of such a possibility. After this, in a section entitled "Historic," Kant describes Emanuel Swedenborg's accounts of his communication with spirits. At the end of the short treatise, he offers a "Theoretical Conclusion" and a "Practical Conclusion." In the former, he declares himself in favor of the empiricist position, and, in the latter, he argues that justifying the existence of disembodied souls is not practically advisable because it promotes impure motivations for moral action:

> What, is it only good to be virtuous because there is another world? Or is it not rather the case that actions will one day be rewarded because they are good and virtuous in themselves? Does not the heart of man contain within itself immediate moral prescriptions? [. . .] Can that person really be called honest, can he really be called virtuous, who would readily abandon himself to his favorite vices, were it not for the deterrence of future punishment? [. . .] For this reason, it seems more consonant with human nature and moral purity to base the expectation of a future world on the sentiments of a nobly constituted soul than, conversely, to base its noble conduct on the hope of another world (*Ak.* II, 372–373).

In sum, the *Dreams* contains two essential features of the skeptical method, namely, the presentation of two opposite claims and a final practical solution. I think that these features do not necessarily reveal a Pyrrhonian crisis but may express Kant's engagement with other skeptical sources, particularly Hume. In fact, Kant's opposition to a morality grounded on the belief in the rewards of a future life as well as his allusion to the "immediate moral prescriptions" of the "human heart," are indications that he had Hume in mind when writing the practical conclusion to this treatise. In a very Humean way, Kant expresses there the inconveniences of basing morality on religion. However, he also seems to begin to consider that it may be fair to do the opposite—to found the religious beliefs in God and the immortality of the soul on morality, which, of course, is very non-Humean. This argument is further developed in Kant's critical stage, as we will see in the next chapter. For now, we can see that two of Kant's skeptical method conditions are present in this early treatise and that the second one does not let us identify Kant's skeptical crisis during this period with Pyrrhonism, as Forster suggests.

My second objection to Forster's claim is that, even though the expression "zetetic method" used in the *Announcement* is, in fact, more Pyrrhonian than Academic,[12] both ancient schools used the same method of opposing pro and con arguments to show the inconsistency of the dogmatists. Moreover, as G. Tonelli argues, the term was used among modern skeptics and anti-skeptics as a synonym for skepticism in general. Various authors, such as Gassendi, Bayle, de Crousaz, and Baumgarten, defined the skeptical schools (or sects) with the terms "skeptic, zetetic,

[12]The term appears in the *Outlines of Pyrrhonism*, where Sextus notes that, "The skeptic school, then, is also called "zetetic" from its activity in investigation and inquiry" (*Pyr.* I, 7).

and aporetic" (Tonelli 1997, p. 70). According to Tonelli, since Kant uses the term 'zetetic' in the *Announcement,* some reflections (*Refl. 4455*, *Ak.* XVII, 558; *Refl. 4450*, *Ak.* XVII, 560; *Refl. 1636*, *Ak.* XVI, 60) and the *Blomberg Lectures* (*VL-Blomberg*, 213), it is possible that the term was widely used in academic circles and he chose it to avoid public censure:

> [...] we should come to the conclusion that, for Kant, "zetetic" meant the same as "sceptical" (in the good sense). If in the *Report of 1765* he avoided the term "sceptical," one should believe that this was in deference to public opinion, which could take offence at it (Tonelli 1997, p. 71).

We can conclude, then, that Kant's use of the term 'zetetic' shows his familiarity with the technical jargon during the eighteenth century to refer to skepticism, but does not necessarily express a Pyrrhonian crisis or phase. Finally, Forster does not offer sufficient evidence of Kant's contact with the sources of Pyrrhonian skepticism. Although it is easy to think that his praise of Pyrrhonism in the *Blomberg Lectures* was a result of his having read Sextus Empiricus, Tonelli's hypothesis, namely that Kant knew about Pyrrhonism only through modern secondary sources, seems to be more probable:

> [...] in all likelihood, Kant's knowledge of ancient skepticism was not based on a first-hand study of ancient Greek texts, but rather on the received image of the Greek skeptics, mainly as it was to be found in the modern skeptics, their opponents, and the historians of philosophy of the times (Tonelli 1997, p. 69).

Indeed, his discussion of Pyrrhonism in the *Blomberg Lectures* points to the hypothesis that Kant may not have read Sextus. There, he does not speak of Sextus directly but only mentions Diogenes Laërtius's description of Pyrrho (*VL-Blomberg*, 214; cf. Guyer 2014, p. 13). Only later, in the *Jäsche Lectures* – the final set of lectures where, in contrast, Kant provides a much more simplified outline of Pyrrhonism – does he mention Sextus: "From these skeptics, nothing has remained for us, however, but the two works of Sextus Empiricus, in which he brought together all doubts." (*VL-Jäsche*, 31). This change is noteworthy, as it seems that Kant's initial contact with Pyrrhonism did not result from having read its primary source.[13] Now, if Kant's acquaintance with Pyrrhonism came from secondary sources, what might these be?

First, we know that Kant used Georg Friedrich Maier's *Auszug aus der Vernunftlehre* (*Excerpts from the Doctrine of Reason*), edited in 1752 in Halle, as a textbook for his *Lectures on Logic*, from the beginning of his career as a *Privatdozent* in 1755 (Young 2004, p. xvii; Warda 1922, p. 52). Gradually, however, he appears to have set aside Maier's text so that, in the *Jäsche Lectures,* which bring together his final years of teaching in the first half of the 1790s, his students no longer needed to read this work (Young 2004, p. xxiii). Hence, it is highly likely that

[13] Sextus Empiricus's *Outlines of Pyrrhonism* does not appear in Warda's list of books owned by Kant. See: Warda 1922.

a significant part of Kant's understanding of Pyrrhonism, at least as presented in the *Lectures*, comes from Maier.

In addition to Maier, Tonelli indicates that Haller's *Prüfung der Secte, die an allem zweifelt*, the abridged German version of de Crousaz's *Examen du Pyrrhonisme ancien et moderne* (de Crousaz 1733), edited in 1733, probably influenced Kant's characterization of skepticism. Kant refers to this text in the *Blomberg Lectures*:

> But if one wants closer instruction in the scepticismus of modern times, then one can read with great profit the writing that Haller published under the title: *Examination of the sect that doubts everything*. Also, the writings of Hume mentioned above [*Philosophical Inquiries*] are actually, in a certain way, to be recommended too (*VL-Blomberg*, 218).

Haller's translation into German of Formey's abridged version of de Crousaz's book was published in 1751.[14] Both Haller and Formey were members of the Prussian Academy in Berlin, which had an anti-skeptical orientation (Laursen 1996, pp. 261–282, 1997, pp. 117–126; Smith 2013, p. 248). Their members translated and criticized skeptical texts, among them Hume's first *Enquiry*, translated by Sulzer in 1755, and Cicero's *Academica*, prepared by de Castillon in 1779. As Laursen argues, while their purpose was to defend Christian faith from the outbreaks of deism and atheism, they had an ambivalent attitude towards skepticism and "to the extent that they actually domesticated many a skeptical point into their own canons of argument, they contributed to the process of assimilating skepticism into the modern world view" (Laursen 1997, p. 155). The combined efforts of these anti-skeptics, then, helped spread the works of ancient and modern skeptical authors, guaranteeing their access to the European intellectual audience.

We should not underestimate the significance that Haller's version of de Crousaz's text could have had for Kant's understanding of Pyrrhonism.[15] De Crousaz's treatise intends to refute the skepticism of Sextus Empiricus and Bayle. The first part gives a general account of Pyrrhonism; the second comprises a translation and refutation of some sections of Sextus's *Outlines* and *Adversus Mathematicos*, and the third is devoted to a rebuttal of Bayle's skeptical views. Some of Kant's remarks about the difference between Pyrrhonism and Academic skepticism may come from the sections of the *Outlines* contained in the second part of de Crousaz's treatise.[16] However, given the anti-skeptical tone of this text, Kant's positive comments about Pyrrhonism probably come from other accounts. Tonelli argues that Kant's most important source of skepticism is Baumgarten's *Philosophia Generalis*, edited by Förster in 1770. According to him, this text's two sections, "*Thetica*" and

[14]Formey's text is a compilation of extracts of de Crousaz's book. This abridged version was entitled, in French, *Le triomphe de l'évidence*, and published in Berlin in 1756.

[15]For a more detailed account of the influence of de Crousaz's in the formation of Kant's critical standpoint, see González 2020.

[16]While Haller's translation of Formey's version de Crousaz's *Examen* into German does not appear in Warda's list of books owned by Kant, de Crousaz's *Observations critiques sur l'abregé de la logique de monsieur Wolff* does. See: Warda 1922, p. 47.

"*Antithetica*," may have been the source for Kant's understanding of the skeptical method (Tonelli 1997, p. 77). Also, Kant seems to have taken the term "dialectic" and the distinction between "phenomena" and "noumena" from this text:

> Kant's interest in the sceptical method certainly would have made him notice these expressions [*noumenon*], in Baumgarten's book (and others). [...] It can probably be assumed that he first had his attention drawn to the term "*noumena*" by Förster's note, considering that the place mentioned above refers to a skeptical use of the term, which is nearer to the Kantian use than to the Platonic, against which both the skeptics and Kant polemicized (Tonelli 1997, p. 73).

Tonelli argues that in the prologue of Baumgarten's book, the editor – Förster – lists three main types of skepticism – namely Pyrrhonism, Academic skepticism, and Cartesian skepticism – and refers to Sextus Empiricus's distinction between *phenomena* and *noumena*. Nonetheless, as Tonelli acknowledges (1997, p. 73), Kant's adoption of these terms does not involve their skeptical meaning. In fact, given their transcendental basis, Kant's appearances or *phenomena* are objectively valid representations, while Pyrrhonian appearances are only subjectively valid impressions. As Makkreel argues:

> Kant seems to make an initial concession to the skeptics when he allows that we are conscious of appearances, not of things in themselves. But for Kant, our consciousness of appearances places us in a position to also know them objectively. Thus, according to the Transcendental Deduction of the *Critique of Pure Reason*, the conditions that serve to unify and formally organize appearances are, at the same time, the conditions that make it possible for us to know objects of experience. Objects so redefined are not merely private mental representations but things that are in principle intersubjectively available through our senses (Makkreel 1998, pp. 101–102).

Makkreel's indication is crucial, in my view. In the Transcendental Aesthetic, Kant warns us against considering as phenomena "that which pertains to them [intuitions] only contingently, because it is not valid for the relation to sensibility in general but only for a particular situation or organization of this or that sense." (A45). Pyrrhonian phenomena are precisely such subjective impressions, which, as Kant says, pertain to the representations only contingently. They depend on the particularities of the subject's organs of perception or the circumstances surrounding them and, for this reason, are only subjectively valid. As a result, Pyrrhonian appearances are not the same as Kant's phenomena, which, depending on the transcendental conditions of experience, are universally valid and constitute objective knowledge.

These modern sources of Pyrrhonism show Kant's interest in the debates between skeptics and anti-skeptics during the seventeenth and eighteenth centuries. Bayle's *Dictionary* has a special place among them since it was a centerpiece of most of these disputes. Let us turn now to Kant's reception of Bayle's influential account of Zeno's paradoxes.

6.3 Kant's Bayle: A Modern Source for Pyrrho and Zeno's Paradoxes

Deeply significant among Kant's modern sources for Pyrrhonism is Pierre Bayle's *Historical and Critical Dictionary* (1687). According to Tonelli:

> Naturally, he [Kant] knew Bayle in the original. Bayle's *Dictionnaire* was known far and wide. In addition to the many editions of his work in the eighteenth century, Frederik II and d'Argens published a selection from the *Dictionnaire* in 1765 [. . .]. Furthermore, Gottsched had translated the whole *Dictionnaire* into German (Leipzig 1741–47), furnishing it with footnotes that were supposed to serve as an antidote to the skeptical poison (Tonelli 1997, p. 82).

From Gotsched's translation, Kant may have had access to Bayle's famous *Dictionary*. Bayle's article "Pyrrho," for instance, probably influenced Kant's conception of Pyrrho's life and skepticism. In several sections of the *Lectures*, Kant claims that Pyrrho would contradict his opponents using the expression *non-liquet*,[17] which is used in Bayle's account, yet neither appears in Diogenes Laërtius nor in Sextus. Bayle says: "Everywhere he found reasons to affirm and reasons to deny; for this reason, he withheld his consent having examined the case for and against, and he reduced his decisions to a *non-liquet*, let it be more widely investigated" (Bayle 2005, p. 99, my translation).[18] And Kant repeats: "Pyrrho had a certain motto for his sect: *Non-liquet.* To anyone who maintained his propositions with the greatest and most settled gravity, he shouted out in opposition, *Non-liquet*, it is not yet settled, and then he began to expound his speech against it" (*VL-Blomberg*, 213).

Kant also reproduces Bayle's opinion that Laërtius' view that Pyrrho suspended judgment to such a degree that, doubting the existence of carriages passing on his path, he risked his life with every step, was just an exaggerated rumor. Bayle claims:

> We must take as a joke in bad taste Antigonus of Carystus's tales that Pyrrho had no preference for anything, that a wagon or a chasm never forced him to take a step back or sideways, and that his friends who followed him frequently saved his life. There is no indication at all that he was crazy to this extent (Bayle 1991, p. 195).

And, similarly in Kant:

> It is said of him, e.g., that he denied the certainty of all experiences, that he, therefore, walked directly into swamps and morasses, that he feared neither the running of horses nor deep chasms or ditches; and add to these disparaging reports that he is said to have lived to be 90 years old, as an obvious proof that all such charges as had been invented about him must be false; for with such boldness he would certainly have lost his life earlier (*VL-Blomberg*, 214).

These passages give us a taste of Kant's engagement with Bayle's *Dictionary.* But Bayle's influence on Kant is still more significant methodologically. As P. Smith

[17]. See: *VL-Blomberg*, 208 and 213, and *VL-Vienna,* 886.

[18]Unfortunately, The English translation does not include the explicit reference to the *non liquet*. See: Bayle 1991, p. 194.

has argued, Bayle's *Dictionary* was a decisive source for Kant's conception of the Antinomies (Smith 2008, pp. 467–476, 2013, pp. 247–263; 2013, pp. 19–30). Smith claims that Bayle's adopted and refined Montaigne's method of arguing on both sides of a question so that the antinomical opposition of not just contrary claims but contradictory ones became the landmark of his skepticism. By opposing contradictory claims, he could reduce to the absurd the opposite one:

> Bayle, for his part, worked with contradictory alternatives (p or ¬p). [...]. Thus, one can argue in favor of a thesis by a *reductio ad absurdum.* Reason proceeds in a negative fashion, and instead of positively arguing for p, establishes the falsity of ¬p. [...] And in the Kantian antinomies, too, reason is only able to attack and destroy, not to defend an opinion; and it too proceeds by a *reductio ad absurdum*, trying to prove the veracity of one thesis by establishing the falsity of its contradictory (Smith 2013, p. 255).

Bayle's logical antinomies, Smith continues, significantly departed from the ancient method (Pyrrhonian or Academic) of *diaphônia*, which could involve more than two points of view or theories positively supported by a set of arguments. Bayle's version of the dialectical opposition, then, "reinterprets equipollence":

> We no longer face two parties with good arguments in their favor, but two parties that have good reasons to refuse the opposed party, but no convincing reasons in their favor. There is here a neutralization of reason [...]. It is precisely this destructive conception of reason what Kant takes from Bayle (Smith 2008, p. 472).

Without a doubt, this is one of the most significant features of Kant's antinomies. A clue to the specific article in Bayle's *Dictionary* that could have inspired Kant's views in this respect is given by his mention of Zeno of Elea in the Critical Solution to the Antinomies. Perhaps to avoid censure, Kant did not want to refer to Bayle in his Antinomies directly. Still, Bayle's article "Zeno" is a clear source, since he treats there the problem of the infinite divisibility of space, time, and matter in antinomical form. Kant references to Zeno, in my opinion, may be interpreted as an oblique defense of Bayle. He says:

> Zeno the Eleatic, a subtle dialectician, who was already severely censured by Plato as a wanton sophist who, to show his art, would seek to prove some proposition through plausible arguments and then immediately to overthrow the same proposition through other arguments just as strong. He asserted that God (presumably for him this was nothing but the world) is neither finite nor infinite, is neither in motion nor at rest, and is neither like nor unlike any other think. To those who judged him, it appeared that he wanted entirely to deny two mutually contradictory propositions, which is absurd. But I do not find that this charge can be justly lodged against him (A502/B530).

Kant associates Zeno with the antinomies and defends him (or Bayle) from the charge of being a "sophist." He argues that Zeno's intention was not to prove the inexistence of movement but only to offer opposing arguments to the theses concerning the infinite divisibility of space, time, and matter, in order to bring about suspension of judgment. He did not seek to destroy scientific systems but only recommended not to pass judgment before thoroughly analyzing the case in question. Kant's positive appraisal of Zeno coincides with his assessment of Pyrrhonism in the *Lectures on Logic*. In conclusion, we can say that Bayle's *Dictionary* exerted a critical influence on Kant's general view of Pyrrho and Pyrrhonism.

Perhaps when Kant finally read Sextus's works – if he ever did – he already had a Baylean conception of Pyrrhonism that did not let him appreciate crucial aspects of the ancient school, such as the broad suspension of judgment and the search for *ataraxia.*

Given that Kant's antinomies seem to find their origin in Bayle's article "Zeno" (Tonelli 1997, p. 81), we should now reconstruct their general context, that is, the modern discussion about Zeno's paradoxes. The debate involved a wide variety of approaches, both skeptical and dogmatic. I will now describe its key elements and pay particular attention to the place Bayle's article had in it.

6.4 Zeno of Elea and the Modern Dispute on the Infinite Divisibility of Space, Time, and Matter

In a series of articles entitled "The History of Zeno's Arguments on Motion: Phases in the Development of the Theory of Limits," F. Cajori gives a historical overview of the modern reception of Zeno's paradoxes (Cajori 1915a, b, c, d). In the seventeenth and eighteenth centuries, the paradoxes were debated in relation to the question of the infinite divisibility of space, time, and matter, which led to the first developments of infinitesimal calculus. Nicholas of Cusa and other Renaissance thinkers had already discussed the issue, but it was Galileo who, in the seventeenth century, gave the initial thrust to the debate in his celebrated work *Discourses and Mathematical Demonstrations Relating to Two New Sciences* (1638). Galileo claimed that Zeno's paradoxes showed how our minds are finite and, hence, cannot grasp the infinite without attributing to it finite properties (Cajori 1915a, p. 78). Almost simultaneously, in 1647, Gregory Saint-Vincent, a Flemish mathematician, put forward a geometric analysis of Achilles and the Tortoise's paradox, calculating the moment and exact location where Achilles would have overtaken the tortoise. This solution was reprised by those who argued that the problems set out in Zeno's paradoxes were solvable by mathematical calculus. They wanted to show to the skeptics that the paradoxes were not irresolvable but instead mere sophistical inventions. We have a direct reference to Saint Vincent's solution in Diderot's 1754 version of the *Encyclopedia.* M. Formey's entry "Motion" discusses the issue thus:

> The most famous of all the sophisms against motion is the one that Zeno called "the Achilles" in order to draw attention to his strength, which he held to be invincible [. . .] But, 1) even if it were true that Achilles never catches the tortoise, it would not follow that motion were impossible, because Achilles and the Tortoise are really moving, as Achilles always approaches the tortoise that is supposedly ahead of him by an infinitely small distance. 2) Zeno's sophism has a direct response. Gregory St. Vincent was the first to demonstrate its falseness and was the one who specified the exact point at which Achilles would reach the tortoise [. . .] (Formey 2016, p. 831; my translation).

Formey calls Zeno's paradoxes "sophisms" and relies on St. Vincent's calculus to show that the argument does not prove the inexistence of motion. Even though like Formey, many considered the matter settled, some philosophers in the seventeenth century persisted in re-opening the discussion and offered diverse solutions. According to Cajori, among the leading philosophers that engaged with the issue were Descartes, Gassendi, and Hobbes (Cajori 1915a, pp. 80–81). The latter, for example, stated the widely-held opinion that the term "infinite" should be understood as a synonym of "indefinite," since it merely indicates the possibility that somebody may indefinitely divide a section of time or space, but not an ontological infinity (Cajori 1915a, p. 81). As we shall see in the next chapter, Kant shares this understanding of infinity in his resolution of the mathematical antinomies.

As already mentioned, however, the most prominent source of these debates in the seventeenth century was not Hobbes but Bayle's *Dictionary*. According to Cajori, Bayle gave the most detailed skeptical treatment of the matter in his entry "Zeno" (Cajori 1915b, p. 109). Bayle's article explains Zeno's arguments, offers Aristotle's objections, and discusses some Cartesian appropriations, such as Malebranche's. Two aspects of Bayle's entry, I believe, are essential for this historical account. The first is his claim that Zeno was not seeking to defend the non-existence of motion and matter, which would involve a sort of negative dogmatism, but that his arguments were rather dialectical. In other words, Zeno aimed at showing how the thesis of the infinite divisibility of space, time, and matter led to the absurdity of denying motion and matter. Bayle states:

> Let us also observe that perhaps our Zeno only maintained that there is nothing while arguing about those principles he wanted to oppose. It may be that people concluded that he taught this positively and absolutely from an *ad hominem* argument he used, although he only offered this as a doctrine that resulted from the hypothesis that he had undertaken to refute (Bayle 1991, pp. 352–353).

The second relevant aspect is Bayle's view that the difficulty involved in these issues should lead us to conclude that our understanding is limited by nature, and we cannot attain knowledge about these matters. As a result, in these and other metaphysical questions, we should accept the dogmas of religion:

> I am even convinced that the exposition of these arguments can be of great service to religion, and I say here about the difficulties concerning motion that which Nicole has said about those concerning infinite divisibility: "The value that can be gained from these speculations is not simply that of acquiring this sort of knowledge, which is pretty sterile in itself; but it is in learning to know the limits of our mind, and making it admit, in spite of itself, that there are things that exist though it is not capable of understanding them. For this reason, it is good that we wear it out with these subtleties in order to check its presumption and to keep it from ever being foolhardy enough to oppose its feeble light to the truths that the Church proposes [...]" (Bayle 1991, p. 372).

Bayle's position reminds us of Galileo's view regarding our inability to understand the notion of infinity. However, the former's argument has undeniable fideistic overtones. He says that, even if Christian dogmas appear to be rationally contradictory, they are no more so than the metaphysical propositions, once we submit them to scrutiny—an argument he had also made in his entry "Pyrrho." Thus, the contradictions of both ancient (Aristotelean) and modern (Cartesian) metaphysics should lead us to accept Christian dogma with yet greater authority, although only by faith and not by reason (Popkin 2003, pp. 283–302). As we will see in the next chapter, Kant does not embrace this fideistic answer to the antinomical problems represented by Zeno. However, it is important to notice here that he was aware that such an answer existed.

On the other side of the dispute, we find some seventeenth-century mathematicians and philosophers who hold the infinite divisibility of time, space, and matter. To my mind, Pascal's account in *Of the Geometrical Spirit* summarizes well this point of view (Pascal 2014, pp. 10–12). He claimed that, even if a motion is remarkably rapid, we can still conceive of another that is even more rapid, and so on to infinity. The same goes for a very slow movement, as an even slower one may always be conceived up to infinity, irrespective of how slow it is. The same thing again happens with numbers and, finally, with space and time—we will always be able to conceive greater and smaller spaces and times, up to infinity. Pascal considered that this argument provided sufficient proof of the infinity of these four entities. However, he acknowledged that some philosophers denied the view:

> I have never known any person who thought that a space could not be increased. But I have seen some, very capable in other respects, who affirmed that a space could be divided into two indivisible parts, however absurd the idea may seem. [...] It is an infirmity natural to mankind to believe that he possesses truth directly; and thence it comes that he is always disposed to deny everything that is incomprehensible to him; while, in fact, he knows naturally nothing but falsehood, and while he ought to receive as true only those things the contrary of which appear to him as false (Pascal 2014, p. 11).

For Pascal, the position denying the infinite divisibility of these entities is based on the false assumption that we can attain truth directly when, in fact, only an indirect method can give us certainty. In other words, one should only take a statement as true when its opposite is shown to be false, that is, when it is reduced to the absurd. And, this is precisely the way Pascal set out to prove his thesis—by showing that the argument denying the infinite divisibility of these entities is false. To do this, first, Pascal said that it is evidently false or absurd to think that, when dividing a space, an indivisible part can be reached, because if something is indivisible, it is necessarily not extended, but we cannot say without contradiction that one part of extension is not extended. Second, he argued that one would have to ask whether the two indivisible parts that make up a divisible space would touch each other or not. And, since parts must be in contact to comprise a whole, we would also have to ask whether they touch only on one side or on all sides. The answers to these questions would lead to contradiction because, if the parts touch on all sides, they constitute a single thing and are not, strictly speaking, parts of a whole. But, if they only touch on one side, then these sides, in turn, have sides, so that they are not indivisible. Finally, Pascal claimed that, since the indivisible is by definition non-extended, to say that a space can be composed of indivisible parts is like saying that it is made out of two "nothingness" of extension. Thus, he concluded that "two negations of extension cannot make an extension" (Pascal 2014, p. 14) and insisted that sustaining this is as absurd as suggesting that two zeros added together can make up a numerical unit (Pascal 2014, p. 17). Using these arguments, Pascal demonstrated by *reductio ad absurdum* the falsity of the theses that deny the infinite divisibility of space, time, numbers, and motion, and the truth of the contrary thesis—namely the infinite divisibility of such entities.

This procedure, also shared by Bayle, is crucial for Kant's conception of the antinomies. However, Kant's assessment of the proofs based on the *reductio ad absurdum* of the opposite thesis differed from Pascal's dogmatic assertions, and neither committed him to Bayle's fideistic conclusions. In the next chapter, we will see in more detail how Kant argues that, although the antinomical issues appear unsolvable given the *reductio ad absurdum* proofs on both sides of the question, they can be unraveled by showing the "illusion of reason" at their basis. But, for the moment, I only want to draw attention to how Bayle's and Pascal's discussions of Zeno's paradoxes are sources of the structure and logical proofs used in Kant's antinomies.

Similar to Pascal's views are those of the philosopher, mathematician, and Bayle's most significant opponent, G. Leibniz. Leibniz's contribution to this dispute was crucial both mathematically and metaphysically. His stance underwent many changes, along with the mathematical problems he was tackling throughout his life, as documented by Samuel Levey (1998, pp. 49–96). Here, it is worth mentioning that at the outset of his career, Leibniz denied that infinitesimals could express anything other than mere "mathematical fictions" (Levey 1998, p. 55). Later, however, while studying infinite numerical series, he returned to the hypothesis of an actual infinite divisibility of matter, based on the concept of infinite convergent series (Levey 1998, pp. 68–79). After this, he held that the possibility that matter is divided into infinite finite parts could be mathematically proved. As Cajori cites, in his correspondence with Foucher (1693), Leibniz claimed:

> Father Gregory St. Vincent has correctly shown, through calculation even of infinite divisibility, the place where Achilles would catch the tortoise ahead of him, according to their proportional speeds. Thus, geometry serves to dispel these apparent difficulties. I am totally in favor of actual infinity since, rather than holding that nature abhors it, as is commonly said, I affirm that it [the infinite] affects nature everywhere to better indicate the perfections of its author. Hence, I believe that there is no part of matter that is not, I say not divisible but really divided and, in consequence, the smallest particle should be considered as a world full of an infinity of different creatures (cited by Cajori 1915b, p. 113; my translation).

The passage shows that Leibniz was aware of Bayle's skeptical characterization of Zeno's arguments and held, like other mathematicians, that is it possible to find by infinitesimal calculus the exact moment at which Achilles overtakes the tortoise. More importantly, however, he also claimed that the infinite divisibility of matter could be mathematically proved. In this way, Leibniz's thought expresses the metaphysical synthesis of Pascal's Saint-Vincent's views.

Now, by the eighteenth century and strongly influenced by Bayle, the English empiricists denied the infinite divisibility of space, time, and matter. In particular, Berkeley attacked infinitesimal calculus in his works *Treatise concerning the Principles of Human Knowledge, Philosophical Commentaries,* and *The Analyst.* Berkeley's arguments are manifold, but they are generally based on his ontological principle *Esse est percipi* (Robles 1993, pp. 29–44). If everything that is, is perception, it cannot be claimed that there are extended objects with infinitesimally divisible or divided parts, as we do not perceive them and their parts. For Berkeley, the term "infinitesimal" lacks any meaning. A term has to refer to an idea to make sense, and an idea is always a sensible mental representation. Given that there are no sensible mental representations of infinitesimal parts of matter, these entities are meaningless (Robles 1993, p. 31). In other words, since extension is just extended perception, it cannot be composed of infinite parts, as infinite parts would be imperceptible and, hence, would not exist. Extension, then, is formed by finite, perceptible parts, which Berkeley calls *minima sensibilia*. According to Robles:

> The MS [*minima sensibilia*] are ultimate sensible elements; they are, literally, perceptual atoms that are to be found in any object of perception. Still, in any perceptual presentation, only a finite number of these atoms are present. [...] If an object of perception is divisible, it will only be so into a finite number of parts since, past a specific limit, no object of perception remains. However, what this means, for Berkeley, is that there is no object, and he is strongly opposed to this leap from being to not being, or [...] to claiming that being, the perceptible, is formed of the non-being, the non-perceptible (Robles 1993, 42; my translation).

It is remarkable, in my view, that both sides of the dispute, the side that asserted the existence of infinitesimals and the side that denied it, alluded to one common argument, namely the impossibility that being arises from non-being. Let us recall that, in support of the thesis of the infinite divisibility of space, Pascal claimed that "two negations of extension cannot make an extension" (Pascal 2014, p. 14). By the same token, Berkeley would probably argue that two "non-perceptions" cannot make a perception in support of the contrary position. As we can see, the symmetry between the arguments on both sides of the issue suggests the equipollence that Kant will develop in his antinomies.

Another fundamental aspect of Berkeley's critique of the infinite divisibility of matter and extension relates to his defense of religion. Berkeley wrote *The Analyst* as a critique of the mathematical dogmatism of his age, arguing that, just like Christian religion holds absurd and mysterious dogmas, mathematics falls into obscurities and contradictory claims. This thesis is undoubtedly influenced by Bayle. As Cajori claims, for Berkeley, "just as mathematics, though involved in obscurities, is esteemed excellent and useful, so articles of Christian faith should be accepted as no less true and excellent" (Cajori 1915c, p. 144). In its day, it seems, this was a recurring fideistic topic.

Finally, strongly influenced by Berkeley, Hume turned his attention to the problem of the infinite divisibility of space and time. I will leave my exploration of this issue to the final section of this chapter, where I examine more in-depth the broader discussion about the role of Humean skepticism in Kant's conception of his critical project.

To conclude this section, I believe we can confidently assert that the modern dispute on Zeno's paradoxes, with their dogmatic and skeptical interpretations, heavily influenced Kant's conception of the Antinomies of Pure Reason. The most important aspects of this influence are, in my opinion, four: i). the skeptical conclusion, most prominently expressed by Bayle, that the contradictions involved in these issues show the limits of our understanding; ii). the structure of thesis and antithesis, also present in Bayle; iii). the use of *reductio ad absurdum* proofs to sustain one side while refuting the opposite one, and vice-versa, which appears both in Bayle and Pascal; and iv). the idea that the infinite extension of time and space should be understood as an indefinite progress of the human mind when thinking these issues, which appears in Hobbes but is also a widespread topic in the seventeenth century.

With this discussion, I close the analysis of the modern sources of Pyrrhonism in Kant. Now, I turn to the Academic sources by detailing Kant's relation to Cicero, his opinion on Academic skepticism in the *Lectures on Logic*, and his relation to Hume's Academic understanding of the Antinomies, particularly of the antinomy of freedom and necessity.

6.5 Kant, Cicero, and Academic Skepticism

Having already delineated Kant's sources for Pyrrhonian skepticism, we must now consider whether Kant had any direct knowledge of the primary sources of Academic skepticism, beginning with a broad overview of his relationship with Cicero.

We know that Kant read Cicero with interest as a student of Latin during secondary school and as a mature man when he was writing his *Groundwork of the Metaphysics of Morals*. In his biography of Kant, M. Kühn claims that, during his youth, Kant was very fond of the classical Roman authors, including Cicero:

> The backbone of the education at the *Collegium Fridericianum* was Latin. Not only did the students spend the most time learning Latin, it was also by far the most important discipline. […] By their third year, students were expected to read Cornelius Nepos; the fourth year consisted of a repetition of all of Nepos, some Cicero, and some poetry. In the fifth class, they read Caesar and more Cicero, and in the sixth year Cicero (*De Officiis,* among other selections), Muretus, Curtius, and Pliny […] Emanuel did well in Latin, and those who knew him then thought that he would make classics his chosen field of studies. […] Kant continued to think highly of the ancients, reading them throughout his life (Kühn 2001a, pp. 48–49).

Beyond this, Kant's first-hand knowledge of Cicero's works is attested by at least twenty references to the Roman philosopher in his *Gesammelte Schriften*, especially in his *Lectures on Logic*. Since Kant seldom referred by name to the authors he read, this piece of information should not be underestimated. Having said this, though, it is also important to notice that Kant's assessment of the Roman philosopher in these references is somewhat ambiguous. To begin with, Kant misses no opportunity to deride Cicero's pompous oratorical style, although he immediately balances his criticism by acknowledging his persuasive power. For example, in the *Hechsel Lectures*, Kant affirms: "Cicero shows an affected taste and art of speaking. One forgets this if the exposition has energy; then we are charmed and transported by the exposition" (*VL-Hechsel,* 115).

As a philosopher, Kant appears not to hold Cicero in very high esteem, but to see him instead as a historian of philosophy:

> Learning that is historical in form but rational in matter is called historical cognition. There are many Ciceros in philosophy, many, indeed, who do not know at all what sort of difference there is between philosophy and the science of philosophizing (*VL-Blomberg*, 52).

Despite these criticisms, he concedes Cicero's ability to make philosophy popular and thinks that, in this sense, he sets an example to be followed:

> To learn true popularity, however, one must read the ancients, e.g., Cicero's philosophical writings, the poets Horace, Virgil, etc., and among the moderns, Hume, Shaftesbury, et al. Men who have all had a good deal of intercourse with the refined world, without which one cannot be popular. [. . .] This truly popular perfection of cognition is, in fact, a great and rare perfection, which shows much insight into science (*VL- Jäsche*, 47).

I have discussed Kant's conception of popularity and its relationship with Ciceronian rhetoric elsewhere (González 2015, pp. 47–60). Still, I wish to stress here that Kant recognizes Cicero as a classical writer that should be followed to cultivate a popular philosophical exposition. Such exposition, for Kant, is not ancillary; rather, it is a perfection of discourse that bears witness to its proximity to truth (*VL- Jäsche*, 47–48).

Unsurprisingly, Kant most positively values Cicero's moral philosophy. He regarded the Roman philosopher as a great exponent of Stoic moral doctrine, and *De Officiis* exerted a substantial influence on him while writing the *Groundwork of the Metaphysics of Morals*. This influence has been the object of discussion among various twentieth-century scholars. The hypothesis that Kant wrote the *Groundwork* as a response to Christian Garve's 1783 translation and commentary of Cicero's *De Officiis* was first presented in 1939 by K. Reich and subsequently examined by various scholars in the second half of the twentieth century (Reich 1939, pp. 446–463; Des Jardins 1967, pp. 237–242; Glei 1999, pp. 49–61; Kühn 2001b, pp. 270–278). In a letter from Herder to Hamman on February 8 of 1784, Herder reports Kant's intention to write a response to Cicero's moral system, as Garve presented it (Reich 1939, p. 447). The discussion, in broad terms, seeks to determine the reach of *De Officiis*'s influence on Kant's *Groundwork*. Specifically, the debate centers on whether the Stoic moral formula *naturae convenienter vivere* – to live suitably in accordance with nature – may have affected Kant's conception of the categorical imperative.

Without delving into the details of the discussion, I will summarize the different opinions concerning the relationship between Kant and Cicero on this issue. Among the commentators, Des Jardins considers that Cicero's writings may have been as crucial for Kant as they were for Hume:

> Today the fame of Cicero has decayed, and the fashion is rather to associate Hume with Kant. But Kant, who has never been regarded as an easy or eloquent philosopher, was nonetheless indebted to Cicero, particularly in ethics. [. . .] The suggestion is, at least, that Cicero provided the context in which Kant expressed his moral philosophy and that Kant, much like Hume, may have been understood by his contemporaries for having announced his philosophy in the vocabulary of *De Officiis* (Des Jardins 1967, pp. 238–242).

Des Jardins makes a case for a strong connection between Cicero and Kant, to the point that he compares Kant's relation to Cicero with Hume's. Against this view, Kühn declares that, while "Kant knew Cicero well," both because of his training in Latin during primary and secondary school, and through Garve's translation of *De Officiis*, the influence has been overstated:

> One has to be suspicious when it is suggested that "Cicero provided the essential context in which Kant expressed his moral philosophy." But we must be even more suspicious when it is claimed that Cicero's *On Duties* played an important role in Kant's formulation of the three versions of the categorical imperative. [...] Such suspicions do not mean, of course, that Cicero and Garve where completely unimportant to Kant when he was writing his prodrome of the metaphysics of morals, even if this seems to be what the majority of Kant scholars believe. [...] The *Foundations* deliver not just Kant's answer to Garve's Cicero but his answer to any kind of Ciceronian ethics (Kühn 2001b, p. 272).

For Kühn, although the historical relationship between Cicero and Kant is clear, the former's influence on the latter's ethics has been overestimated. He believes that the differences between Cicero's highly eudaemonistic and communitarian moral theory and Kant's universalist and deontological ethics are more significant than the similarities.[19] Kühn's analysis is, in my view, correct, except that he forgets Cicero's universalistic aims. His rejection of a link between the two theories may come from *De Officiis*'s pedagogical tone and defense of Roman civility, which somehow blurs the universal import of Panaetius's ethics. Also, Kant's derogatory views on Cicero, already mentioned, may have led Kühn to conclude that the link was not sufficiently robust.[20] And, in fact, acknowledging them, Kühn claims that "Kant saw Cicero more as a follower than an original philosopher" (Kühn 2001b, p. 271). He even cites the paragraph in the *Jäsche Lectures* where Kant asserts: "When philosophy subsequently passed from the Greeks to the Romans, it was not extended; for the Romans always remained just disciples" (*VL-Jäsche*, 31). Thus, Kant's negative comments on Cicero may have guided Kühn's assessment of their connection in ethics. Nonetheless, I also believe that Des Jardins's view that Cicero's influence on Kant is comparable to his influence on Hume is exaggerated. In this sense, Kühn correctly calls attention to Kant's criticism of Cicero. We should, then, balance these extreme evaluations through a closer examination of Kant's comments, not just about Cicero, but also about Academic skepticism.

Even if Kant was familiar with Cicero's works, he probably was not directly acquainted with the *Academica*. Just as he did not read Sextus Empiricus but learned about Pyrrhonism from secondary sources, we can expect that his evaluation of

[19] "A Ciceronian ethics that remains founded on common life, expressed by such concepts as honor (*honestas*), faithfulness (*fides*), fellowship (*societas*), and seemliness (*decorum*) was too superficial and unphilosophical for Kant. For this reason, Kant rejected Cicero and all those trying to develop a Ciceronian ethics" (Kühn 2001b, p. 276).

[20] Kühn forgets the link between *On Duties* and *The Republic* which Glei points out in "Lux Regiomontana," and explains why Ciceronian ethics is universalistic and not based on communitarian restrictions. In my view, the most solid argument concerning Cicero's influence on Kant's *Groundwork* is offered by Glei (1999, pp. 49–61).

Academic skepticism did not come from his reading of Cicero's epistemological treatise but from modern appropriations. We must bear in mind that the Prussian Academy published De Castillon's French translation of the *Academica* only in 1779. Although Kant may have had access to this text after its publication, his conception of the *Critique of Pure Reason,* with his adoption of the skeptical method, had already taken place, so it is doubtful that this text had any impact on him.

Furthermore, as we have already mentioned, in the *Lectures on Logic,* Kant sees Academic skepticism as a radical kind that extends doubt to all cognition by holding the principle that nothing is certain. Kant defines academic doubt as "dogmatic doubt:"

> Dogmatic doubt is always accompanied with persuasion of the certainty that nothing at all can be established or maintained concerning this or that cognition. Hence, it consists in nothing but judging that one can never attain complete certainty with a cognition, and that all enquiry, furthermore, is thus always conducted in vain and for nothing (*VL-Blomberg*, 209).[21]

Kant's poor opinion of the Academy probably comes from modern accounts of ancient skepticism, which reproduced the opening paragraphs of the *Outlines of Pyrrhonism.* As I argued before, most likely the source here is de Crousaz's *Examen du Pyrrhonism*, which included fragments of the original work by Sextus. The exact description of Academic skepticism as a sort of negative dogmatism appears in Kant's portrayal of Arcesilaus and Carneades. About Arcesilaus, he writes:

> But Arcesilaus set up a skeptical school. This is Academic skepticism. He said that everything is so uncertain that it is even uncertain whether this is uncertain; this was called a real purgative of reason, so that by seeking to purge reason, he also got rid of reason itself (*VL-Vienna*, 886).

And he attributes the same principle to Carneades:

> Carneades, the founder of the last Academic school, deviated from Pyrrho's doctrines very much and actually maintained that one could not attain a proper certainty, or come to certain truth, in any matter (*VL-Blomberg*, 215).

In other passages of the *Lectures,* Kant's characterizations of Carneades are also derogatory. Following the anecdotes from classical doxography – particularly concerning Carneades's participation in the Embassy to Rome in 155 BCE, when he gave successive speeches for and against justice – Kant describes him as a sophist practicing a shameless and fallacious dialectics: "The third school of Carneades consisted of disgusting sophists and dialecticians" (*VL-Vienna*, 886). Given these opinions, it appears that Kant did not obtain his views on Academic skepticism from

---

[21] "[. . .] they doubted things, etc., but nevertheless believed that they had a right that every man should give approval to their proofs and accept their doubts. Thus, they contradicted themselves, for they said that everything is uncertain, without distinction, and nonetheless they maintained their propositions and attributed to them an infallible certainty. They had, e.g., the proposition: everything is uncertain, and in such a way that it is even uncertain that everything is uncertain." (*VL-Blomberg*, 216).

the *Academica* or other, more balanced, sources. It seems that he took them rather from anti-skeptical treatises, who derided Academic skepticism for their irreligious consequences.

Now, if this is the case, why should we think that Kant's skeptical method could have been influenced in any way by Academic skepticism? Does it not make more sense to consider that it proceeded from the kind of skepticism he praises, namely Pyrrhonism, probably in its Baylean adaptation? Indeed, if his opinions about Cicero as an orator are quite negative and as a philosopher not at all enthusiastic, why insist on finding an Academic influence on Kant's skeptical method?

I suggest that part of the answer to these questions may be found in an analysis of Kant's account of the concept of suspension of judgment in his *Lectures on Logic*. There he assesses positively a limited suspension, more akin to the Academics' approval of impressions for practical purposes than to the Pyrrhonists' broad *epochê* and acceptance of appearances and the laws and customs of one's community to act. Given this conception of suspension of judgment, we should seriously explore whether Kant was acquainted with the general spirit of the New Academy's skeptical position, either through his reading of other Ciceronian treatises or, indirectly, via Hume. I hope to make these connections clearer in the remaining parts of this chapter.

6.5.1 Academic Suspension of Judgement in the *Lectures on Logic*[22]

When Kant writes more extensively about skepticism's fundamental concepts in the *Lectures on Logic*, he reveals a greater affinity with the views of the New Academy than with those of Pyrrhonism. This affinity is especially notorious in his evaluation of the suspension of judgment valuable for (critical) philosophy. In the *Vienna Lectures*, Kant defines suspension of judgment as an attitude by which one withholds reason's natural inclination to take as true a provisional judgment: "this is the *suspensio judicii*, the intention not to let a provisional judgment become a determining one" (*VL-Vienna*, 860). This definition places suspension of judgment in the context of scientific inquiry broadly construed. According to Kant, this suspension may take one of two forms:

> We can divide this withheld approval in two. *Suspensio judicii indagatoria* is when one still intends to seek out the grounds for the determinate judgment. This is actually a postponement of judgment, in order to inquire more about this in time; or there is *renunciatio judicii*, in that we renounce our judgment and give up all hope of determining anything. Skepticism renounces all judgments in regard to various things (*VL-Vienna*, 860).

For Kant, the first type of suspension – *suspensio indagatoria* – is helpful because it allows the careful examination of grounds of a judgment before adopting it. This is

---

[22] See note 7.

the sort of suspension that, for him, a cautious philosopher or scientist exercises. To avoid taking a mere hypothesis for a conclusive judgment, the enquirer postpones her assent while examining the grounds of the judgment in greater detail. The second type – the *renunciatio judicii* – is defined by Kant as a more radically skeptical attitude, which abandons the possibility of ever attaining truth from the very beginning of the search. This is, of course, the kind of suspension suggested by a negative dogmatism, and which, as we have seen, Kant attributes to the Academics.

In the posterior *Dohna-Wundlacken Lectures*, Kant's description of the difference between the two kinds of suspension has some variations:

> *Suspensio judicii* consists in this, that when all the grounds on both sides are equally sufficient, none is predominant. It can be (I.) *ob indifferentiam* – [2.] *ob aequilibrium.* To hold oneself in *suspensio judicii* by choice (this is the mean between holding-to-be-true and rejecting – it remains only a problematic judgment) shows a very great mind and is extremely hard on account of the fact that the inclination toward immediate judgment of the understanding interferes. *Ob aequilibrium*, to be and to remain inconclusive, gives evidence of weak minds (*VL-Dohna-Wundlacken*, 736).

Here, Kant characterizes suspension of judgment as the finding, upon skeptical examination, that two opposing judgments have equally insufficient grounds. This is no longer the sort of inquiry in which belief is suspended while a hypothesis is verified. Instead, we see here an explicit reference to the skeptical method of opposing two claims of equal weight to evaluate their rational grounds. That the *Dohna Wundlacken Lectures* date from the first years of the 1790s, while the *Vienna Lectures* are from the 1780s, may explain the more detailed elaboration of the distinction in the former set.

Kant's terminology in these passages is noteworthy. An *ob indifferentiam* suspension, on the one hand, is that in which the enquirer voluntarily chooses to be indifferent to the matter in question—she decides to exert a method to discipline her reason not to follow her natural tendency to judge. An *ob aequilibrium* suspension, on the other hand, is the attitude that results from rational indecisiveness—an inclination to constantly change one's point of view in the face of the difficulties imposed by scrutiny. Even though the use of the term *aequilibrium* appears to suggest the more positive Pyrrhonian idea of maintaining a certain kind of "balanced" mental attitude – that is, of sustaining the state of *epochê* for attaining psychological peace of mind – it does not have this meaning in Kant. And, given Kant's earlier characterization of the two forms of skepticism – the Academic as a negative dogmatism, and the Pyrrhonian as a valuable device to continue the inquiry – it is more accurate to infer that he correlated the former kind of suspension, *indagatoria* or *ob indifferentiam,* with Pyrrhonism. The second kind, suspension *ob aequilibrium,* seems to be associated with a hesitant attitude and not with any type of suspension proposed by the ancient skeptical schools. Kant may have had in mind Descartes's mention of skeptical irresolution, which he sought to remedy with the second maxim of his provisional morals (*AT.* VI., 24). In this way, the suspension *indagatoria,* for Kant, would represent the type of attitude expressed by Pyrrho's *non-liquet,* while the suspension *ob equilibrium* would describe an attitude of constantly hesitating and ending inquiry before arriving at any conclusion.

However, other sections of the *Dohna-Wundlacken Lectures* suggest that Kant was familiar with the suspension of judgment that Carneades proposed, whereby it is possible to continue investigating a belief while *approving it for practical purposes*. Kant says:

> In *suspensio judicii* there lies some freedom. One can also give his approval and yet, in judging, not make a decision – which still awaits further grounds – where one accepts the opposite as still possible. I am free when I accept a proposition in a practical respect for the good that can follow from it. One can say, then, that there are things in matter of religion, e.g., our eternal continued existence, which we accept just for this good, for this life is not at all adequate to the idea of the highest good (*VL-Dohna-Wundlacken*, 736).

There are a few notable points in this fragment. First, the substance and tone of the passage differ significantly from the previously-discussed ones where Kant speaks about skepticism in the context of scientific inquiry. Here, Kant no longer refers to the suspension of hypotheses until they are empirically verified, but instead talks about the belief in an immortal soul—a belief that, evidently, cannot be considered an empirical hypothesis. Second, Kant says that one can suspend judgment in religious matters while approving the claim in question for practical purposes. This change of register is significant because, as we have seen, his skeptical method involved not only the examination of the grounds for opposite theses but also a practical verdict or practical resolution of the issue. Third, Kant attributes greater rational freedom to this way of proceeding. This is a new element of the discussion since, in previous accounts, Kant had only spoken of the skeptics' attitude of caution and careful examination. But here, by contrast, the emphasis is placed on one's autonomy or capacity to determine oneself to act according to a particular belief, even if this belief proves insufficiently grounded in the final examination.

Of course, dating from the 1790s, the *Dohna-Wundlacken Lectures* are well into Kant's critical stage and reconstruct many aspects of his treatment of metaphysics in the first *Critique*. However, the same argument about the practical advisability of approving a metaphysical claim like the soul's immortality already appeared in the Practical Conclusion of the *Dreams of a Spirit-Seer*. Thus, Kant's contact with Academic skepticism and their narrow suspension of judgment may have occurred in the pre-critical period, via Hume. In any case, the novelty of this paragraph and its resemblance with the New Academy's view on suspension of judgment and "approval" of persuasive impressions should be taken into account.

Now, one may object that the approval Kant recommends in this paragraph is not Academic but Pyrrhonian since, as we saw, Sextus also advised accepting impressions that are relevant for action. To avoid inactivity, the Pyrrhonists only suspend judgments regarding things as they are, not as they appear (*Pyr.* I, 19–20). However, the kind of judgment Kant suggests approving for practical purposes in this passage does not express subjective impressions, like the Pyrrhonian appearances, but instead states a metaphysical claim, namely that the soul is immortal. Kant claims that we can approve this judgment given our natural desire to obtain the greatest good – i.e., the convergence of morality and happiness – and the impossibility of attaining it during our finite lives. We should also note that this metaphysical claim does have a rational basis, albeit only from a practical point of view.

Would the Pyrrhonists accept metaphysical assertions like this in order to act? As I argued in the second chapter, on theological or religious matters, the Pyrrhonists suspend judgment and apply the practical criterion of undogmatically following the laws and customs of their community. But accepting these practices does not mean that they have chosen to believe in them—indeed, they have only decided to *abide* by them. Their acceptance of social norms is not the product of rational scrutiny, like Kant proposes. In fact, approving the immortality of the soul would not be a Pyrrhonian way of resolving the practical problem that, according to Kant, arises from the incompatibility between searching for happiness and conducting oneself morally in a finite lifetime. The Pyrrhonists would see this opposition between happiness and morality as an indication that it is best to suspend judgment on the matter entirely. Thus, they would not approve the judgment "the soul is immortal" for the reasons put forward by Kant, but would only observe their community's religious conventions on the matter. For example, they would attend funeral rites to ask the Gods to show goodwill towards immortal souls, if their community saw it fit. Still, they would do this to conform to social mores and not because they approved the underlying theological claims.

As I will discuss in the next chapter, Kant thinks that adopting the belief in the immortality of the soul, albeit only as an incentive for living a moral life, is a way of practically resolving the issue. The value of this approval derives from the freedom implicit in acting according to a rationally examined belief, despite the insufficiency of its theoretical grounds. This is an autonomous choice, an expression of the practical freedom of a rational individual. The Pyrrhonian attitude would, in Kant's eyes, represent a heteronomous choice, though it would at least have the virtue of avoiding metaphysical dogmatism. The freedom that Kant sees in approving a metaphysical claim for practical purposes is, then, very unlike the Pyrrhonists' acceptance of appearances and religious traditions.

In conclusion, based on Kant's discussion of suspension of judgment in the *Lectures on Logic,* we can claim that Kant was probably acquainted with the New Academy's narrow suspension and solution to the *apraxia* problem, even if he did not identify it with the ancient school. He mistook Academic skepticism for a negative dogmatism while deemed Pyrrhonian skepticism a merely cautious way of conducting empirical research. These misinterpretations may be founded on the second-hand sources Kant used for his account of the ancient skeptical schools. However, his view that a metaphysical claim can be adopted for practical purposes is properly understood as Academic and may have come from Hume, who, as we have seen in previous chapters, was more directly influenced by Cicero's skeptical treatises and had a more nuanced view of the differences between the ancient skeptical schools. This being said, I will now discuss the role that Hume's Academic skepticism may have played in shaping Kant's understanding of the skeptical method and the antinomies. I will do so by framing the discussion within the broader debate about whether it was Hume or the antinomies that caused Kant's "awakening" from his "dogmatic slumber."

6.6 Humean Academic Skepticism in Kant's Awakening from his Dogmatic Slumber

The relationship between Kant's critical project and Humean skepticism was studied intensively in the twentieth century, particularly in regard to Kant's response to Hume's problem of causality. Kantian scholars, especially in the analytic tradition, exhaustively examined and assessed this relationship and argued that Kant's Transcendental Analytic of the *Critique* gives a response to "the skeptic," particularly in the Deduction of the Categories and the Analogies of Experience (Guyer 2008, pp. 71–160; Stepanenko 2013, p. 17). Despite this interest in Kant's connection with Humean skepticism, most scholars overlook that Kant also retrieved Hume's skeptical engagement with traditional metaphysics. The debate about the cause of Kant's "awakening" from his "dogmatic slumber" stands out as an instance of this omission.

The problem revolves around Kant's two testimonies on the matter. In one of them, Kant attributes the cause of his awakening to Hume, while, in the other, he says that it was his encounter with the cosmological or antinomical issues that brought it about. The first testimony appears in the *Prolegomena*:

> I freely admit that the remembrance of David Hume [*Erinnerung des David Hume*] was the very thing that many years ago first interrupted my dogmatic slumber and gave a completely different direction to my researches in the field of speculative philosophy (*P. Ak*. IV, 260).

The second can be found in the 1798 letter to Garve:

> It was not the investigation of the existence of God, immortality and so on, but rather the antinomy of pure reason – "The world has a beginning; it has no beginning, and so on, right up to the fourth: There is freedom in man, vs. there is no freedom, only the necessity of nature" – that is what first arose me from my dogmatic slumber and drove me to the critique of reason itself, in order to resolve the scandal of ostensible contradiction of reason with itself (*C. Ak*. XII, 257–258).

The disparity between the two assertions has led to different views on the matter. First, we have the traditional interpretation, according to which the two accounts are incompatible. Among the commentators who share this view, the boldest suggestion, undoubtedly, comes from Lewis White Beck. He claimed that only Hume's influence was decisive and that the statement in the letter to Garve should be considered a mere "memory lapse" from a somewhat senile Kant (Beck 1978, p. 119, n. 23). After Beck, most Kant scholars have identified the relationship between Kant and Hume exclusively in terms of the problem of causality, largely ignoring Kant's allusion to the antinomy in the letter to Garve (Beck 1978, pp. 118–119, n. 22; Anderson 2010, pp. 81–120; Kühn 1983, pp. 175–193).

This interpretation is for the most part based on the English translation of the German expression *Erinnerung des David Hume* as the "suggestion," "objection," or "critical observation"[23] of David Hume, instead of as the "memory," "recollection,"

[23] Beck (1978) began the tendency to translate this section with the noun "suggestion." Kühn (1983) accepted the term, while Anderson (2010) preferred "objection" or "critical observation."

or "remembrance" of David Hume.[24] The term's choice is significant because translating *Erinnerung* as "suggestion" or "objection" makes it easier to interpret Kant as referring to Hume's skeptical treatment of the principle of causality—the "objection" of David Hume is construed here as his critical treatment of causality. In this line of interpretation, Kant's reference, in his letter to Garve, to the antinomy of pure reason is easily dispatched by arguing that Hume's problem of causality ultimately led him to the cosmological issues mentioned there. I find this answer unsatisfactory, because even if the problem of causality is, in fact, related to the cosmological matters – particularly in the third antinomy of freedom and necessity – the issues do not entirely overlap, and Hume afforded them different skeptical treatments. Thus, in my view, these commentators do not do justice to Kant's relationship to Hume's skepticism by circumscribing it to the question of causality.

However, there is a second interpretive tradition maintaining that in both testimonies, Kant refers to a single intellectual event—namely his encounter with Hume's formulation of the antinomy of reason (Kühn 1983, pp. 175–193; Ertl 2002, pp. 617–640; Kreimendhal 1990; Falkenstein 1995, pp. 63–79). These commentators acknowledge the variety of Hume's skeptical arguments and develop a more complex version of the relationship between Hume's skepticism concerning metaphysics and Kant's critical project. To better appreciate their views, however, we must delve more into the intricacies of the debate.

Both interpretative camps face the problem of determining which of Hume's texts produced Kant's awakening and when it occurred exactly or approximately. Here is a brief sketch of the most important positions:

1. Traditional interpretations suggest that Kant would have had contact with the cosmological issues around 1765 through sources other than Hume, while his first encounter with the latter would only have taken place around 1772. This contact would have occurred through the German translation of a text by Jeames Beattie – Hume's most vigorous opponent – that included fragments of Hume's *Treatise*, specifically those containing Hume's critique of the principle of causality (Smith 1950, p. xxxviii; Forster 2008, p. 15; Wolff 1960, pp. 117–123). Kant's reading of these passages would be the decisive stimulus for his critical project.
2. A more heterodox interpretation maintains that the two testimonies are compatible, and considers two possible dates for Kant's encounter with Hume and the antinomy of pure reason—1769 or sometime between 1760 and 1765. First, some scholars, principally L. Kreimendhal, hold that there is conclusive textual evidence to situate the moment of Kant's awakening in 1769 (Kreimendhal 1990; Kühn 1983, pp. 175–193). The most crucial piece of evidence for this interpretation is Reflection No. 5037, in which Kant affirms:

[24]I use G. Hartfield's translation here. In an explicative footnote, Hartfield says that the term *Erinnerung* can also mean "suggestion" or "critical observation": "The German word *Erinnerung* can mean a "memory" or "remembrance" (as shown here), or it can mean a "reminder," "admonition," or "warning." Kant used the term both ways [. . .]. Thus, his words here need not imply a specific act of remembering Hume's work, but may simply be invoking Hume's admonition or warning about the use of the causal concept in traditional metaphysics." (*P*. n. 9, p. 10.)

> Initially, I saw this doctrine as if in twilight (*wie in einer Dämmerung*). I tried quite earnestly to prove propositions and their opposite, not in order to establish a skeptical doctrine, but rather because I suspected I could discover in what an illusion of the understanding was hiding. The year 69 gave me a great light (*gab mir großes Licht*). (*Ak.* XVIII, 69)

Based on this fragment, the best date to situate the beginning of Kant's critical project would be the year 1769.[25] Kreimendhal also claims that the text involved in Kant's awakening was probably J. G. Hamman's translation of the conclusion of the first book of the *Treatise*, which Kant may have read in 1769, before its publication in Germany (Kreimendhal 1990, p. 5).[26] Other commentators argue that the text involved in Kant's awakening is Hume's first *Enquiry* (Falkenstein 1995, pp. 63–79; Anderson 2010, pp. 81–121), whose German edition, by Sulzer, was already in circulation by 1755. For them, Kant's encounter with this text, and his awakening, would have taken place between 1760 and 1765 (Kühn 1983, pp. 180–181).

Against this latter position, some scholars have objected that in the *Reflections* and the *Correspondence*, Kant refers explicitly to the year 1769 as the one that gave him "great light," while his reading of the *Enquiry* happened almost five years before that (Falkenstein 1995, p. 66). Thus, accepting that the *Enquiry* caused the awakening would involve dating the event much earlier than Kant's testimonies indicate. Moreover, in this scenario, it would be difficult to explain Kant's strange "return" to metaphysics in his *Inaugural Dissertation* of 1770, where he accepted the possibility of attaining rational knowledge of metaphysical entities. Had Kant already experienced his awakening, it is not easy to understand why he would have fallen back into metaphysical dogmatism in 1770. Finally, for some commentators, Hume's treatment of the problem of causality in the *Enquiry* is not sufficiently "skeptical" to have produced Kant's awakening. They, consequently, propose the alternative that it was his later reading of the *Treatise* that led to his critique of metaphysics (Smith 1950, p. xxxvii; Kühn 1983, p. 181).

For my part, I find the hypothesis that Kant's reading of the *Enquiry* produced his awakening to be entirely plausible, and I will defend it in what follows. The objections mentioned above can be easily resolved by having in mind, first, that even if the "great light" that sparked Kant's criticism occurred in 1769 or later in 1772, Kant's early reading of the *Enquiry*, between 1760 and 1765, still could have produced his awakening. The event could have happened by his "remembering" – as the term *Erinnerung* more naturally implies – of his earlier reading of Hume's text.[27] After all, philosophers often re-read the texts that greatly impress them, especially if

---

[25] This date is also supported by the fact that, in correspondence from 1783 with Garve and Mendelssohn, Kant states that the development of the first Critique took him approximately 12 years (*C. Ak.* XX, 338; XX, 345).

[26] Kühn claims that Kant read this text later, once it was published, in 1771 (Kühn 1983, pp. 185–186).

[27] The hypothesis is reinforced by Kant's own testimony in his 1764 essay, *Inquiry Concerning the Distinctness of the Principles of Natural Theology and Morality* (*Ak.* II, 273–301), according to which he already had a command of the antonymic problems by that period (Falkenstein 1995, p. 69).

they seek to comprehend better the issues involved. Indeed, there is no reason to think that Kant did not read the *Enquiry* on different occasions throughout his "silent decade." N. K. Smith contemplates this hypothesis: "[. . .] we cannot be absolutely certain that it was not a re-reading of the *Enquiry* or a recalling of its argument that suggested to Kant the central problem of his critical philosophy." (Smith 1950, pp. xxxv–vi).

Second, Kant's falling back into metaphysics in the *Dissertation* of 1770 has been explained as part of his gradual process of conceiving the *Critique*. As Hinske argues, Kant was still uncertain in 1770, even after his awakening, about metaphysics' actual scope; hence he temporarily adopted the traditional position, whereby knowledge of metaphysical entities is possible. However, at this stage of his thinking, Kant had already begun to envisage the general principles of his critique. In the *Dissertation,* he suggested an opposition or antinomy between the principles of understanding – oriented toward the sensible realm – and reason – seeking to grasp an intelligible world (Hinske 1965, pp. 485–496). Therefore, although Kant had not yet posited the illegitimacy of speculative claims, he had recognized the antonymic structure of metaphysical problems and was on the road to their skeptical or non-dogmatic solution.

Third, it is important to note, as Kühn does (1983, p. 180), that in the preface to his edition of the *Enquiry,* Sulzer expresses his wish that the arrival of Hume's text in Germany "would awaken German philosophers from their idle slumber."[28] The coincidence between Sulzer's phrasing and Kant's repeated allusions to his "awakening from [a] dogmatic slumber" is noteworthy. Indeed, why would Kant have repeated the formulation almost literally, if not because he felt that it contained the proof that the *Enquiry* had the effect desired by Sulzer? Finally, I think that the commentators who dismiss the *Enquiry* because its tone is not as explicitly skeptical as the *Treatise*'s are mistaken. In this respect, I agree with Falkenstein, who regards Hume's metaphysical skepticism as much more solidly and eloquently expressed in the *Enquiry* than in the *Treatise* (Falkenstein 1995, p. 77). One of the reasons commentators think that the *Enquiry* is less skeptical than the *Treatise* is that, as I have shown in previous chapters, Hume's skepticism in the *Enquiry* is decidedly Academic and, hence, more mitigated. Still, this mitigated skepticism is consistently and successfully applied in the *Enquiry* to the issues of miracles, liberty and necessity, divine providence, and the infinite divisibility of time and space, as we will see shortly. In conclusion, Kant's awakening could be caused by his reading of the *Enquiry* and his recalling, in the next decade, Hume's skeptical approach to traditional metaphysics.

Before moving on to an analysis of these passages of the *Enquiry* that I believe most prominently influenced Kant's antinomies, I want to survey yet another interpretation. For some scholars, the crucial text in Kant's awakening is the

[28]The phrase in German reads: "*Ich hoffe, dass die Bekanntmachung dieses Werkes sie [die deutschen Philosophen] aus ihrer müßigen Ruhe ein wenig aufwecken, und ihnen eine neue Tätigkeit geben werde*" (Cited by Kühn 1983, p. 180).

Conclusion to Book I of the *Treatise*, translated by Hamman and published in 1771 with the title *Nachtgedanken eines Zweiflers* ("Night Thoughts of a Skeptic"). This name certainly recalls the melancholy tone of the piece, where Hume laments the skeptical nature of his philosophical reflection, his irresolvable doubts, and their disturbing emotional consequences. Commentators who see in this text the origin of Kant's critical turn suggest that Hume's depiction of the conflict between imagination and reason conveys an antinomy of reason. Let us remember the passage in question:

> 'Tis this principle [of the mind or imagination], which makes us reason from causes and effects; and 'tis the same principle which convinces us of the continu'd existence of the external objects, when absent from the senses. But tho' these two operations be equally natural and necessary in the human mind, yet, in some circumstances, they are directly contrary, nor is it possible for us to reason justly and regularly from causes and effects, and at the same time believe the continu'd existence of matter. How then shall we adjust those principles together? Which of them shall we prefer? Or in case we prefer neither of them, but successively assent to both, as is usual among philosophers, with what confidence can we afterwards usurp that glorious title when we thus knowingly embrace a manifest contradiction? (*THN*. I, 266).

The passage complains about the tendency of reason to become involved in contradictions or antinomies. The human mind, says Hume, possesses a principle that leads us, on the one hand, to infer causal relations and, on the other, to believe in the existence of external objects. However, in some circumstances, these two inferences contradict each other, and the same principle leads us alternatively to belief and doubt. Thus, doubt and belief, negation and assent, come from the same natural operations of our mind. A little later in the same section, Hume offers what appears to be his last word on the matter, namely that human understanding has a fateful tendency to fall into contradictions when it submits itself to philosophical speculation: "For I have already shown that the understanding when it acts alone, and according to its most general principles, entirely subverts itself and leaves not the lowest degree of evidence in any proposition, either in philosophy or common life." (*THN*. I, 267–268). Hume's assessment in these passages coincides with the Kantian idea that when reason surpasses the limits of experience, it becomes entangled in insurmountable contradictions. Thus, it is possible that these fragments of Hume's *Treatise* significantly contributed to Kant's awakening.

However, we should note that nowhere in the Conclusion to the first book of the *Treatise* does Hume specifically refer to the cosmological issues. And, since Kant insists in his letter to Garve that it was the cosmological problems – "The world has a beginning; it has no beginning, and so on, right up to the fourth: There is freedom in man, vs. there is no freedom, only the necessity of nature" (*C. Ak*. XII, 257–258) – that made him adopt the critical standpoint, I find that the section of the *Treatise* translated by Hamman does not completely explain Kant's awakening from his dogmatic slumber.

I do want to note that Hume did address the issue of the infinite divisibility of space and time in the *Treatise*, in a section before the Conclusion—section two of the first book, entitled "Of the Ideas of Space and Time." However, I do not believe that this text can be related to Kant's awakening. Let me explain why. This section may have reached Kant's hands by way of Beattie's translated excerpts. There, Hume is mainly concerned with identifying the difficulties that the mathematical notion of infinite divisibility brings for the physical conception of space and time. Doubtlessly following Berkeley's lead on the matter, Hume offers proofs he considers sufficient to determine the falsity of the mathematical thesis of the infinite divisibility of space and time.

In general, Hume's argument is grounded on the empiricist premise that ideas, as copies of impressions, are suitable representations of objects, and, consequently, the characteristics attributed to ideas can be extended to their corresponding objects. From this premise, Hume argues, it clearly follows that "whatever appears to be impossible and contradictory upon the comparison of these ideas, must be really impossible and contradictory, without any farther excuse or evasion." (*THN*. I, 29). Still, we know that our mind's capacity is limited and, therefore, it cannot adequately grasp the idea of the infinite. When our mind seeks, for example, to divide the content of its ideas, it invariably arrives at an end in this division; that is, it "advances to a minimum"—an idea that allows no further subdivision (*THN*. I, 28). Then, Hume concludes that "no finite extension is capable of containing an infinite number of parts; and consequently [. . .] no finite extension is infinitely divisible" (*THN*. I, 30). Clearly, Hume alludes here to Berkeley's *minima sensibilia*.

A second argument, more specifically targeted at the infinite divisibility of space and time, runs as follows: an essential property of time is the succession of its parts; in other words, time is the successive order of moments. Hence, moments in time cannot co-exist or be simultaneous. As a result, time must be composed of indivisible moments since otherwise, there would be an infinite number of co-existing moments or parts of time. But if infinite co-existing moments or parts constituted the same moment, the nature of time as a successive order would be contradicted:

> 'Tis certain, then, that time, as it exists, must be compos'd of indivisible moments. For if in time we cou'd never arrive at an end of division and if each moment, as it succeeds another, were not perfectly single and indivisible, there would be an infinite number of co-existing moments or parts of time; which I believe will be allow'd to be an arrant contradiction (*THN*. I, 31).

This argument is akin to one of the Aristotelian objections to Zeno's paradoxes (*Physics*, 281^{a}-10-20) recorded by Bayle:

> Since it is thus impossible that Monday and Tuesday exist together, and since it must be the case necessarily that Monday cease before Tuesday begin to exist, there is no other part of time, whatever it may be, that can co-exist with another. Every one has to exist alone. Everyone must begin to exist, when the other ceases to do so. Every one must cease to exist, before the following one begins to be. From which it follows that time is not divisible to infinity, and the successive duration of things is composed of moments properly so-called, of which each is simple and indivisible, perfectly distinct from the past and the future, and contains only the present time (Bayle 1991, pp. 353–354).

The refutation of the infinite divisibility of time also applies to the infinite divisibility of space. Hume's argument in this respect is very schematic. He simply says: "The infinite divisibility of space implies that of time, as is evident from the nature of motion. If the latter, therefore, be impossible, the former must be equally so" (*THN* I., 31).

As can be appreciated from these fragments, while some of Hume's arguments are similar to Berkeley's, others seem to be taken from Bayle's account. However, the arguments he takes from Bayle correspond to the *reductio ad absurdum* of the rationalistic side, that is, they are the arguments that Berkeley also adopted for his purposes. Hume does not retrieve from Bayle the skeptical tone of his article. In other words, for the young Hume of the *Treatise*, the issue is not an irresolvable antinomy. He simply assumes the empiricist position against the infinite divisibility of these entities and sets out the relevant arguments to confirm this position against the opposite view. P. Smith claims:

> As in the case of the objections against the senses, Hume believes that he has a solution for Bayle's problem. Far from seeing it as a supposedly irresoluble conflict, Hume shows confidence in his own capacity to find a solution not yet envisaged by anyone. The doctrine of colored or tangible points, according to Hume, solves the difficulties highlighted by Bayle, and allows to deny the infinite divisibility of space. Hume calls this solution a "system about space and time" composed of two parts that mutually reinforce each other; which include decisive arguments in favor and the refutation of the objections against it. In this way, Hume believes he is capable of finding the solution for another difficulty posed by the skeptical method (Smith 2011, p. 77).

Indeed, Hume's treatment of this problem in the *Treatise* is far from skeptical. The clearest indication that he is proceeding in a dogmatic way is his claim that the issue of the infinite divisibility of space and time is entirely resolvable and that his refutation of the mathematical theses is conclusive:

> But here we may observe, that nothing can be more absurd than this custom of calling difficulty what pretends to be a demonstration, and endeavoring by that means to elude its force and evidence. 'Tis not in demonstrations as in probabilities, that difficulties can take place, and one argument counterbalance another, and diminish its authority. 'Tis a demonstration, if just, admits of no opposite difficulty; and if not just, 'tis a mere sophism, and consequently can never be a difficulty (*THN.* I, 32).

There are no signs of skepticism in these claims. By all appearances, Hume sees this problem not as an antinomy but only as a sophism. As he asserts, there are no gradations of probability in this matter, but only demonstrative arguments and definitive conclusions. In my view, the fact that Hume's treatment of this problem in the *Treatise* is much less skeptical than in the *Enquiry* (or not skeptical at all) should cast doubt on the interpretation that Hume's *Treatise,* via Bettie, caused Kant's awakening by turning his attention to the cosmological problems of the antinomies. As we will now see, in the *Enquiry,* we find an unambiguously skeptical version of the cosmological issues, including the infinite divisibility of space and time. I believe, then, that it was the skeptical approach to the antinomies in the *Enquiry* that decisively influenced Kant's conception of the Antinomies.

In section XII of the *Enquiry*, Hume reviews the diverse forms of skepticism and their main topics. In the context of this discussion, we find important allusions to the cosmological issues. First, Hume calls attention to the infinite divisibility of extension:

> No priestly dogmas [...] ever shocked common sense more than the doctrine of the infinite divisibility of extension, with its consequences; as they are pompously displayed by all geometricians and metaphysicians, with a kind of triumph and exaltation. A real quantity, infinitely less than any finite quantity, containing quantities infinitely less than itself, and so on, in infinitum, this is an edifice so bold and prodigious, that it is too weighty for any pretended demonstration to support, because it shocks the clearest and most natural principles of human reason. But what renders the matter more extraordinary, is that these seemingly absurd opinions are supported by a chain of reasoning, the clearest and most natural; nor is it possible for us to allow the premises without admitting the consequences (*EHU*, 156).

At the beginning of the passage, we have the impression that Hume continues to hold the view that the problem is easily resolvable, given his claim that these abstract notions "are a scandal to common sense," for contradicting "the clearest and most natural principles of human reason." However, a few lines after, he acknowledges that these arguments "are supported by a chain of reasoning, the clearest and most natural," so that we cannot deny their logical validity once we accept their premises. Unlike the Hume of the *Treatise,* who found it possible to settle the matter, this Hume appears to be less convinced that the arguments for the infinite divisibility of space and time can be decisively refuted. No matter how "absurd" the opinions of those who hold them may be, they are founded on logically valid inferences. Immediately after this, Hume argues that the same kind of opposition occurs as regards the infinite divisibility of time:

> The absurdity of these bold determinations of the abstract sciences seems to become, if possible, still more palpable with regard to time than extension. An infinite number of real parts of time, passing in succession, and exhausted one after another, appears so evident a contradiction, that no man, one should think, whose judgment is not corrupted, instead of being improved, by the sciences, would ever be able to admit of it. Yet still reason must remain restless, and unquiet, even with regard to that scepticism, to which she is driven by these seeming absurdities and contradictions. How any clear, distinct idea can contain circumstances contradictory to itself, or to any other clear, distinct idea, is absolutely incomprehensible, and is, perhaps, as absurd as any proposition, which can be formed. So that nothing can be more sceptical, or more full of doubt and hesitation, than this scepticism itself, which arises from some of the paradoxical conclusions of geometry or the science of quantity (*EHU* XII, 157).

In this section, Hume is perplexed by the contradictions the understanding falls into when examining the infinite divisibility of time. While, again, the idea of an infinite temporal succession goes against common sense, mathematicians and metaphysicians of all times have never ceased to try to demonstrate them with their abstract reasoning. Thus, although Hume considers it "absolutely incomprehensible" that a clear and distinct idea might bring about such insurmountable contradictions, he accepts that this is the case and that it causes much "restlessness and unquiet." Now, we are indeed facing a skeptically conceived antinomy, such as Bayle put forward.

One in which the contending thesis and antithesis are based on equally clear and distinct, but contradictory, ideas.

Although someone could object that Hume is referring here to the same antonymic tendency of reason he had already outlined in the Conclusion to the first book of the *Treatise*, I believe that the differences between both treatments are substantial. First of all, in the *Enquiry,* this tendency is discussed directly in relation to the cosmological issue of the infinite divisibility of time and space, while in the *Treatise,* it was treated with regard to the causal inference and the continued existence of external objects—problems that, for Kant, are solved through the doctrine of the categories, in the Second Analogy of Experience and the Refutation of Idealism, respectively. Second, here Hume describes the mind's unavoidable inclination to become entangled in metaphysical contradictions as having the logical (Baylean) form of an antinomy, which he had not done in the earlier account. These factors, in my view, contribute to a more solidly established Humean skepticism concerning traditional metaphysics in the *Enquiry*. With this in mind, it seems reasonable to think that, having read the *Enquiry* several times throughout his silent decade, Kant would have encountered Hume's approach to the antinomical problems in these passages and they would have triggered his awakening.

Now, there is yet another matter linked to Kant's letter to Garve that we should take into account. He specifically mentions the third antinomy, although he says it is the fourth—whether there is freedom or only natural necessity. Hume devoted the third part of the second book of the *Treatise* and the eighth section of the *Enquiry* to his treatment of this metaphysical problem. I ignore whether Beattie included among the translated extracts of Hume this section of the *Treatise,* but, since undoubtedly Kant had access to the problem in the *Enquiry*, I will refer to the latter in what follows.

6.6.1 Hume's Antinomy of Freedom and Cicero's *De Fato*

Just as Hume's *Dialogues concerning Natural Religion* are modeled on Cicero's *De Natura Deorum*, his discussion of the metaphysical problem of freedom seems to draw extensively from Cicero's *De Fato*. In this work, Cicero explores the debate among Stoics, Epicureans, and Academics about whether there is freedom or the universe is determined by natural causes. He defends Carneades's claim that, since some events are in our control, the human will is free, and there is no fate (*Fato*, XIV, 31). Given the very complex nature of the discussion, I will not describe it in detail but only point to the similarities in Cicero's and Hume's accounts of the issue.

Hume begins section VIII of the *Enquiry* by claiming that the traditional metaphysical opposition between freedom and necessity should be understood as a merely terminological dispute. For him, if metaphysicians had agreed from the start what they understand by the concept of freedom, they would have realized that freedom and natural causality have the same meaning, and the disputes on the matter would never have arisen:

> From this circumstance alone, that a controversy has been long kept on foot, and remains still undecided, we may presume that there is some ambiguity in the expression, and that the disputants affix different ideas to the terms employed in the controversy (*EHU*, XVIII, 80).

As Peter Jones has pointed out, this is a prevalent Ciceronian topic that can be found not only in *De Fato* but also in *De Finibus* and other works.[29] The line of *De Fato* he mentions appears in XIX. 44. Cicero discusses the dispute on freedom and fate between the Stoics, particularly Chrysippus, and the skeptics, notably Carneades. The Stoics argued that there are natural causes but not fate, that is, there is human freedom, while the Academics claimed that things in our control are by definition not necessitated by fate, even if they do involve natural causes. For Cicero, both parties affirm the same—that there is natural causality and human freedom.

Without doubt, the idea that the metaphysical disputes concerning freedom and necessity are due to terminological misunderstandings is far from Kant's stance on the issue. However, I believe that Hume's adoption of the Ciceronian topic may have contributed to the Kantian idea that there is an "illusion of reason" at the heart of the antinomies, which is possible to reveal through critical investigation. Hume's argumentative strategy aims at uncovering the cause of the metaphysical dispute. In the same way, Kant's view is that there is an underlying issue in the antinomies that explains why nobody has been able to demonstrate the truth of one of the positions and the falseness of the opposite one.

Hume also insists that the discussion on freedom and necessity is less a speculative or metaphysical problem than a practical one. In this, again, he is following Cicero, who claims that this field of inquiry belongs to ethics (*Fato*, I, 1). His Academic skepticism commits him to the view that the problem of freedom has a solution but only in the practical realm (*EHU*, XVIII, 81). And, as we will see in the next chapter, Kant also adopts this feature of Academic skepticism.

After setting out the problem, Hume puts forward his thesis. There has always been an underlying agreement regarding necessity and freedom since, ultimately, all human beings accept that psychological motives do cause their actions. In other words, the terminological issue disappears if we acknowledge that both "necessity" and "freedom" can be understood in terms of "causality" (*EHU*, XVIII, 81). Then, Hume goes on to describe the first side of the dispute. If we understand necessity as causality, we can correctly attribute necessity to both the works of matter and human action. Given that causal inferences are based on the observation of natural events in constant conjunction, and human actions and motives also exhibit such constant conjunction (hate is usually accompanied by violent actions, compassion by benevolent ones, etc.), we should attribute as much necessity to the latter as to the former:

[29] "Cicero shows a tendency, later displayed by Hume himself, to dismiss apparently substantive disagreements as 'merely verbal' disputes" (Jones 1982, p. 39).

> It is universally acknowledged that there is a great uniformity among the actions of men, in all nations and ages, and that human nature remains still the same, in its principles and operations. The same motives always produce the same actions: The same events follow from the same causes. [...] So readily and universally do we acknowledge a uniformity in human motives and actions as well as in the operations of body. (*EHU*, XVIII, 83–84)

Having examined the first position, Hume proceeds to consider the reasons on the other side of the debate—the side defending freedom. He claims that the cause of the doctrine of freedom is "a false sensation or seeming experience which we have, or may have, of liberty or indifference, in many of our actions" which makes us believe that the relationship between motives and actions is not as strong as that of natural causes and effects (*EHU*, XVIII, 94, note). In other words, given that we do not have a robust inner sensation of the bond between our motives and our actions, we falsely believe that we are free or that there is no causality whatsoever in our will.

After this, Hume turns to the practical consequences of the two doctrines. Despite their internal cogency, he says, both are often attacked for their harmful effects on morality and religion. Hume begins by examining the common accusation to the doctrine of necessity that there would be no moral responsibility if human actions were the result of natural causes. He answers that far from being detrimental to morality, this doctrine gives a foundation to our moral intuitions and legal practices, since both the education of children and the enforcement of laws are based on the belief that rewards and punishments are efficient motives for the will (*EHU*, XVIII, 97–98). On the contrary, Hume suggests, denying causality would be detrimental to moral responsibility because, if human actions and motives were not causally connected, we would not be able to claim guilt or innocence:

> According to the principle, therefore, which denies necessity, and consequently causes, a man is as pure and untainted, after having committed the most horrid crime, as at the first moment of his birth, nor is his character anywise concerned in his actions; since they are not derived from it, and the wickedness of the one can never be used as a proof of the depravity of the other (*EHU*, XVII, 98).

In conclusion, understanding necessity as it should – namely as causality or the constant conjunction of human motives and actions – safeguards rather than endangers morality.

In turn, the doctrine of freedom, Hume argues, if adequately understood as causality, also admits the connection between motives and human actions that makes moral judgment possible.[30] Thus, both doctrines elude the common criticism:

> It will be equally easy to prove, and from the same arguments, that liberty, according to that definition above mentioned, in which all men agree, is also essential to morality, and that no human actions, where it is wanting, are susceptible of any moral qualities, or can be the objects either of approbation or dislike (*EHU*, XVIII, 99).

Finally, Hume argues that the two doctrines have no negative consequences for religion. In this regard, the most significant criticism against the doctrine of necessity

[30] A similar argument appears in *Fato*, V. 11.

is that if natural causes wholly governed human wills, everything in the universe would be determined, and our actions would have been decided by God from the very beginning of time. But, then, either vicious human actions would not be evil – since they would be the effects of divine will and this will is good by definition – or, if they were evil, only God should be blamed for them. That is to say, if we accept necessity, we reject either morality or religion (*EHU*, XVIII, 100).

Hume, then, reminds us how philosophers in antiquity and modernity have attempted to solve this dilemma through all sorts of theodicies (*EHU*, XVIII, 101), which ultimately converge in the same argument: God is not guilty of the evils of the world because what appears to us as evil is, in fact, a means for a greater good, according to his general plan or providence. Nonetheless, he claims, this argument does not prevent us from attributing moral responsibility to human beings, because our moral judgments depend on natural principles:

> But though this topic be specious and sublime, it was soon found in practice weak and ineffectual. [...] The mind of man is so formed by nature that, upon the appearance of certain characters, dispositions, and actions, it immediately feels the sentiment of approbation or blame; nor are there any emotions more essential to its frame and constitution (*EHU*, XVIII, 101–102).

Since morality is founded on natural sentiments, no metaphysical consideration would stop us from feeling outraged by the evil we see in others or from having remorse for the harm we cause to others. Why, then, concern ourselves with such reflections? Moreover, attempting to reconcile moral evil with divine benevolence is beyond our capacity and, therefore, philosophy should limit itself to practical matters fundamental for human life:

> Happy, if she [philosophy] be thence sensible of her temerity, when she pries into these sublime mysteries; and leaving a scene so full of obscurities and perplexities, return, with suitable modesty, to her true and proper province, the examination of common life; where she will end difficulties enough to employ her enquiries, without launching into so boundless an ocean of doubt, uncertainty, and contradiction! (*EHU*, XVIII, 103).

At the end of the section, Hume invites philosophers to set aside metaphysical speculation and accept a theoretical and a practical claim—that human actions spring from natural causes and these are sufficient grounds to determine moral responsibility.

As we will see in more detail in the next chapter, Hume's Academic position heavily influenced Kant. First, like Hume, Kant defined freedom in the third antinomy in terms of causality (A444/B472-A445/B473). However, for Kant, freedom is not "natural causality" but "spontaneous causality"—a kind of cause that does not depend on a previous cause and so on *ad infinitum*, as natural causality does. Thus, the antinomy remains for Kant a real problem and not simply a question of terminology, easy to solve when one ascribes the same meaning to the two opposite terms. Kant, in fact, redefines the antinomy: the first side of the debate holds the possibility of spontaneous causes or freedom without denying natural causes, and the opposite side advances the claim that everything is sheer natural causality. In other words, Kant preserves Hume's definition of freedom as causality and takes out from the discussion the view that freedom is mere hazard. And he uses Hume's description of the doctrine necessity as sheer natural causality as the position on the other side of the antinomy.

Second, like Hume, Kant believes that besides a theoretical solution to the conflict, there should be a practical one (A532/B560-A534/B562). As we will see in the next chapter, Kant's theoretical solution to the third antinomy allows for the acceptance of the two kinds of causality, but he also emphasizes the illegitimacy of metaphysical speculation and the need to restrict the problem of freedom to the practical realm. As a result, Kant defines "practical freedom" as freedom of the will, which is the basis for moral judgments, while "transcendental freedom" remains a theoretically problematic concept. However, unlike Hume, Kant's "practical freedom" is a faculty of human reason, so that the motives that determine our actions are not natural, but rational:

> Freedom in the practical sense is the independence of the power of choice from necessitation by impulses of sensibility [...]. The human power of choice is indeed an *arbitrium sensitivum*, yet not *brutum*, but *liberum*, because sensibility does not render its action necessary, but in the human being there is a faculty of determining oneself from oneself, independently of necessitation by sensible impulses (A534/B562).

Despite the differences, I believe we can confidently say that Hume's cosmological discussions in the *Enquiry*, particularly his portrayal of this debate, must have been an essential stimulus for Kant's awakening. Hume makes here a superb exposition of what we understand as Kant's skeptical method: he sets the problem as an antinomy of reason, with theses and arguments on both sides of the question, examines the cause of the contradiction, and reaches a practical verdict to settle the question.

This last feature of Kant's skeptical method is explicit in Hume's treatment of the problem of liberty and necessity when he argues that philosophical speculation should be circumscribed to the practical sphere. As I have previously discussed, Hume wholeheartedly adopted this aspect of Academic skepticism. It is possible that the same practical answer appeared at the end of *De Fato*, just as it did at the end of *De Natura Deorum*. Unfortunately, given *De Fato*'s fragmentary extant form, we cannot know whether Cicero used the same argumentative strategy or not. Yet, since *De Fato* was planned as a sequel to *De Natura Deorum* (*Fato*, I. 1), we can suspect that Cicero proceeded accordingly. In any case, we can see in Hume's discussion of freedom and necessity an excellent example of Kant's skeptical method. Probably, then, this work contributed crucially to his awakening from his dogmatic slumber.

I began this chapter with Kant's characterization of Humean skepticism concerning causality as the sort of skepticism that should be avoided given that it involved an external censorship of reason. However, as Kant knew well, Hume's skepticism was not circumscribed to the problem of causality, and his treatment of traditional metaphysics, in particular, exhibited all the marks of Academic skepticism. For this reason, against mainstream interpretations that see Kant's engagement with Humean skepticism mainly through the problem of causality, I hold that Kant's awakening is a Humean uprising against traditional metaphysics, and that it was made possible by the antinomical problems. In the next chapter, I will go back to the

Academic features of Kant's treatment of the Antinomies of Pure Reason. Since one of the essential elements of Kant's skeptical method is the practical verdict given after the skeptical examination, I will focus on Kant's practical resolution of the Antinomies and interpret it as an Academic answer to the *apraxia* problem.

* * *

As a conclusion, I want to summarize the main points of this chapter. I have argued that Kant's skeptical method has both ancient and modern skeptical influences. While some commentators have argued that the ancient school that most influenced Kant is Pyrrhonism, I have contended that Kant neither seems to have been very acquainted with Sextus Empiricus's writings nor with the principal tenets of the school; hence, his sources for this kind of skepticism are either modern skeptics or anti-skeptics, such as the members of the Prussian Academy of Sciences. One crucial modern source is Bayle, who modified the traditional skeptical method of arguing on both sides of a question into the antinomical one of opposing two contradictory theses. I have argued that this way of proceeding can be observed in the modern disputes on Zeno of Elea's paradoxes regarding the infinite divisibility of space, time, and matter, both in Bayle's account and in Pascal's dogmatic solution to the question, among others. Based on Cajori's reconstruction of the discussion, I have given an overview of how modern rationalists, empiricists, and skeptics argued on both sides of the problem, giving rise to Kant's mathematical antinomies. I have also examined the influence of Cicero's *De Officiis* on Kant's *Groundwork* and his comments on both Cicero and Academic skepticism in his *Lectures on Logic*. Despite his negative apprisal of both, Kant's notion of the type of suspension of judgment that is critically useful is very close to the Academics'. Finally, I have explored the scholarly discussion about whether Hume or the Antinomies produced Kant's "awakening from [his] dogmatic slumber." I argued that the two possibilities are compatible, and Hume's treatment of the problem of freedom and necessity in the *Enquiry* – in turn influenced by Cicero's *De Fato* – most probably produced the decisive impulse for Kant's critical project. In this way, Hume's Academic treatment of metaphysical issues most probably provided the key elements of Kant's skeptical method.

Bibliography

Anderson, Abraham. 2010. The Objection of David Hume. In *Rethinking Kant*, ed. P. Muchnik, vol. 2, 81–121. Newcastle upon Tyne: Cambridge Scholars Publishing.

Aristotle. 1984. Physics. In *The Complete Works of Aristotle*, ed. Jonathan Barnes, Vol. I, 315–446. Princeton: Princeton University Press.

Bacon, Francis. 2001. *The Advancement of Learning*. New York: The Modern Library.

Bayle, Pierre. 1991. *Historical and Critical Dictionary. Selections*. Trans. Richard H. Popkin. Indianapolis/Cambridge: Hackett Publishing Company.

———. 2005. *Dictionnaire historique et critique*. Vol. XII. Paris: Elibron Classics.

Beck, Lewis White. 1978. *Essays on Kant and Hume*. New Haven: Yale University Press.

Berkeley, George. 2014. *Treatise Concerning the Principles of Human Knowledge*. Cambridge: Cambridge University Press.
Cajori, Florian. 1915a. The History of Zeno's Arguments on Motion: Phases in the Development of the Theory of Limits. *The American Mathematical Monthly* 22 (3): 77–82.
———. 1915b. The History of Zeno's Arguments on Motion: Phases in the Development of the Theory of Limits. *The American Mathematical Monthly* 22 (4): 109–115.
———. 1915c. The History of Zeno's Arguments on Motion: Phases in the Development of the Theory of Limits. *The American Mathematical Monthly* 22 (5): 143–149.
———. 1915d. The History of Zeno's Arguments on Motion: Phases in the Development of the Theory of Limits. *The American Mathematical Monthly* 22 (6): 179–186.
Cicero, Marcus Tullius. 1950. *Tusculanae Disputationes (Tusculan Disputations)*. Trans. J.E. King. Cambridge/London: Harvard University Press.
———. 1997. *De Officiis* (*On Duties*). Trans. Walter Miller. Cambridge/London: Harvard University Press.
———. 2000. *De Natura Deorum (On the Nature of the Gods)*. Trans. H. Rackham. Cambridge/London: Harvard University Press.
———. 2001a. *De Fato* (*On Fate*). Trans. H. Rackham. Cambridge/London: Harvard University.
———. 2001b. *De Finibus (On Moral Ends*). Trans. Julia Annas. Cambridge: Cambridge University Press.
———. 2006a. *Academica (On Academic Scepticism)*. Trans. Charles Brittain. Indianapolis/Cambridge: Hackett Publishing Company.
———. 2006b. *De Divinatione (On Divination)*. Trans. D. Wardle. Oxford/New York: Oxford University Press.
de Crousaz, Jean Pierre. 1733. *Examen du pyrrhonisme ancien et moderne*. La Haye: Hachette livre.
DesJardins, Gregory. 1967. Terms of *De Officiis* in Hume and Kant. *Journal of the History of Ideas* 28 (2): 237–242.
Ertl, Wolfgang. 2002. Hume's Antinomy and Kant's Critical Turn. *British Journal for the History of Philosophy* 10 (4): 617–640.
Falkenstein, Lorne. 1995. The Great Light of 1769: A Humean Awakening? Comments on Lothar Kreimendahl's account of Hume's influence on Kant. *Archiv für Geschichte der Philosophie* 77 (1): 63–79.
Formey, Henri Samuel. 2016. Mouvement. In *Encyclopédie ou Dictionnaire raisonné des sciences, des arts et des métiers, Vol. 10*, ed. Denis Diderot and Jean D'Alembert, 830–840. ARTFL Encyclopédie Project, University of Chicago. http://encyclopedie.uchicago.edu.
Forster, Michael N. 2008. *Kant and Skepticism*. Princeton: Princeton University Press.
Glei, Reinhold F. 1999. Lux Regiomontana. Der kategorische Imperativ in Ciceros *De Officiis*. *Bochumer Philosophisches Jahrbuch für Antike und Mittelalter* 4 (1): 49–61.
González, Catalina. 2011. Pyrrhonism vs. Academic Skepticism in Kant's *Critique of Pure Reason*. *Philosophy Today* 55: 225–230.
———. 2015. Kant, Cicero, and 'Popularity' in the Lectures on Logic. In *Kant's Lectures/Kant's Vorlesungen*, ed. B. Dörflinger, R. Louden, et al., 46–59. Berlin: De Gruyter.
———. 2020. The Prussian Academy's Struggle Between Dogmatism and Skepticism: An Antecedent to the Kantian Critique. In *Critique in German Philosophy*, ed. María del Rosario Acosta and Collin McQuillan. New York: SUNY Press.
Guyer, Paul. 2003. Kant on Common Sense and Scepticism. *Kantian Review* 7: 1–37.
———. 2008. *Knowledge, Reason, and Taste: Kant's Response to Hume*. Princeton: Princeton University Press.
———. 2009. Kant and Skepticism (Book Review). *The Philosophical Review* 118 (3): 384–389.
———. 2014. *Kant*. London: Routledge.
Hinske, Norbert. 1965. Kants Begriff der Antinomie und die Etappen seiner Ausarbeitung. *Kant-Studien* 56 (3–4): 485–496.
Hobbes, Thomas. 1839. *Elements of Philosophy (Vol. I)*, ed. W. Molesworth. London: John Bohn.
Hume, David. 1896. *A Tratise of Human Nature*, ed. L.A. Selby-Bigge. Oxford: Clarendon Press.

———. 1902. *Enquiries Concerning the Human Understanding and Concerning the Principles of Morals*, ed. L.A. Selby-Bigge. Oxford: Clarendon Press.

———. 2007. *Dialogues concerning Natural Religion and Other Writings*, ed. Dorothy Coleman. Cambridge: Cambridge University Press.

Janitsch, Julius. 1879. *Kants Urteile über Berkeley*. Strassburg: Kaiser-Wilhelm-Universität Strassburg.

Jáuregui, Claudia. 2008. *Sentido Interno y Subjetividad. Un análisis del problema del autoconocimiento en la filosofía trascendental de Kant*. Buenos Aires: Prometeo Libros.

Jones, Peter, 1982. Hume's Sentiments. Their Ciceronian and French Context. Edinburgh: Edinburgh University Press.

Kant, Immanuel. 1992. *Lectures on Logic* (*VL*). Trans. J. Michael Young. Cambridge: Cambridge University Press.

———. 1996. *Critique of Practical Reason (KpV)*. In *Practical Philosophy*. Trans. Mary J. Gregor, 133–271. Cambridge/New York: Cambridge University Press.

———. 1998. *Critique of Pure Reason (KrV)*. Trans. Paul Guyer and Allen Wood. Cambridge/New York: Cambridge University Press.

———. 1999. *Correspondence* (*C*). Trans. Arnulf Zweig. Cambridge/New York: Cambridge University Press.

———. 2000. *Critique of the Power of Judgment (KU)*. Trans. Paul Guyer. Cambridge: Cambridge University Press.

———. 2002a. Announcement of the Programme of Lectures for the Winter Semester of 1765-66. In Theoretical Philosophy 1755–1770. Trans. D. Walford and R. Meerbote, 290–300. Cambridge/New York: Cambridge University Press.

———. 2002b. Dreams of a Spirit-Seer Elucidated by Dreams of Metaphysics. In Theoretical Philosophy 1755–1770. Trans. D. Walford and R. Meerbote, 301–359. Cambridge/New York: Cambridge University Press.

———. 2002c. Inquiry Concerning the Distinctness of the Principles of Natural Theology and Morality. In Theoretical Philosophy 1755–1770. Trans. D. Walford and R. Meerbote, 243–275. Cambridge/New York: Cambridge University Press.

———. 2004. *Prolegomena to Any Future Metaphysics* (*P*). Trans. G. Hartfield. Cambridge/New York: Cambridge University Press.

———. 2005. *Notes and Fragments*. Trans. C. Bowman, P. Guyer, and F. Rauscher. Cambridge/New York: Cambridge University Press.

Kreimendahl, Lothar. 1990. *Kant-Der Durchbruch von 1769*. Köln: Jürgen Dinter.

Kühn, Manfred. 1983. Kant's Conception of 'Hume's Problem'. *Journal of the History of Philosophy* 21 (2): 175–193.

———. 2001a. *Kant: A Biography*. Cambridge: Cambridge University Press.

———. 2001b. Kant and Cicero. In *Kant und die Berliner Aufklärung*, ed. V. Gerhardt, R.-P. Horstmann, and R. Schumacher, 270–278. Berlin/Boston: De Gruyter.

Laursen, John C. 1996. Swiss Anti-skeptics in Berlin. In *Schweizer im Berlin des 18. Jahrhunderts*, ed. M. Fontius and H. Holzhey, 261–282. Berlin: Akademie Verlag.

———. 1997. Cicero in the Prussian Academy: Castillon's translation of the *Academica*. *History of European Ideas* 23 (2–4): 117–126.

Laursen, John C., Richard H. Popkin, and Peter Briscoe. 1997. Hume in the Prussian Academy: Jean-Bernard Mérian's 'On the Phenomenalism of David Hume'. *Hume Studies* 23 (1): 153–162.

Lazos, Efraín. 2014. *Disonancias de la Crítica*. Variaciones sobre cuatro temas kantianos. México: Universidad Nacional Autónoma de México.

Leibniz, Gottfried W. 1875. *Die philosophischen Schriften (Vol. 1)*, ed. C. I. Gerhardt. Berlin: Weidmannsche Buchhandlung.

Levey, Samuel. 1998. Leibniz on Mathematics and the Actually Infinite Division of Matter. *The Philosophical Review* 107 (1): 49–96.

Luiz, Eva. 2011. Bacon's Doctrine of the Idols and Skepticism. In *Pyrrhonism in Ancient, Modern, and Contemporary Philosophy*, ed. D. Machuca, 99–129. Dordrecht: Springer.
Makkreel, Rudolf A. 1998. Kant's Responses to Skepticism. In *The Skeptical Tradition Around 1800: Skepticism in Philosophy, Science, and Society*, ed. Jean Van der Zande and Richard H. Popkin, 101–109. Dordrecht: Kluwer Academic Publishers.
Manzo, Silvia. 2009. Probability, Certainty, and Facts in Francis Bacon's Natural Histories. A Double Attitude Towards Skepticism. In *Scepticism in the Modern Age: Building on the Work of Richard Popkin*, ed. J.R. Maia Neto, G. Paganini, and J.C. Laursen, 123–137. Leiden: Brill.
———. Reading Scepticism Historically. Scepticism, Acatalepsia and the Fall of Adam in Francis Bacon. In *Academic Scepticism in the Development of Early Modern Philosophy*, ed. S. Charles and P. Smith, 81–102. Dordrecht: Springer.
Moller, Sofie C. 2013. The Court of Reason in Kant's *Critique of Pure Reason*. *Kant Studien* 104 (3): 301–320.
Pascal, Blaise. 2014. *Of the Geometrical Spirit*. Trans. O. Williams Wight. California: Create Space Publishing Platform – The Perfect Library.
Popkin, Richard H. 2003. *The History of Scepticism*. New York: Oxford University Press.
Reich, Klaus. 1939. Kant and Greek Ethics (II.). *Mind* 48 (192): 446–463.
Robinson, Lewis. 1924. Contributions a l'histoire de l'évolution philosophique de Kant. *Revue de Métaphysique et de Morale* 31 (2): 269–353.
Robles, Jose Antonio. 1993. *Las ideas matemáticas de George Berkeley, obispo de Cloyne*. México: Universidad Nacional Autónoma de México.
Santozki, Ulrike. 2004. Die *Bedeutung antiker Theorien für die Genese und Systematik von Kants Philosophie: eine Analyse der drei Kritiken*. Philipps-Universität Marburg, Diss.
Sextus, Empiricus. 2000. *Pyrrhoniae Hypotyposes (Outlines of Scepticism)*, ed. Julia Annas and Jonathan Barnes. Cambridge: Cambridge University Press.
———. 2006a. *Adversus Mathematicos VII (Against the Logicians)*. Trans. Robert G. Bury. Cambridge: Harvard University Press.
———. 2006b. *Adversus Mathematicos IX (Against the Physicists; Against the Ethicists)*. Trans. Robert G. Bury. Cambridge: Harvard University Press.
Smith, Norman Kemp. 1950. *A Commentary to Kant's Critique of Pure Reason*. New York: Humanities Press.
Smith, Plínio J. 2008. La *Critique de la raison pure* face aux scepticismes cartésien, baylien et humien. *Dialogue* 47: 463–500.
———. 2011. Cómo Hume se volvió escéptico. *Revista Internacional de Filosofía* 52: 71–84.
———. 2013. Kant's Criticism and the Legacy of Modern Scepticism. In *Scepticism in the Eighteenth Century: Enlightenment, Lumières, Aufklärung*, ed. Sébastien Charles and Plínio J. Smith, 247–263. Dordrecht: Springer.
Stepanenko, Pedro. 2005. Kant y los diversos rostros del escéptico. *Ideas y Valores* 129: 35–46.
———. 2013. *Categorías y Autoconciencia en Kant*. México: Universidad Autónoma de México.
Tonelli, Gorgio. 1997. Kant and the Ancient Sceptics. In *Scepticism in the Enlightenment*, ed. Richard H. Popkin, Ezequiel de Olaso, and Giorgio Tonelli, 69–98. Dordrecht: Kluwer Academic Publishers.
Warda, Arthur. 1922. *Immanuel Kants Bücher*. Berlin: Martin Breslauer Verlag.
Wolff, Robert P. 1960. Kant's Debt to Hume Via Beattie. *Journal of the History of Ideas* 21 (1): 117–123.
Young, J. Michael. 2004. Translator's Introduction. In *Lectures on Logic*, ed. Immanuel Kant, xv–xxxii. Cambridge/New York: Cambridge University Press.

Chapter 7
Kant's Critical Solution to the Antinomies as an Academic Response to the *Apraxia* Objection

Abstract As we saw in previous chapters, the debates among ancient skeptics and dogmatists often ended in the same argument against skepticism—the famous *apraxia* objection. With some variations, the argument held that the skeptics either maintained some beliefs despite their alleged suspension of judgment or suspended judgment in all matters but were, in consequence, unable to act. Given Hume's and other modern philosopher's versions of this traditional argument, we might expect that Kant was aware of it. And, if this is the case, it is possible that the inclusion of a final practical verdict in his "skeptical method," aimed at answering the objection. With this in mind, in this chapter, I begin by briefly describing Kant's account of the cosmological problems, as they appear in in the Transcendental Dialectic of the Critique of Pure Reason (*KrV*). After this, I explain Kant's theoretical solution to these problems, according to which Transcendental Idealism is the correct metaphysical viewpoint to reveal the illusion at the basis of the opposition between theses and antitheses. I move, then, to detail why this solution is actually insufficient from a practical point of view—in other words, how no matter how cogent metaphysically, it does not withstand the *apraxia* objection. Finally, I analyze Kant's understanding of the regulative principle of the idea of the wold and the beliefs in freedom, the existence of God, and the immortality of the soul, to show how his notion of a belief that we adopt for practical purposes may be interpreted as a Academic response to the *apraxia* objection, that is, an adaptation of the Academic practice of "approving" persuasive impressions to orient the skeptic's behavior.

As we saw in previous chapters, the debates among ancient skeptics and dogmatists often ended in the same argument against skepticism—the famous *apraxia* objection. With some variations, the argument held that the skeptics either maintained some beliefs despite their alleged suspension of judgment or suspended judgment in all matters but were, in consequence, unable to act. Hume mentioned the objection multiple times in his works, using very similar terms to those employed by Aristotle and Plutarch. Let us recall, for example, what Cleanthes told Philo in the first section of the *Dialogues*:

> Whether your scepticism be as absolute and sincere as you pretend, we shall learn by and by, when the company breaks up: We shall then see, whether you go out at the door or the window; and whether you really doubt, if your body has gravity, or can be injured by its fall;

C. González Quintero, *Academic Skepticism in Hume and Kant*, Synthese Library 449, https://doi.org/10.1007/978-3-030-89750-5_7

> according to popular opinion, derived from our fallacious senses and more fallacious experience (*DNR* I, 5; 132).

Given Hume's and other modern philosopher's versions of this traditional argument, we might expect that Kant was aware of it. And, if this is the case, it is possible that the inclusion of a final practical verdict on the matter examined in his skeptical method aimed at answering the objection. Although this claim may be questioned on the grounds that Kant never directly speaks of the *apraxia* objection, I believe I can show how having in mind this discussion is helpful to interpret Kant's application of his "skeptical method" to metaphysical issues and his practical response to them. Since Kant maintained that practical reason held primacy over theoretical reason (*KpV*, *Ak.* V, 119–121), it is unsurprising that he concluded his critique of metaphysics by arguing that even if theoretical knowledge about certain aspects of reality is not possible, it is practically advisable to rationally and non-dogmatically adopt some metaphysical beliefs. In fact, Kant offers, besides a theoretical resolution of the antinomies, a practical one, in the Canon of Pure Reason of the first *Critique* and the Postulates of Practical Reason of the second *Critique*, where he discusses these beliefs. Notably, his resolution of the third antinomy of freedom and natural necessity in terms of the distinction between "transcendental" and "practical freedom" indicates that he was aware that an answer to the question of how to act should be provided while the metaphysical concept of freedom was under critical examination. Finally, Kant makes an important distinction among "knowledge," "belief," and "opinion," which, I think, encapsulates the Academic view that, even though assent to theoretical claims should be suspended, some beliefs may be approved to orient everyday life. Building on these arguments and others, in this chapter, I advance an interpretation of Kant's critical solution to the antinomies as a veiled or indirect response to the *apraxia* problem. Since, as far as I know, no commentator has made a similar claim, I will mainly focus on Kant's text and provide a relatively novel interpretation, unlike in previous chapters where I engaged more extensively with secondary literature.

To support this interpretation, I will begin by briefly describing Kant's account of the cosmological problems in the Transcendental Dialectic of the *Critique of Pure Reason* (*KrV*).[1] After this, I will explain his theoretical solution to these problems, according to which Transcendental Idealism is the correct metaphysical viewpoint to reveal the illusion at the basis of the opposition between theses and antitheses. Then, I will detail why this solution is actually insufficient from a practical point of view—

[1] I am not concerned here with assessing the logical validity of the proofs of the antinomies, as a large part of the secondary literature on the Dialectic does, especially in the Anglo-American analytical tradition. By 1940, Milton Fried was already claiming: "...The proofs of the antinomies are nothing more than a philosophical *tour de force*, exhibiting, at their best, a series of linguistic maneuvers." (Fried, 1940, p. 207). Following his lead, most commentators criticize Kant by arguing that the proofs of the antinomies lack validity, since they are constructed on ambiguous concepts or their conclusions do not follow from their premises. Russell, Fried, and Bennett are among the commentators who hold the former opinion; while Strawson and Smith sustain the latter. Allison (2004, pp. 82–108) offers a good synthesis of these accounts. Likewise, a clear response to Strawson's appraisal can be found in Baldacchino (1984). In my opinion, these critiques lack argumentative soundness, because they do not take into account the historical sources of the antinomical problems.

in other words, how no matter how cogent metaphysically, it does not withstand the *apraxia* objection. Finally, I will analyze Kant's understanding of the beliefs in freedom, the existence of God, and the immortality of the soul, to show how his notion of a belief that we adopt for practical purposes may be seen as an adaptation of the Academic practice of "approving" persuasive impressions to orient the skeptic's behavior. By the chapter's conclusion, I hope to have convincingly established that Kant's views on freedom and religion can be fruitfully interpreted as an Academic response to the *apraxia* objection.

7.1 Skepticism, Natural Dialectic, and Cosmology

As noted in the preceding chapter, Kant's critical adoption of the skeptical method aimed to internally interrogate reason's speculative impulses and reveal the illusion it falls into when dealing with metaphysical matters. Kant argued that when our reason strays from the solid ground of empirical cognition, it becomes lost in a sea of abstract thoughts and illusory inferences:

> But this land [of experience] is an island, and enclosed in unalterable boundaries by nature itself. It is the land of truth (a charming name), surrounded by a broad and stormy ocean, the true seat of illusion, where many a fog bank and rapidly melting iceberg pretend to be new lands and, ceaselessly deceiving with empty hopes the voyager looking around for new discoveries, entwine him in adventures from which he can never escape and yet also never bring to an end (A236/B295).

As the metaphor of the island suggests, metaphysics is for Kant a realm of illusory claims—mere mirages of reason with the appearance of logical cogency and objective validity. This tendency of reason to go beyond the limits of experience – the adventurous desire to know metaphysical entities – is natural and unavoidable, so that, no matter how we try to escape reason's speculative illusions, we always fall back into them, because they are grounded on necessary and purposefully founded principles (Guyer 2014, p. 145). Even if it can never completely dispel reason's natural tendency to metaphysics, the skeptical method serves the purpose of acknowledging and checking this inclination, showing its illegitimate grounds, and offering practical orientation.

Kant's view of how this occurs is quite compelling. Reason naturally seeks "the unconditioned for conditioned cognitions of the understanding, with which its unity will be completed" (B364). In other words, objects of experienced are conditioned by material, spatial-temporal, and causal relations, that is, *a priori* laws the understanding imposes on reality. However, following its architectonical impulses, reason seeks for the "unconditioned" that unifies these objects, namely, some sort of entity that is not dependent on these relations and represents the *whole* of possible experience. As Guyer portrays it:

> Kant's view is that reason inevitably leads us to form ideas of such "unconditioned" realities, that we cannot think of such things as if they were given by sensibility because everything given by sensibility is inherently "conditioned" [. . .], and so we inevitably think of anything "unconditioned" as if it were an object that lies beyond and transcends the limits of sensibility (Guyer 2014, p. 149).

This ascent of reason, from the conditioned in experience to the metaphysically unconditioned, occurs by way of prosyllogisms (A331/B388). Unlike ordinary or deductive syllogisms, in which, from the logical combination of a universal condition (the major premise) and a particular one (the minor premise), a logically valid a conclusion is inferred, in prosyllogisms, reason proceeds inversely: from the conclusion and the minor premise to the major premise or universal condition. In this type of inference reason moves, first, from a particular object of experience to the universal condition that subsumes it, next, from that principle to an even more encompassing ultimate principle, and so on, until it surpasses all limits of experience and enters the realm of metaphysical illusion (Guyer 2014, p. 150). For example, starting from determinate judgments of experience like "bodies are alterable" and "bodies are composites," reason ascends by prosyllogisms to the universal principle of experience that subsumes both, namely, that "everything composite is alterable." However, it does not stop here. Instead, it continues until it infers an unconditioned principle that encompasses the totality of composites, namely, the idea of the world (B387). This construct of reason expresses the "unconditioned" that corresponds to the totality of both the phenomenally conditioned (all objects of possible experience) and their phenomenal conditions (their general principles of relation) (A322/B379).

In its search for the unconditioned, reason produces three main ideas: (1) the idea of the *soul*, by which it conceives the "unconditioned unity of the thinking subject;" (2) the idea of the *world* or *cosmos*, by which it thinks the "absolute or unconditioned unity of the series of phenomena;" and (3) the idea of *God*, by which it conceives the total or unconditioned "unity of all objects of thought in general", that is, of everything conceivable by human reason (A334). In addition, reason produces the idea of freedom, which is a sort of sub-product of the idea of the world, taken as nature, and refers to the unconditioned spontaneity of a cause (A446/B474). Kant calls this ascending tendency of reason a *natural dialectic*:

> Hence there is a natural and unavoidable dialectic of pure reason, not one in which a bungler might be entangled through lack of acquaintance or one that some sophist has artfully invented in order to confuse rational people, but one that irremediably attaches to human reason, so that even after we have exposed the mirage, it will still not cease to lead our reason on with false hopes, continually propelling it into momentary aberrations that always need to be removed (A298/B354-55).

Kant's term "dialectic" may have been retrieved from Bacon's use, for whom it refers to a deceitful argument that, despite having the appearance of a logically valid inference, leads to error and confusion. However, for Kant, since this dialectic is a natural tendency, reason's illusory constructions are not mere sophistry. The ideas of the world, freedom, the immortal soul, and God, and the dialectical arguments supporting them, are unavoidable and grounded on solid principles of reason. This is why, despite our critical efforts to determine the point at which reason illegitimately surpasses the limits of experience and begins to be lost in metaphysical subtleties, we will always contrive these ideas and become entangled in their conflicting arguments.

The spontaneous or natural character of this dialectic is, in my view, an interesting Kantian recognition that, regardless of our efforts to undermine metaphysics critically or skeptically, our mind will always persevere in its attempt to make such claims. In fact, these metaphysical arguments are not only formulated philosophically but – perhaps more prominently – are expressed in popular sayings, common riddles, and children's charades, such as the traditional English song, "*The green grass grows all around*," or the Spanish recitation, "*La lora está en la casa, la casa está en la esquina, la esquina está en el barrio…etc.*"[2] These examples express our reason's movement from very particular objects of experience to more encompassing ones, and finally to the metaphysical idea of the "world." Kant's aim is to articulate this natural tendency of reason philosophically, in response to a long tradition of metaphysical disputes involving them, and to understand how these audacious arguments are the result of an illusion of our human faculties.

Unlike the other two rational ideas – *God* and the *soul* – the particular illusion that occurs when reason conceives the unconditioned through the idea of the *world* has the form of an antinomy. This means that, whenever reason tries to determine the totality of phenomena, it becomes entangled in a conflict between a thesis and an antithesis of equal logical validity.[3] Kant calls this conflict the Antinomies of Pure Reason (A418/B446):

> Now one can think of this unconditioned either as subsisting merely in the whole series, in which thus every member without exception is conditioned, and only their whole is absolutely unconditioned, or else the absolutely unconditioned is only a part of the series, to which the remaining members of the series are subordinated, but that itself stands under no other condition. In the first case, the series is given *a parte priori* without bounds (without a beginning), i.e., it is given as infinite and at the same time whole […]. In the second case, there is a first [member] in the series, which in regard to the past time is called the beginning of the world […] (A417/ B444-A418/B446).

Let us pause here for a moment. In the case of the idea of the world or cosmos, the illusion of reason involves a sort of "dialectical reasoning," both in the logical, Baconian sense of a fallacious inference, and in the ancient sense of an argument affirming two contradictory claims. Kant considered that when reason thinks the unconditioned whole of phenomena – the idea of the world – it engages in an internal conflict, in which arguments for and against a particular configuration of the world stand in opposition. As he affirmed in the first Prologue to the *Critique*, this conflict is the very essence of metaphysics: "The battlefield of these endless controversies is metaphysics" (Aviii). It is important to see here how Kant incorporated the idea of a dialectical conflict of reason in his assessment of metaphysics. As I argued before, this depiction already appeared in Hume, who saw in reason's struggle with itself the source of the skeptic's melancholy and ostracism, and thought it could only be overcome by adopting the mitigated skepticism of the Academy. Now Kant appeals to a similar conflict of reason to illustrate the illegitimate character of metaphysics. And, like Hume, Kant too will provide a solution based on Academic skepticism, as we will see later.

---

[2] *Lit.* 'The parrot is in the house, the house is on the corner, the corner is in the neighborhood…'

[3] This is not to say that the metaphysical arguments that support the idea of the *soul* (the Paralogisms) and of *God* (the proofs of God's existence, in the Ideal of Pure Reason) are not fallacious or illusory, only that they have a different logical structure.

This dialectical conflict arises when we think of the world as an unconditioned whole, which can be conceived as either a finite or an infinite set of phenomena. The two configurations give rise to the thesis and antithesis of Kant's four antinomies. The first antinomy discusses whether the world as a whole is finite or infinite, both in its temporal and spatial regards; the second, whether matter is finitely or infinitely divisible; the third, whether there is only an infinite chain of causes and effects or there are also spontaneous causes (namely, freedom); and, the fourth, whether or not the world has a necessary being as a first cause.[4] The first two problems constitute what Kant calls the *mathematical antinomies* because the cosmos is understood there in relation to its magnitude, that is, as a mathematical whole. The latter two are entitled *dynamical antinomies* since, in this case, the world is considered with regard to its causality, that is, as a dynamical whole or *nature* (B446-B448).

Since every antinomy has a thesis and an antithesis, the side of the theses represents the view that the cosmos is a finite set, while the side of the antitheses, that it is an infinite one. In this way, the theses hold, first, that there is a temporal and spatial limit to the world's extension; second, that there is a limit to the divisibility of its matter (namely, a simple part); third, that there is a limit on the causes that produce its phenomena or, put differently, that there are spontaneous or initial causes within it (i.e., freedom); and fourth, that there is a limit to its contingency, namely a necessary being, which is its first cause. By contrast, the antitheses hold, first, that the world's extension in space and time has no limit but is infinite; second, that there is no limit to the divisibility of its material entities (there are no simple parts but a division *ad infinitum*); third, that the chain of natural causes and effects in the world is not limited (that is, there is no freedom, only natural causes *ad infinitum*); and fourth, that there is no necessary being as its cause, that is, the world is absolutely contingent and infinitely so (A418/B446). All of the above can be summarized in the following way:

Thesis	Antithesis
First Antinomy	
The world has a beginning in time and a limit in space.	The world has neither a beginning in time nor a limit in space.
Second Antinomy	
Everything in the world is a composite of simple parts.	Nothing in the world is a composite of simple parts.
Third Antinomy	
Causality in agreement with the laws of nature is not the only one; there is also *freedom.*	Causality in agreement with the laws of nature is the only one (there is *no freedom).*
Fourth Antinomy	
An absolutely necessary being exists, which is the cause of the world (as a part of it or outside it).	An absolutely necessary being, which is the cause of the world, does not exist (neither as a part of it or outside it).

[4]The four pure concepts, of sensibility and understanding, to which the antinomic problems correspond, are: space, time, causality, and necessity or contingency (A412-A415).

Each side of the antinomy can be logically proven through a *reductio ad absurdum* of the other; that is, the theses are demonstrated by showing that the antitheses are either impossible or logically contradictory and *vice versa*. For example, the thesis of the first antinomy, which states, in part, that the world has a beginning in time, can be proven by demonstrating that the antithesis "the world has no beginning in time," involves an ontologically impossible situation, namely that an infinite series of events, which are a condition of the present moment, has in fact occurred. But, since the series is infinite, it cannot have been completed at a determinate time, so the antithesis is absurd (A426/B454). Equally, the corresponding antithesis regarding the world's infinite time can be proven by demonstrating the thesis' internal contradiction, namely, that the statement, "the world has a beginning in time" presumes a time when the cosmos did not exist, a time when there were no ontological conditions, or, as it were, an "empty" time. But, such a time could not give rise to a new series of events (A427/B455), or, simply put, it is absurd to think that something arises out of nothing. Consequently, the opposite statement, i.e., that the world does not have a beginning in time, is true.

In the second antinomy, the thesis and antithesis are concerned with whether matter is finitely or infinitely divisible. The thesis holds that simple parts exist, while the antithesis maintains the infinite divisibility of material entities. Once again, the thesis and antithesis of the second antinomy are not "positively" demonstrated but proven indirectly by showing how the opposite thesis necessarily leads to a contradiction.

The proof of the third antinomy's thesis argues that if we accept the antithesis, namely that natural causes wholly determine the world, then, each cause also has its own cause, and so on *ad infinitum*. But, if this is the case, no particular event can occur because each event depends on its own cause, and so on, without end. In this way, the antithesis is reduced to the absurd, and the truth of the thesis, namely that there are spontaneous causes in the world, or, in Kant's terms, that there is "transcendental freedom," is logically proven (A444/B472). The proof of the antithesis proceeds in the same way that the proof of the first mathematical antinomy. If there is transcendental freedom in the world – that is, if there are spontaneous causes – then there was a time when no previous ontological determinations existed, or, again, a kind of "empty" time. But an empty time is contrary to causality since there cannot be something that arises from nothing. Hence, the thesis of the third antinomy is also reduced to the absurd (A445/B473).

Finally, in the fourth antinomy, causality is considered in connection with the concepts of necessity and contingency. In this way, nature is thought of as originating in a necessary being or as absolute contingency or change. This antinomy is closely related to the so-called "cosmological proof of the existence of God" (A603/B631) that Kant surveys in the Ideal of Pure Reason (A603/B631-A620/B648). The thesis holds that a necessary being, which is the cause of the world, exists and is either a part of the cosmos or an external entity, while the antithesis states that no such being exists, neither as a part of the world nor as an external cause (A452/B480-A460/B488).

As we saw in Chap. 6, the origin of this method of antinomy and its particular application to the cosmological problems can be traced back to Bayle's elaboration of the ancient practice of discussing on both sides of the question (Smith 2008, p. 469). As P. Smith claimed:

> [. . .] it seems that Kant had taken as a model the skepticism of Bayle, rather than that of the ancient skeptics. The reason for this is that Bayle introduced some modifications to ancient skepticism that Kant also wanted to adopt. [. . .] The structure and form of his antinomical problems lead me to think that Bayle was an important source for Kant (Smith 2008, p. 469).

I believe that this procedure was also linked to how seventeenth-century mathematicians and rationalists, particularly Pascal, used the *reductio ad absurdum* to demonstrate the infinite divisibility of space, time, and matter. Interestingly, they were unaware that the empiricists that opposed them could use the same proof to demonstrate that space, time, and matter are not infinitely divisible. In my view, Kant's genius lies in understanding the fallaciousness of the proofs and showing how the metaphysical illusion resides in the impossibility of obtaining empirical verification on either side of the question (Guyer 2014, pp. 159–160). In other words, Kant recognized that the same type of logical proof could be used by both sides of the dispute for the same purposes and with identical results, precisely because they lack the only positive demonstration that could end the dispute—an empirical cognition of the world.

After describing the four antinomies, Kant argued that two attitudes naturally ensue from this dialectical reasoning—"skeptical hopelessness" and "dogmatic stubbornness." The first attitude arises when one realizes that it is not possible to rationally choose either of the opposing statements and is forced to suspend judgment on the matter. By contrast, "dogmatic stubbornness" involves deciding to hold one of the two sides of the antinomy on mere dogmatic grounds, that is, "without giving a fair hearing to the grounds for the opposite" (B 434). Both attitudes are harmful for reason, although the former is somewhat more comfortable and, therefore, Kant calls it the "euthanasia of pure reason" (B434). This implies that by suspending judgment on these matters, reason meets a "good death"—a death accompanied by the ancient Pyrrhonians' desired *ataraxia*. Evidently, there is a great deal of sarcasm here. This time, however, Kant seems to have acknowledge the end of Pyrrhonian *epochê*.

Kant did not recommend either of these attitudes. He argued that, without his critical approach, we would be forced to continue to dispute metaphysical issues endlessly. Thus, instead of suspending judgment or arbitrarily adhering to our favorite metaphysical position, Kant thought we should solve reason's natural dialectic by critically showing the origin of the problem, that is, by applying the skeptical method to reveal the illegitimate grounds at the basis of the antinomies. I shall now discuss Kant's theoretical solution to the antinomies and its practical implications.

7.2 Kant's Theoretical Solution to the Antinomies

Since the antinomies' theses or antitheses cannot be verified through experience, it is impossible to solve theoretically the cosmological problems detailed in the previous section. In Kantian terms, without intuitions that guarantee the correspondence of a concept to an object of experience, we cannot cognitively determine whether the world is finite or infinite in any of its quantitative or qualitative aspects (Guyer 2006, p. 159). How, then, can these problems be resolved from the theoretical point of view?

To help us with this problem, we may want to go back to Kant's distinction between the "skeptical method" and "skepticism." Surely, we cannot simply adopt skepticism about these matters because that would amount to suspending judgment and forsaking any possible theoretical resolution of the antinomies. Doing so would entail accepting the "euthanasia of reason" and abandoning the effort to discover the causes of the rational illusion that leads us to these metaphysical problems. But, if we adopted another means to resolve the issue, namely, if we used the "skeptical method," we would fare better, because this method exposes the rational illusion at the basis of antinomical reasoning:

> This method of watching or even occasioning a contest between assertions, not in order to decide it to the advantage of one party or the other, but to investigate whether the object of the dispute is not perhaps a mere mirage at which each would snap in vain without being able to gain anything, even if he met with no resistance – this procedure, I say, can be called the **skeptical method**. [. . .] This skeptical method, however, is essentially suited only to transcendental philosophy (A 423/B451-A424/B452).

In contrast to skepticism, the skeptical method explains the grounds of the cosmological illusion (A420/B448). It does so, first, by examining the foundational inference that guides it, namely, "If the conditioned is given [*gegeben ist*], then the whole series of conditions for it is also given; now objects of the senses are given [*gegeben sind*] as conditioned; consequently, etc." (A497/B525). For Kant, although this appears to be a logically valid inference, it involves an epistemological fallacy. The error lies in overlooking that only the conditioned, the intuitive content of any possible experience, can "be given." The entire series of conditions of a particular experience – the world as the totality of phenomena – cannot be "given" because it is not the object of any possible experience. Therefore, the inference must be corrected by stating, instead, "if the conditioned is given [*gegeben ist*], then, through it, a regress in the series of all conditions for it is given to us as a problem [*aufgegeben sei*]" (A498). While it is difficult to translate the German play of words, the meaning of the inference is relatively straightforward. Kant affirms that whenever an object of experience is given to us (in other words, whenever we perceive a thing), a conceptual return to its conditions is rationally *presented* to us, such that, our reason seeks to represent the complete series of conditions that made it possible. The grounds for this rational process is that "the concept of the conditioned already entails that something is related to a condition, and if this condition is once again conditioned, to a more remote condition, and so through all the members of the series" (A498/B526). In other words, when we perceive an object empirically, our reason begins a process of inferring the various conditions that produced it. For example, when I perceive a car, I may think about it being made in a manufacturing plant. This can lead me to think, in turn, about the conditions needed to produce a manufacturing plant, namely materials, human endeavor, and so on, until, by following the chain of causes, I arrive at the idea of the entire universe. However, we cannot *perceive* a given object's complete series of conditions, because this whole chain of causes cannot be the content of any possible experience. And, the fact that we cannot perceive the complete series of conditions of a given object should be enough, says Kant, to show the disputants on both sides of the antinomies that neither claim is actually justifiable and, hence, the objects of their debates are mere illusions of reason:

> When such a fallacy has been shown to ground the common argument (for the cosmological assertions), the demands of both disputing parties could rightfully be dismissed as being based on no well-grounded title. [...] Thus no means is left for ending the dispute in a well-grounded way and to the satisfaction of both sides, unless through the fact that they can do such a fine job of refuting each other they are finally won over to the view that they are disputing about nothing, and that a certain transcendental illusion has portrayed a reality to them where none is present (A501/B529).

Still, given that there are two kinds of antinomies (mathematical and dynamical), one more step needs to be provided to solve them theoretically. Kant proceeds then to examine the two types of antinomies separately. In the case of the mathematical antinomies, he argues that there is actually no transcendental opposition between thesis and antithesis since it could be the case that both assertions were false. In fact, given that we do not have an intuition corresponding to the world, we can assert neither that the world is finite or has a finite magnitude, nor that it is infinite or has an infinite magnitude (A501/B529-A502/B530).

As scholars of Kant have remarked, the core of the solution to the mathematical antinomies is that it indirectly proves transcendental idealism (Allison 2004, pp. 96–113; Guyer 2006, pp. 164–165; Smith 1950, pp. 503–504). The argument goes as follows: the position described in the premise of the erroneous logical inference "if the conditioned is given, the series of conditions is also given" is *metaphysical realism*—namely the belief that phenomena and noumena are the same or that the world is such as we perceive it. However, since metaphysical realism makes no distinction between phenomena and noumena, it fails to distinguish between the aspects of reality that can be given to experience and cognized, from those that cannot and thus are unknowable. In other words, metaphysical realism ontologically conflates particular objects of experience with the totality of these objects—the idea of the world. This confusion is the cause of the antinomies and of reason becoming entangled in metaphysical disputes. *Transcendental idealism*, on the contrary, resolves the problem by showing how both theses and antitheses of the mathematical antinomies rest on a fallacious ground:

> If one regards the two propositions, "The world is infinite in magnitude," "The world is finite in magnitude," as contradictory opposites, then one assumes that the world (the whole series

> of appearances) is a thing in itself. For the world remains, even though I may rule out the infinite or finite regress in the series of its appearances. But if I take away this presupposition, or rather this transcendental illusion, and deny that it is a thing in itself, then the contradictory conflict of the two assertions is transformed into a merely dialectical conflict, and because the world does not exist in itself (independently of the regressive series of my representations), it exists, neither as an in itself infinite whole nor as an in itself finite whole (A505/B33).

Again, the point is that, if we consider that phenomena and noumena are the same (as metaphysical realism does), then, there is no way to distinguish between what is given in experience and what is not. This is what Kant conveyed with the assertion that, in this case, "one assumes that the world (the whole series of appearances) is a thing in itself" (A505/B33). But once we accept *transcendental idealism*, the illusion fades away, because maintaining the distinction between phenomena and noumena prevents us from attributing magnitude to the world-in-itself. And, since finity and infinity are only determinations of a phenomenenical magnitude, once we take the point of view of transcendental idealism, we see how it is not necessary to determine whether the world-in-itself is finite or infinite.

According to H. Allison, Kant's argumentative strategy can be understood as a *modus tollens*. If the world is a thing in itself (the point of view of metaphysical realism), it is an entity in space and time, and if it is an entity in space and time, it is finite or infinite. But, given that the proofs of the mathematical antinomies are mutually contradictory, the assertion "the world is finite or infinite" is false. Hence, it is also false to say the world exists in itself:

> The logical form of this argument is a *modus tollens* combined with an immediate inference. The denial of the consequent ("the world is either finite or infinite") is used, in turn, to deny the antecedent ("the world is a whole existing in itself"), which is then taken to entail the conclusion: "appearances in general are nothing outside our representations." This latter statement is the thesis of transcendental idealism (Allison 2004, p. 388).

Allison's explanation is interesting, but loses sight of the fact the argument has a dialectical or skeptical form. Immediately after his solution to the mathematical antinomies, Kant discussed Zeno's strategy of opposing two judgments to demonstrate the invalidity of the *presupposition* they share (A502/B530). Allison himself acknowledges this point (Allison 2004, pp. 386–387), still, he fails to see that, strictly speaking, we are dealing here with dialectical, not logical, reasoning: "If two mutually-opposed judgments presuppose an inadmissible condition, then despite their conflict (which is, however, not a real contradiction) both of them collapse, because the condition collapses under which alone either of them would be valid" (A503/B531). Thus, Kant's argument is not a *modus tollens*, but rather the result of applying the guidelines of the skeptical method—that whenever two opposing arguments have equal weight, their grounds should be considered invalid, and judgment should be suspended.

However, Kant has not yet finished giving us his theoretical solution to these antinomies. To explain Zeno's dialectical principle further, Kant used the Aristotelean distinction between contrary and contradictory terms. He gave the example of the assertion "every body either smells good or does not smell good," which necessarily excludes a third option, namely, that some bodies do not smell at all, since the predicates "smells good" (*riecht gut*) and "does not smell good" (*riecht nicht gut*) are contrary but not contradictory. But, the judgment "every body is either good-smelling (*wohlriechend*) or not good-smelling (*nicht wohlriechend*)," does not omit any logical possibility because the term "not good-smelling" includes the options of smelling bad and having no smell at all (*gar nicht riechen*), and, therefore, these claims are, in fact, contradictory (A503/B531). Similarly, Kant argued, if we speak of the world as "finite or infinite," we overlook a third possibility, namely, that it has no magnitude at all. However, if we say that the cosmos is either "infinite or not infinite," then the possibility that it has no magnitude is included in the latter term. Thus, for Kant, the mathematical antinomies express a "dialectical opposition" rather than an "analytical opposition"—that is, an opposition between contrary but not between contradictory claims (A504/B532).[5]

This distinction becomes crucial for Kant to solve the mathematical antinomies theoretically. If adopting transcendental idealism allows us to say that we can attribute no magnitude to the world as a thing-in-itself, because the concept of magnitude can only be predicated on phenomena, then both the thesis ("the world is finite") and the antithesis ("the world is infinite") may be false. This move reveals how Zeno's dialectical framework, via Bayle's method of antinomy, informed Kant's skeptical method of opposing thesis and antithesis to examine their corresponding foundations. In conclusion, Kant solves the problem theoretically by revealing the dialectical structure of the antinomies.

Now, Kant's solution to the dynamical antinomies is very different from this. Concerning them, Kant argued that both thesis and antithesis could be *true*. The logical inference made in these antinomies differs considerably from the one made in the mathematical ones, because, in the latter case, said Kant, "the concept of understanding" grounding these ideas contains "[...] a synthesis of homogeneous things, which is presupposed in the case of every magnitude [...]." (A530/B558). In contrast, in the dynamical antinomies, reason operates "a synthesis of things not homogeneous" (A530/B558). In other words, since in the mathematical antinomies the grounding concept is that of *magnitude*, and this concept involves the same phenomenal synthesis for both claims – that the cosmos is finite and that it is infinite – both theses and antitheses can be shown to be false if we simply reject that the world, as a thing-in-itself, has any magnitude at all—if we reject the underlying presupposition. Nevertheless, in the case of the dynamical antinomies, the grounding concepts are causality and freedom (for the third antinomy) and necessity and contingency (for the fourth antinomy), that is, two different metaphysical concepts, involving a heterogeneous synthesis. Another way to put it is by saying that the sort of metaphysical realism underlying the dynamical antinomies

---

[5]I want to thank the late Professor Antonio Aguirre for explaining this to me long time ago when I was still a graduate student.

holds, on the side of the theses, that there are two compatible, though metaphysically distinct, causalities (natural causality and freedom) and entities (a necessary being and a contingent being), while the antitheses claim that there is only one kind of metaphysical causality (natural causality) and entity (a contingent being). (Beade 2010, p. 203). In this case, transcendental idealism solves the antinomy by showing that both theses may be true if we think of the world both as a phenomenon (an object of experience, ruled by natural causality and contingence) and a noumenon (an object of thought, governed by freedom and necessity). Given this, the dynamical antinomies involve both "phenomenal" causality and contingency, and "intelligible" spontaneity and necessity. Kant claims:

> [. . .] the dynamic series of sensible conditions, on the contrary, allows a further condition different in kind, one that is not part of the series but, as merely intelligible, lies outside the series; in this way, reason can be given satisfaction, and the unconditioned can be posited prior to appearances without confounding the series of appearances, which is always conditioned, and without any violation of principles of the understanding (A530-31/B558-59).

However, from the point of view of transcendental idealism, we cannot affirm that there is freedom and a necessary being as the cause of the world. We can only think of the world as possibly having them, not as positively doing so. Moreover, we cannot further characterize these entities, that is, determine them in judgment and assert them dogmatically.

Besides this, since the third antinomy is mainly concerned with "causality," which had already been shown in the Transcendental Analytic to be a category of experience (A189/B233-A211/B256), we must accept that phenomena are bound by natural causes according to the laws of the understanding (Carnois 1987, pp. 8–23). For this reason, the thesis and antithesis of this antinomy are worded only to assert or deny that there may *also* exist non-phenomenal causality—namely, freedom (A543/B571). Given that transcendental idealism is only committed to the existence of a transcendental, yet unknowable, realm correlative to phenomena (Allais 2015, pp. 3–36; Langton 1998, pp. 15–47),[6] the dynamical antinomies are complementary because they affirm the possibility of both phenomenal and noumenal causes—in the third antinomy, natural causality and freedom, and, in the fourth, a merely contingent world and a necessary being as its ultimate cause. Adopting transcendental idealism produces no contradiction and opens the possibility to conceive theses and antitheses as complementary (A532/B560). The solution to the dynamical antinomies offers, then, what Baldacchino calls the "positive version" of transcendental idealism—a perspective in which phenomena may be not only subjected to natural causes in the field of experience but also to non-natural causes, even if unknowable, in the noumenal realm (Baldacchino 1984, pp. 91–97).

---

[6] I agree with both commentators on their metaphysical reading of transcendental idealism as holding the existence of things in themselves, particularly with Allais's moderate metaphysical stance.

In the "Commentary on the Antithesis of the Fourth Antinomy," Kant illustrated this compatibility with the example of two astronomers who, given their particular geographical locations, make apparently opposed, but, in reality, complementary inferences:

> M. de Mairan took the controversy between two famous astronomers arising from a similar difficulty in the choice of standpoint to be a sufficiently strange phenomenon that he wrote a special treatise about it. One inferred, namely, that the moon turns on its axis because it constantly turns the same side toward the earth; the other, that the moon does not turn on an axis, just because it constantly turns the same side towards the earth. Both inferences were correct, depending on the standpoint taken when observing the moon's motion (A461/B489).

To summarize, the advantage of transcendental idealism is that it allows the critical philosopher to embrace both perspectives concerning the pair causality-freedom and necessity-contingency. Like the observations of the astronomers, both views may be correct because they express different positions about a single object—the world as a phenomenon and as a thing-in-itself. Refering to the solution of the third antinomy, L. Parra has argued that Kant's critical project can be understood as "the introduction of a perspectivism in anthropological matters." He explains:

> Thus, even if it is unavoidable to evaluate our actions from a natural causal perspective, it is also possible – even necessary – to do it from another point of view, namely that of freedom. [...] In fact, Kant asserts that, when the difference of perspectives from which a human action may be examined is established, the appearance of contradiction vanishes (Parra 2017, pp. 39–41; my translation).

Indeed, we can take the theoretical resolution of the dynamical antinomies as the adoption of this perspectivism. Nevertheless, transcendental idealism's dual point of view brings about some other difficulties, especially in the practical realm. First, we have to keep in mind that, as Kant claimed, this dialectic is a natural tendency of reason. Thus, even if we adopt transcendental idealism and recognize the grounds of the metaphysical illusion, the antinomies will rise again whenever reason follows its internal drive to think about the unconditioned totality of phenomena. Second, this theoretical resolution of the antinomies, particularly of the dynamical ones, does not seem to give a practical answer to the parties involved in the dispute, but simply tells them that we can accept that there may be both natural causality and freedom, and both a necessary first cause of the universe and absolute contingency.[7] But, practically speaking, it appears as though Kant were advocating for the suspension of judgment on the matter. Otherwise put, given that the theoretical solution to the antinomies, particularly of the dynamical ones, does not provide actual orientation on practical issues as urgent as to decide whether there is freedom

[7] There is one more issue, debated by commentators, about the third antinomy, that is, whether Kant is discussing a cosmological or a practical concept of freedom. While I agree that the antinomy, as Beade argues, involves both the cosmological and the practical concept of freedom (Beade, 2010, pp. 204–205), I believe that Kant does distinguish between a theoretical and a practical solutions to the antinomy, when he asserts that "practical freedom," not "transcendental freedom," can be proven from the *factum* of moral law.

or not, his solution is as liable to the *apraxia objection* as would be suspending judgment on the matter. Given these problems, we have to take a step back and discuss how Kant's theoretical solution to the Antinomies can be compared to a skeptical suspension of judgment. I will do that in the next section of this chapter, to prepare the discussion about Kant's practical solution to the antinomies.

7.3 Practical Limitations of the Antinomies' Theoretical Solution

Let us consider in what sense this theoretical solution to the antinomies differs from the skeptical suspension of judgment and whether or not it allows Kant to escape the *apraxia* objection. We saw already that the attitude resulting from the natural dialectic of reason is *perplexity*. Without critically unveiling the illusion at the foundation of the antinomies, the opposition of theses and antitheses leaves reason disoriented, incapable of continuing with its speculative efforts, which are, nonetheless, essential for its vocation as a faculty of principles (B356). Moreover, as we saw, in a veiled reference to Pyrrhonian suspension of judgment, Kant claimed that this could result in a "euthanasia of reason" (B434) and, continuing with his metaphor of the skeptics as "nomads," he said that it did not provide a "dwelling place for permanent residence, for the latter can only be found in complete certainty, whether it be of the cognition of the objects themselves or of the boundaries within which all of our cognition of objects is enclosed" (A761/B789). Suspension of judgment, then, is an insufficient response to metaphysics.

In contrast, Kant saw his theoretical resolution to the antinomies as a way to provide a "dwelling place" for reason. In Kant's estimation, transcendental idealism, with its distinction between phenomena and noumena, actually supplied this starting point for reason's inquiries. His idea was that accepting transcendental idealism entails focusing our rational efforts on investigating particular phenomena rather than seeking to grasp ultimate metaphysical principles—a goal that, although speculatively less ambitious, is scientifically more productive. In other words, transcendental idealism's limitation of reason's endeavors to the realm of experience helps the faculty to refrain from venturing into the dark ocean of metaphysical speculation and to exercise sovereignty in its legitimate domain:

> But it will be asked: What sort of treasure is it that we intend to leave to posterity, in the form of a metaphysics that has been purified through criticism but thereby also bought into a changeless state? On a cursory overview of this work, one might believe that one perceives it to be only of negative utility, teaching us never to venture with speculative reason beyond the boundaries of experience; and in fact, that is its first usefulness. But this utility soon becomes positive when we become aware that the principles with which speculative reason ventures beyond its boundaries do not, in fact, result in extending our use of reason, but rather, if one considers them more closely, inevitably result in narrowing it [...] (BXXIV–XXV).

Through the adoption of transcendental idealism, the theoretical resolution of the antinomies provides a restriction of reason's speculative impulses and a precise determination of the terrain in which knowledge is possible. Before continuing with Kant's argument, I want to note that we can see in this solution a very Academic gesture—the rejection of metaphysical speculation and restriction of philosophical inquiry to human affairs was, indeed, at the core of Hume's Academic skepticism. For Hume, however, this entailed a radical abandonment of metaphysical investigation and an embracement of the study of human nature. For Kant, in contrast, this limitation can only occur through adopting a particular metaphysical framework – namely transcendental idealism – only on whose basis can reason direct its efforts to human matters.

Now, one could still object that this conscious limitation of the field of theoretical knowledge may be scientifically valuable but does not supply specific practical guidelines. In other words, it could still be argued that Kant's proposal fails to offer adequate direction for human action *vis-à-vis* issues that he himself accepted to be unavoidable and primordial, such as freedom and the existence of God. It may even be objected that, insofar as transcendental idealism accepts that we cannot know things-in-themselves, Kant's theoretical solution remains too similar to a skeptical suspension of judgment in these matters. After all, it is hard to see how arguing that there is an inaccessible realm of reality is not actually inviting people to suspend judgment on such a realm.

To be fair, Kant did recognize that metaphysical beliefs, especially those examined in the third and fourth antinomies, have great practical significance, so much so that he argued that practical reason has primacy over theoretical reason in these matters (*KpV*, Ak. 119–123). However, I think he acknowledged that transcendental idealism could not provide a practical solution to the antinomies. Or, to put it more directly, I maintain that Kant knew that the antinomies' theoretical solution was insufficient insofar as it did not answer the *apraxia* objection.

Some examples may help to clarify this problem. One practical issue closely related to the cosmological problems is the methodological question of how to conduct scientific investigations about the world as the totality of phenomena. Let us illustrate this with the case of an astronomer who wants to solve the question of the temporal magnitude of the world by empirically determining the exact time of the Big Bang. For such a scientist, Kant's theoretical solution to the first antinomy – according to which, let us recall, adopting transcendental idealism would allow to conclude that both the temporal finitude and infinitude of the world may be false – would, in practical terms, differ very little from the suspension of judgment resulting from the natural dialectic of reason. In fact, this scientist would find no practical difference between adopting transcendental idealism and adopting a radical sort of skepticism—both viewpoints would carry the same consequence of bringing her investigation to a complete halt. The same could be said of a quantum physicist interested in learning about the behavior of sub-atomic particles. Kant's solution to the second antinomy suggests that matter is neither finitely nor infinitely divisible because, as a thing-in-itself, we cannot know whether or not it has a magnitude. Consequently, our physicist would have to abandon all experimental work on these particles. Again, I fail to see how this differs from suspending judgment and stopping inquiry altogether.

A second practical problem, perhaps even more pressing, concerns Kant's solution to the third antinomy. It deals with how to assign moral responsibility, given that transcendental idealism allows for the possibility that both the thesis, that there is both freedom and natural causality, and the antithesis, that there is no freedom but mere natural causes *ad infinitum*, are true. Faced with this answer, I dare say, we would not be able to assign moral responsibility at all. For example, a person who wanted to resolve the issue of morally judging a murderer from the point of view of transcendental idealism, would have to accept both that the accused could have acted freely and, thus, was responsible for the crime, *and* that she could have been compelled to commit the crime by empirical causes *ad infinitum*, and, hence, was not accountable for her actions. In such a case, transcendental idealism would leave the question of the perpetrator's guilt or innocence completely unsettled. Consequently, while transcendental idealism may be helpful for a philosopher or social scientist who wants to better understand the issue of human freedom and responsibility by approaching it from the points of view of freedom an natural causality, it would make it impossible for an ordinary person to make an everyday-life moral judgment.

Analogously, the theoretical solution to the fourth dynamical antinomy brings about the problem of how to judge and act on matters of religion. What would be the right attitude to adopt in this regard, by a person who has considered Kant's claim that the thesis and antithesis of the last antinomy may both be true from the point of view of transcendental idealism? Would this person not feel compelled to adopt the same sort of agnosticism produced by the natural dialectic of reason? Or, in other words, would this actually provide any practical orientation in religious matters other than abstaining oneself from acting? Granted, this theoretical solution is essential for Kant's view that theological claims do not have the status of knowledge, and dogmatic metaphysics should be rejected. But, from the practical standpoint, it seems again to bring about the same results of suspending judgment on the matter and to offer no answer to the *apraxia* objection.

At this point of my argument, scholars of Kant may be pretty irritated. Some may think that transcendental idealism does not fail to provide practical orientation but instead recommends a form of "epistemic humility" (Langton 1998, p. 2) that leads to a more balanced view of metaphysics. Others, like P. Smith, may contend that Kant's theoretical solution does settle the question altogether. He argues:

> Once the contradiction [of the antinomies] disappears, the ensuing problems find a solution. In the first place, we can now attribute a truth-value to each proposition so that there is no more epistemological problem. For example, it is true that, in the domain of phenomena, all actions are causally determined and that, in the realm of things-in-themselves, freedom is effective. If everything is determined causally in the field of phenomena, then it is possible to have an empirical science of phenomena and of human behavior. In this way, scientific knowledge [...] is preserved. This also offers a solution to the moral problem since determinism is limited to the realm of phenomena, and freedom is viewed as possible in the realm of things-in-themselves (Smith 2008, p. 476; my translation).

I agree with the first part of his assessment, namely, that Kant's transcendental idealism warrants a deterministic science of phenomena. However, I find his last view puzzling, since accepting that freedom is possible in the realm of things-in-themselves – that is, in an uncognizable domain – certainly does not settle the discussion but rather leaves it open. Besides, Kant's resolution of the third antinomy stresses that it may be true that *both* there is natural causality and freedom, and *only* natural causality. In other words, the thesis and antithesis of this antinomy are either that metaphysical compatibilism is the case or that metaphysical determinism is, respectively. But Kant does not accept the thesis and reject the antithesis, which would allow us to, at least, affirm compatibilism. Instead, he solves the antinomy by asserting that both metaphysical positions may be true from the point of view of transcendental idealism. This does not seem to be particularly helpful when it comes to judging someone's actions morally. Therefore, the question remains about how to act concerning freedom and causality. Of course, Kant responded to these issues extensively in the first and second critiques, so that my assessment only holds for his *theoretical* solution of the antinomies. I will turn to their *practical* solution in due course, but before I do, allow me to pause and observe Kant's historical characterization of these antinomical problems, and their relation to what he calls, the practical and speculative interests of reason.[8]

7.4 Practical Interest, Speculative Interest, and Popularity

Before offering a practical solution to the antinomies, Kant considered the factors that lead reason to sustain either the thesis or the antithesis of the antinomies, namely the "interests of reason in these conflicts" (A462/B490-A475/B503). Kant defined an interest of reason as: "a principle that contains the condition under which alone its exercise is promoted" (*KrV, Ak.* 119). In other words, it is a principle that fosters or encourages the activity of a particular faculty. In this case, Kant argued that reason has practical and speculative interests, which are principles that promote assenting to one side or the other of the antinomies.

In my view, this section offers an important alternative version of the antinomic conflict of reason because both it situates the dispute in the context of the history of modern philosophy and shows how these problems are tightly connected to fundamental human preoccupations. Kant begins the section by claiming that, so far, the conflict has been presented only through "dry formulae," that is, only in its logical form and without taking into account its "empirical aspects" (A462/B490). These

---

[8] While this is not the only way of establishing a relationship between the theoretical and practical aspects of Kant's theory, it is the one chosen by him after setting out the antinomies in the first *Critique*. Kant also seeks to establish this link in the *Critique of Judgment* (*KU*), through the concept of reflective judgment. I will leave aside this problem, since I am only concerned here with Kant's adoption of the skeptical method. For a discussion on the topic see: Kleingeld 1998, pp. 311–339; Nuzzo 2005; Parra 2007, pp. 239–286.

empirical aspects are the interests that reason possesses to support or reject the metaphysical claims in question. In fact, Kant said, only by considering these aspects can "the full splendor of reason's assertions shine forth" (A463/B491). In other words, considering the practical motivations behind each side is the only way to completely understand what is at stake in the antinomies:

> Since in this case, we would consult not the logical criterion of truth but merely our interest, our present investigation [. . .] will have the utility of making it comprehensible why the participants in this dispute have sooner taken one side than the other, even if no superior insight into the object has been the cause of it, and it likewise explains still other ancillary things, e.g., the zealous heat of the one side and the cold assurance of the other, and why they hail the one party with joyful approval and are irreconcilably prejudiced against the other (A465/B493).

The two positions in conflict are "dogmatism" and "empiricism," which hold the theses and antitheses, respectively. According to Kant, the dogmatic side has a "practical interest" since its theses rationally support the principles of morality and religion, namely the belief in freedom and the belief in the existence of God as the ultimate cause of the universe (A466/ B494) (Beade 2010, p. 199). This dogmatic side also has a "speculative interest," since it claims that it is possible to have rational knowledge of the totality of phenomena, that is, of the world. In other words, this side considers the transcendental idea of the world as actually constituting *a priori* knowledge of the whole chain of conditions that make particular objects of experience possible (A467/B495). This is, in a word, the "optimistic" side of the dispute, as it contends that reason has no limits and that it can attain knowledge beyond experience.

In contrast to the dogmatic side, Kant argued that the empiricist side is inimical to the practical interests of reason, since it does not offer support to moral and religious precepts and even demands that reason call them into question (A468/B496).[9] Despite this, empiricism also possesses a speculative interest, albeit very different from that of the dogmatic side, because it advocates disciplining reason to ensure that it is constantly reminded of not surpassing the domain of experience. Kant, unsurprisingly, praised this attitude, as it proved consistent with his critical efforts:

> On the contrary, however, empiricism offers advantages to the speculative interests of reason, which are very attractive and far surpass any that the dogmatic teacher of the ideas of reason might promise. For with empiricism, the understanding is at every time on its own proper ground, namely the field solely of possible experiences, whose laws it traces, and by means of which it can endlessly extend its secure and comprehensible cognition (A468/B496).

[9]This is, in fact, a particularly bold and prejudiced claim by Kant. As we have seen in previous chapters, the empiricist critique of the metaphysical concept of freedom and the proofs of the existence of God have an enormous practical interest—namely, acknowledging the harm that religion and superstition have caused to human society, and freeing individuals from the yoke of religious authority so that they may live their life in the framework of a civil morality.

Despite Kant's sympathetic assessment of the empiricists' interest to discipline reason, he recognizes that the dogmatic side enjoys greater "popularity" given its practical and speculative interests. The idea of a spatial-temporal limit or ultimate cause of the world furnishes reason with a sort of speculative resting place, an ultimate ground on which to build the edifice of both science and philosophy (A474/B502). Thus, the dogmatic side's tendency to systematicity and utility in advancing the practical interests of reason account for its popularity. And, of course, the lack of practical interest, together with the demand to restrict inquiry to the field of experience, explains empiricism's poor popularity (A474/B502).

At this point, it is worth noticing that Kant's assessment of empiricism and empirical dogmatism seems to correspond with his previous comments on the skeptical method and skepticism. When empiricism is employed solely to acknowledge the limits of human reason, it is positively valued and analogous to the skeptical method. But, when it becomes dogmatic, when it claims that all metaphysical beliefs lack any validity whatsoever, Kant censures it, as he had done with skepticism, and identifies it with a negative dogmatism:

> Of course, if the empirical philosopher with his antithesis had no other intention than to strike down the impertinent curiosity and presumptuousness of those who so far mistake the true vocation of reason that they make most of insight and knowledge just where insight and knowledge really cease [. . .] then his principle would be a maxim for moderating our claims [. . .]. But if empiricism itself becomes dogmatic in regard to the ideas (as frequently happens) and boldly denies whatever lies beyond the sphere of its intuitive cognitions, then it itself makes the same mistake of immodesty, which is all the more blamable here, because it causes an irreparable disadvantage to the practical interests of reason (A470/B498-A471/B499).

In conclusion, the antinomic disputes are essentially traversed by practical and speculative interests, and historically represented by philosophical dogmatism and empiricism, the former of which, given the practical interest of reason, is for Kant more suitable for the orientation of everyday life than the latter.

Now, is this section on the interests of reason, Kant's actual answer to the *apraxia* objection? Is he saying that, in order to act, we should simply adopt the beliefs of the dogmatic side of the dispute? If this were the case, Kant would be justifying metaphysical dogmatism on its utility. If guided by the practical interests of reason, Kant recommended adopting the dogmatic side of the dispute without modifying its claims or, at least, giving them a non-dogmatic assent, the point of his critical efforts would be missed. Moreover, he would not have adopted a skeptical method at all, but simply have decided that one side of the dispute wins given the practical and speculative interests it furthers. And, in this way, he would be betraying his own project by allowing, as it were, metaphysical dogmatism to slip inside the house through the back door after he threw it out of the front door. Fortunately, this is not Kant's final position. His ultimate proposal is evidently more nuanced, and genuinely retains the Academic spirit of theoretically maintaining doubt while non-dogmatically approving claims to orient action, as we will see in the remaining part of this chapter.

7.5 Practical Solution to the Antinomies

Up to this point, we have discussed how Kant's theoretical solution to the antinomies through transcendental idealism, although essential for revealing the illusion of reason involved in the natural dialectic, is insufficient from a practical point of view. We have also seen that merely considering the speculative and practical interests of reason in each side of the dispute would lead us to give a dogmatic answer to the *apraxia* objection and abandon Kant's critical efforts. What then should the scientist who wants to determine empirically the temporal extension of the cosmos or the behavior of sub-atomic particles, the person who faces the problem of morally judging someone who has committed a crime, and the individual who wants to engage in religious practices, believe to be consistent with Kant's critical project?

Kant provides a very precise explanation of the practical attitude that should be taken in each scenario. First, in response to how to continue the theoretical investigation about cosmological issues, he recommends taking the idea of the world as a *regulative principle* for empirical research—that is, investigating the world *as if* it had an infinite magnitude. Second, Kant proposes accepting a practical conception of freedom while taking the idea of transcendental freedom as merely problematic. And, finally, he argues that some religious beliefs, particularly the beliefs in the existence and providence of God and in the immortality of the soul, may be practically adopted, as they are rational consequences of our moral commitments and incentives for moral action. Let us begin to explain these practical solutions with the regulative use of the idea of the world.

7.5.1 *Regulative Use of Cosmological Ideas*

We have previously noted that a scientist seeking to understand the magnitude of the world would have to suspend her inquiries, should she accept Kant's theoretical solution to the mathematical antinomies and adopt transcendental idealism. Indeed, if we cannot attribute magnitude to the world because, as a thin-in-itself, it remains wholly unknown, it seems that empirically attempting to determine the exact time of the Big Bang or studying sub-atomic particles make little or no sense at all. However, Kant did not believe this was the case. For him, even if it is impossible to cognize the world as a thing in itself, this rational idea has great scientific utility, provided we accord to it an *immanent* and not a *transcendent* meaning:

> [...] it is not the idea itself but only its use that can be either extravagant (transcendent) or indigenous (immanent), according to whether one directs them [the ideas of reason] straight away to a supposed object corresponding to them, or only to the use of the understanding in general regarding the objects with which it has to do (A643/B671).

Taking the idea of the world *immanently* means using it as a general point of reference for scientific research rather than as a cognizable ultimate reality. The immanent idea of the world allows the scientist to make increasingly broader hypotheses about interrelated phenomena while refraining from asserting how the cosmos as a whole really is. As a mere point of reference, the idea of the world can order specific knowledge based on general principles, while knowing that an absolute, ultimate principle is, in reality, unattainable. For Kant, this idea is a kind of "*focus imaginarius*" that orients empirical investigation toward that systematic theoretical body of knowledge that we call "natural science" (A644/B672). In fact, given that our understanding's cognitive attempts are limited and can only determine particular objects in their specific causal chains, we would not have a system of natural laws if reason did not connect and order these specific cognitions with general principles. It is only because of this architectonic tendency of reason, guided by the idea of the world, that we can structure particular cognitions into a whole, using hierarchical and subordinating principles. Otherwise, our knowledge of the world would be a bare enumeration of physical events, bound by a great diversity of specific empirical laws (Grier 2006, p. 205).

At first glance, however, we could ask Kant how his use of the idea of the world as a mere regulative principle – an "as if" useful to guide empirical research – is different from adopting one of the two dogmatic positions. Would it not be better to allow the scientist to adopt the dogmatic side of the mathematical antinomies – i.e., that there is a limit of the world in time and space – given that Kant accepts that this view furthers the speculative interest of reason? Or, should not the scientist accept the disciplining force of the empiricist side, and adopt the belief expressed in the antitheses of the antinomies—namely that the world is infinite both extensively and intensively? The answer is that doing this would amount to giving the idea of the cosmos a "constitutive use," which, in turn, would renew the dialectic of reason with its unsolvable disputes. Instead, the merely "regulative use" of the idea of the world retains the disciplining efforts of reason and constantly warns the scientists against thinking that the world, as a thing-in-itself, can be known, but, at the same time, serves as an incentive to research into ever broader chains of empirical conditions. In other words, scientists should use the idea of the world merely as an ultimate point of reference and always keep in mind that they will never apprehend it as an object of knowledge. Kant explained:

> Accordingly, I assert: the transcendental ideas are never of constitutive use, so that the concepts of certain objects would thereby be given, and in case one so understands them, they are merely sophistical (dialectical) concepts. On the contrary, however, they have an excellent and indispensably necessary regulative use, namely that of directing the understanding to a certain goal respecting which the lines of direction of all its rules converge at one point, which, although it is only an idea (*focus imaginarius*) [...] nonetheless still serves to obtain for these concepts the greatest unity alongside the greatest extension (A644/B672).

Kant offers numerous examples of this regulative use of the idea of the world in science: making hypotheses, organizing concepts into types and species, reducing causal principles to a specific number of chemical agents, etc. (A646/B674-A668/B696). Each of these activities expresses the movement of reason towards more general principles, whose highest unity is represented by the idea of the world. Reason, says Kant, is at once economical and systematic in that it seeks to explain nature with the least possible number of principles, while simultaneously encompassing the greatest possible number of natural laws and specific events (A652/B680). Thus, the idea of the world provides an essential service to science as a unifying referent.

Given the above, the scientist who studies the cosmos should position herself in a middle ground between the points of view of dogmatism and empiricism. On the one hand, she should follow the disciplining effort of empiricism, considering the world *as if* it had an infinite magnitude, so that her ascent to ever-broader phenomenal conditions be not mistaken for an access to the unconditioned. On the other hand, though, she should never deny that, as a thing-in-itself, perhaps the world haves an ultimate limit, which can be neither empirically known nor rationally determined. Kant explained:

> [...] (in cosmology) we have to pursue the conditions of the inner as well as the outer appearances of nature through an investigation that will nowhere be completed, **as if** nature were infinite in itself and without a first or supreme member – although, without denying, outside of all appearances, the merely intelligible primary grounds for them, we may never bring these grounds into connection with explanations of nature because we are not acquainted with them at all (A672/B700).

Kant's insistence on using this idea of reason for merely regulative, and not constitutive, purposes is rooted in his conviction that, since the dialectical illusion is so prominent, metaphysical dogmatism would quickly permeate both the spheres of science and common sense unless we maintain a critical point of view. The tenacity of the natural dialectic of reason makes it even more important for scientists to be attentive to the blurry limits of their task. Only in this way can they, on the one hand, follow reason's tendency to seek general principles by aiming at the *focus imaginarius* that is the idea of the cosmos, and, on the other, keep in mind the merely rational status of this idea in order to avoid falling into metaphysical quandaries. Therefore, the scientist who seeks to determine the specific time of the Big Bang or the behavior of sub-atomic particles should not interrupt her inquiry. Instead, she should continue it by indefinitely pursuing the chain of natural phenomena under observation and using as a regulative principle the idea of the world, conceived *as if* infinite in time and space, but never dogmatically affirmed as such. In sum, with his notion of the world as a regulative principle, Kant has given the scientist a clear answer to the *apraxia* objection.

7.5.2 Transcendental Freedom and Practical Freedom

Since the theoretical solution to the dynamical antinomies – according to which, from the point of view of transcendental idealism, both thesis and antithesis may be true – also leaves us without practical guidance, Kant needed to resolve the issue of how to attribute moral responsibility in the practical realm. What should be done, for example, when judging an individual who has committed a crime? Should this person be deemed free and also determined by natural causes *ad infinitum*? And, as I said before, would not this amount practically to suspending judgment about human freedom and moral responsibility? Or, alternatively, should we accept the thesis of the third antinomy because it furthers the practical interest of reason, even if doing this entails a return to metaphysical dogmatism?

Immediately after presenting his theoretical solution to the dynamical antinomies, Kant discusses the practical problem of freedom (A532/B560-A558/B586). To do so, he draws an important distinction between the concepts of "transcendental freedom" and "practical freedom." The former, he says, refers to the capacity of a particular being or substance to "begin a state from itself, the causality of which does not, in turn, stand under another cause determining it in time in accordance with the law of nature" (A533/B561). In other words, "transcendental freedom" should be understood as spontaneous causality—a cause that is not the effect of a previous one, and so on. Although freedom is a sort of causality, it is opposed to natural causality or a "caused causality, that is, a cause invariably preceded by a prior cause in time *ad infinitum*." (A533/B561). By contrast, Kant defines "practical freedom" in anthropological terms: "Freedom in the practical sense is the independence of the power of choice from necessitation by impulses of sensibility" (A534/B562). Human beings are free in that they are not "pathologically necessitated" like non-human animals, thus making their own choices. Put differently, whereas non-human animals can only follow their natural instincts, human beings can will themselves to act independently of natural causes and solely according to their practical reason. In this context, then, freedom is no longer understood from a cosmological point of view, which concerns speculative reason, but rather from a human standpoint and only involves a description of the faculties that determine action. Kant argues:

> But of reason one cannot say that before the state in which it determines the power of choice, another state precedes in which this state itself is determined. For since reason itself is not an appearance and is not subject at all to any conditions of sensibility, no temporal sequence takes place in it even as to its causality [...]. Reason is thus the persisting condition of all voluntary actions under which the human being appears (A553/B581).

Of course, we can question the assertions that practical freedom is the capacity to act according to reason and escapes all natural determinations or natural causes. As we saw in Chap. 5, Hume's discussion of freedom and necessity significantly influenced Kant's view on freedom as a sort of causality. However, Hume's position was that freedom is identical to natural causality. For him, even when we think we are motivated to act by reason alone, we are in truth acting from natural causes, namely passions, prejudices, education, etc. So, following Hume, we could question Kant's view that practical maxims and imperatives are rational and do not proceed from phenomenal causes *ad infinitum*.

Kant's answer to this possible criticism is that moral imperatives are not empirical and, hence, not determined by natural causes because they are the product of *pure reason*. This means that moral action is founded on moral imperatives or imperatives of duty, which commands us to act by following a universally valid moral law and regardless of our empirical inclinations. In his view, the pure or intelligible character of practical moral reason is proved by the fact that duty usually opposes all possible empirical grounds for action. Clearly, for Kant, when we say that someone "*ought to* act in this way," we mean that this person can and should act independently of sensible determinations or empirical causes:

> The *ought* expresses a species of necessity and a connection with grounds which does not occur anywhere else in the whole of nature. In nature the understanding can cognize only what exists, or has been, or will be. It is impossible that something in it ought to be other than what, in all these time-relations, it in fact is [...]. Now this "ought" expresses a possible action, the ground of which is nothing other than a mere concept, whereas the ground of a merely natural action must always be an appearance (A547/B575).

In other words, for Kant, the fact that moral judgments are normative, that they express the demand that people act the way they ought to, while refraining from acting out of natural determinations, necessarily involves considering reason as a spontaneous causality (Korsgaard 1996, pp. 7–48). I cannot delve here into the problem of Kant's moral normativity but, for our purposes, we should bear in mind that, for Kant, judgments expressing moral duties presuppose human freedom.[10] In the Canon of Pure Reason, he claims:

> [...] but these considerations about that which in regard to our whole condition is desirable, i.e., good and useful, depend on reason. Hence this also yields laws that are imperatives, i.e., objective laws of freedom, and that say what ought to happen, even though perhaps it never does happen, and that are thereby distinguished from laws of nature, which deal only with that which does happen (A802/B830).

Commentators debate whether this concept of "practical freedom" is compatible with the dialectical idea of "transcendental freedom." The discussion focuses on a passage of the Canon where Kant says that practical freedom can be *empirically proved*:

> Practical freedom can be proved through experience. For it is not merely that which stimulates the senses, i.e., immediately affects them, that determines human choice, but we have a capacity to overcome impressions on our sensory faculty of desire by representations of that which is useful or injurious even in a more remote way [...] (A802/B830).

The concept of practical freedom in this passage seems to directly contradict the idea of transcendental freedom, which is, by principle, not empirically verifiable. For this reason, some commentators have argued, first, that both concepts are at odds with each other, and, second, that the concept of practical freedom in the Canon is a remnant of Kant's pre-critical conception of freedom (Guéroult 1954, pp. 331–357; Schönecker 2005, pp. 106–165; Allison 1990, pp. 54–70). Others have answered this interpretative challenge by understanding practical freedom as a kind of transcendental freedom and showing how the "empirical proof" Kant speaks of in this passage should be broadly construed as a moral proof (Esteves 2014, pp. 336–371; Kohl 2014, pp. 313–335; Josifovic 2015, pp. 4483–506). I tend to side with the latter interpretation. Still, I believe that the reason these commentators find a problem between both conceptions of freedom is that they have lost sight of the fact that, with his concept of "practical freedom," Kant is not answering a theoretical issue but merely a practical one. Thus, even if speaking of an "empirical proof," Kant is not talking here of a theoretical proof, rather he is referring to a practical demonstration that involves experience. In other words, once we interpret the issue with the ancient

---

[10] Korsgaard's account of Kantian normativity is widely known. For a lesser known, but equally interesting, Spanish and Latin American analysis of Kantian ethics, see: Aramayo and Rivera, 2017.

apraxia objection in mind, the difficulties become insubstantial, because the real question is how the distinction between "transcendental freedom" and "practical freedom" solves, from a practical point of view, the perplexity caused by the theoretical solution of the third antinomy.

The answer to this question is that "practical freedom" allows us to accept that we are free, in the restricted way of being able to determine our actions rationally, independently from our natural drives, and following the exigencies of morality. This concept of freedom is merely practical and can only be attributed to human beings. In other words, we cannot judge any phenomenon in the cosmos as freely produced, except for human action. This means that the spectrum of what we can accept as "free" is quite limited. We cannot resort to free causes to explain natural phenomena beyond the sphere of human behavior. And, indeed, Kant insists that, in the theoretical realm, we always abide by the *problematic status* of the idea of "transcendental freedom," which means that, although this idea is logically possible, its reality cannot be empirically proven (A558/B586). Thus, we should never conflate the domains of "practical freedom" and "transcendental freedom," for doing this would amount to theoretically hypostasizing a concept that is only practically valid.

With this distinction in mind, it is now easier to see how Kant answered the question about how to attribute moral responsibility. He directly addressed the problem using the example of a person who lies for empirical motives:

> [...] one may take a voluntary action, e.g., a malicious lie, through which a person has brought about a certain confusion in society; and one may first investigate its moving causes, through which it arose, judging on that basis how the lie and its consequences could be imputed to the person. With this first intent, one goes into the sources of the person's empirical character, seeking them in a bad upbringing, bad company, and also finding them in the wickedness of a natural temper insensitive to shame, partly in carelessness and thoughtlessness; in so doing one does not leave out of account the occasioning causes. [...] Now even if one believes the action to be determined by these causes, one nevertheless blames the agent [...]. This blame is grounded on the law of reason, which regards reason as a cause that, regardless of all the empirical conditions just named, could have and ought to have determined the conduct of the person to be other than it is (A555/B583).

As we can appreciate, here Kant provides a clear answer to the *apraxia* objection. This answer is that even if we can consider natural causes when attributing responsibility, the founding principle of any moral judgment is the belief that human beings are free because their rational capacity gives them moral imperatives that are contrary to their natural inclinations.

However, it could be objected that Kant's argument is circular, stating something like "given that when we judge morally, we presuppose freedom, we should presuppose freedom when we judge morally." This objection would raise the question of whether Kant is giving us a real practical orientation or is just leaving the moral problem to practical common sense. But, does common sense actually presupposes practical freedom? Despite the normativity of our moral language, I think we tend to assign moral responsibility by appealing to natural causes. In fact, we often affirm that a particular person could not help acting in a certain way, given her natural inclinations, and that she should be considered innocent or guilty depending on which inclinations she followed. Precisely for this reason, Hume claimed that the doctrine of necessity did not contradict morality but actually founded it (*EHU*, VIII, 97). Moreover, common sense can become easily entangled in the metaphysical antinomy, holding alternately that human beings are free and naturally determined. Thus, without any practical guideline, we could arbitrarily affirm and deny freedom, sometimes considering natural causes and sometimes seeing practical reason as the basis of moral behavior.

Now, in the second *Critique*, Kant affirms that practical freedom is morally *proven*, which could be taken as a step towards metaphysics. He argues that our consciousness of the moral law is a kind of "factum" of reason that proves the reality of freedom:[11]

> [. . .] the moral principle [. . .] serves as the principle of the deduction of an inscrutable faculty which no experience could prove but which speculative reason had to assume as at least possible [. . .], namely the faculty of freedom, of which the moral law [. . .] proves not only the possibility but the reality in beings who cognize this law as binding upon them (*KpV*, *Ak.* V, 47).

Here, we should note that Kant goes beyond saying that practical freedom should be *presupposed* whenever we judge morally—he affirms that it is *real.* However, we must bear in mind that in the second *Critique*, Kant no longer speaks from a theoretical point of view but only from a practical one. Thus, *stricto sensu*, only from this practical point of view is freedom a reality. As Beck famously claimed: "The moral law is the *ratio cognoscendi* of freedom, and freedom is the *ratio essendi* of the moral law." (Beck 1960, p. 27). In other words, we know that we are practically free because the moral law compels us to act according to it, that is, independently of our natural drives. This awareness of the moral law, for Kant, proves the reality of freedom as a capacity of reason to cause action: "[. . .] the moral law thus determines that which speculative philosophy had to leave undetermined, namely the law for a causality, the concept of which was only negative in the latter, and thus for the first time provides objective reality to this concept" (*KpV*, *Ak.* V, 47)

It is undoubtedly odd to attribute to a capacity an objective reality, particularly since Kant also argues that we will never find in experience an actual enactment of freedom.[12] However, the point is that, since "the moral law is given" to us, freedom

---

[11] The proof that is offered in the *KpV* is not directly based on experience, as Kant construed it in the Canon, but we could consider the former as an elaboration of the latter (Schönecker 2005, pp. 106–165; Esteves 2014, pp. 336–371; Kohl 2014, pp. 313–335; Josifovic 2015, pp. 4483–506).

[12] Since reality or actuality was defined in the Postulates of Empirical Thinking of the first *Critique* as "that which is connected with the material conditions of experience (of sensation)" (A218/B265), it is not clear what the conditions of experience are to which freedom is related in these passages. Given that the moral law is a pure or *a priori* law that does not possess any empirical content, we cannot, *stricto sensu*, make this claim. Thus, we cannot affirm that Kant has provided empirical proof of freedom, and we have to keep in mind that the Postulates of Empirical Thinking, as the name indicates, give only the conditions for a theoretical, determinate judgment on the reality of phenomena. Thus, they can neither afford empirical proof of an idea, such as freedom, nor be applied to explain the *practically objective reality* of freedom. For a good discussion of what is the exact meaning of freedom's and the Postulates of practical reason's *objective reality*, see: Zimmermann 2016, pp. 131–155.

is also "firmly established of itself" (*KpV*, *Ak.* V, 47). Once again, Kant warns us not to move illegitimately from the practical to the theoretical level. We must always keep in mind that the reality attributed to freedom is not theoretical but only practical, which means that we can and must affirm freedom only with a view to acting and judging other humans' actions.

This point becomes clearer if, as Kant recommended, we understand this capacity in terms of "autonomy." Autonomy refers to a will's "lawgiving of its own" in opposition to "heteronomy," which is the determination of the will by natural inclinations or other external factors (*KpV*, *Ak.* V, 33). The practical emphasis becomes manifest when we speak of human beings as autonomous or capable of rationally determining their own behavior. This concept takes us immediately to a point of view that is bereft of any metaphysical connotation. Nobody would ask us to theoretically prove autonomy, even though we consider it "real" in a practical sense – namely we expect rational human beings, as long as they are adults and not mentally impaired, to give justifications for their actions. Autonomy, in this context, means that human beings can articulate their motives into "reasons" or practical rules of conduct – either maxims or imperatives – that explain their behavior. Indeed, we educate children to develop this faculty whenever we ask them to tell us their reasons for their behavior. We also feel admiration for those we believe possess it and refer to them as people with a good or virtuous character, capable of good deeds not only because they have the right feelings but also because they have the right reasons and are able to act according to them. And we tend to have remorse when we recognize having acted heteronomously, that is, driven by mere inclinations or by the reasons others impose on us, etc. Hence, if we interpret practical freedom as autonomy, it ceases to have the metaphysical, ontological, and even theological underpinnings of the traditional term.

In conclusion, staying within the practical point of view limits the tendencies of speculative reason to seek an ontological characterization of freedom. Kant argues that, to avoid *apraxia* in moral matters, we can rationally believe that human beings are practically free, namely, that they can determine their will to act out of moral maxims and laws their reason gives to themselves. In this way, we can carry out moral lives, without theoretical certainty but guided by the practical conception of freedom or autonomy.

7.5.3 Beliefs in God and the Immortal Soul

Let us now go back to the theoretical solution to the fourth antinomy. This antinomy brings us face to face with the problem of how to conduct ourselves in matters of religion.[13] To decide whether to engage in religious practices, it seems that we

---

[13] Kant faces this problem again in the refutation of the proofs of the existence of God in the Ideal of Pure Reason of the first *Critique* (A567/B595-A641/B669). However, for reasons of space I shall not discuss this aspect of the Transcendental Dialectic here.

should first make a more fundamental decision—whether to maintain or deny the existence of God as a necessary being and first cause of the universe. However, since the theoretical solution to this antinomy accepts the possibility that both thesis and antithesis are true, it is not clear what religious view we should adopt in practice. Once again, the question can be formulated in terms of the *apraxia* objection. How can we decide whether to engage or not in religious practices if we do not take a clear position on the existence and providence of God? Moreover, should we accept the dogmatists' side of the dispute, given that, as Kant claimed, this position is more favorable to this interests of reason?

Kant discusses the problem in the Canon of Pure Reason of the first *Critique* and the Postulates of Pure Reason of the second *Critique*. His answer involves a practical justification of the ideas of God and the immortality of the soul. Although the idea of the soul as an immortal entity is not related to the antinomic conflict, which only concerns the world, it also results from reason's natural dialectic when it seeks to grasp the absolute and unconditioned synthesis of the thinking subject (A340/B398). Attempting to metaphysically determine the soul as a simple and, consequently, unperishable substance, once again, reason hypostatizes a mere idea. Kant devotes the Paralogisms of Pure Reason of the first *Critique* to assess this problem (A341/B399-A405/B432). Unfortunately, I cannot summarize here this section without making this chapter unnecessarily long, so I will only say that, just as in the case of the idea of the world, Kant argues that the idea of the soul should be taken as a *focus imaginarius*, in order to think the subject as a complete unity. Thus, while the soul's immortality cannot be held theoretically, it has a heuristic and practical utility.

In the Canon of Pure Reason, Kant goes back to mentioning the interests of reason by arguing that reason is practically interested in freedom, the existence of God, and the immortality of the soul (A798-B826). Having already seen his practical solution to the problem of freedom, we could guess that he would recommend, once again, maintaining these ideas as beliefs from the practical point of view. However, unlike human freedom, which was proven as real by the *factum* of the moral law, the beliefs in God and the immortality of the soul do not have such secure practical grounding. According to Kant, we can accept that, given that the moral law commands us to act in a certain way, we must be able to do so; that is, we must be practically free. This argument is Kant's famous "ought implies can" principle (A548/B576). However, it is less clear how we can assert the existence of God and the immortality of the soul at the practical level. Thus, Kant had to give us a more extensive explanation of the practical necessity of these beliefs by connecting them with morality and happiness. But to understand this, we have to take a closer look at Kant's discussion of the "highest good" and the "moral world" in the Canon.

Kant claims that the moral law necessarily leads us to conceive the idea of a "moral world." Given their normative character, moral laws do not determine the world as it is, but only as it *ought to be* (A808/B836). But, since they are necessary and universal imperatives, we cannot reasonably deny the possibility that they may produce a moral state of affairs: "For since they [the moral principles] command that these actions ought to happen, they must also be able to happen" (A807/B835). In other words, we cannot simultaneously say that we *ought to* act morally and that it is impossible to achieve a moral arrangement of the world because this would entail a practical contradiction. Indeed, morality only makes sense if we think that a moral world, in which such moral imperatives are enacted, is possible:

> I call the world as it would be if it were in conformity with all moral laws (as it can be in accordance with the freedom of rational beings and should be in accordance with the necessary laws of morality) a moral world. This is conceived thus far merely as an intelligible world, since abstraction is made therein from all conditions (ends) and even from all hindrances to morality in it (weakness or impurity of human nature). Thus far, it is therefore a mere, yet practical, idea, which really can and should have its influence on the sensible world, in order to make it agree as far as possible with this idea (A808/B836).

This moral world is an idea analogous to that of the world but with a practical foundation and significance. While the idea of the cosmos was a *focus imaginarius* – a mere *as if* that provided a formal reference to give unity to natural laws in scientific research – the idea of the moral world is a kind of *corpus mysticum*. By analogy with the medieval use of the term, whereby the church would become the *corpus mysticum* of Jesus Christ, Kant uses it to refer to the absolute unity of human wills around a common purpose—in this case, the realization of the moral law. For him, this idea of a multitude of wills comprising a unity, once again, has an "objective reality," although only from a practical point of view (Zimmermann 2016, pp. 131–155). Moreover, it is a necessary idea, in the sense that we have to believe that it is achievable, or we would have to deny the imperative character of morality. Another way to put this argument is that if we accept that we not only *should* but *must* act morally, then we have to admit that a global moral society is a real possibility. Otherwise, we would be endorsing a practical obligation that has no possible outcome, and this amounts to falling into a practical contradiction (Tomasi 2016, p. 123).[14]

But this idea of the moral world introduces yet another question, namely, whether we can *hope* not only to partake in a moral world but also in a world where we can attain happiness: "if I do what I should, what may then I hope?" (A805/B833) Kant maintained that human beings have two ultimate ends, namely, morality and happiness. As we experience all too often, these ends are opposed to each other such that our efforts to be morally upright usually lead to unhappiness and *vice versa*. Nevertheless, since we cannot renounce either one, we constantly strive to balance them. Kant portrays the relationship between these two ends as synthetic (*KpV*, Ak. V, 112), which means that none of the two concepts is included in the other, but they require a third term to relate them. For this reason, as B. Recki argues, he is unsatisfied with the ancient accounts of the highest good:

> By rejecting the idea of an analytic connection between happiness and virtue, Kant critically refers to the "attempts at coalition that have hitherto been made" (*KpV*, 5:112), in which virtue and happiness are to be made congruent with one another by identification or deduction. [. . .] As Kant explains, while Epicurus identifies virtue with happiness (or, more specifically, conceives of virtue as an effect of happiness), the Stoics identify happiness with virtue (or, more specifically, conceive of happiness as a consequence of virtue) (Recki 2016, p. 76).

---

[14] Although Tomasi refers here to the psycho-theological proof of the existence of God in the *Critique of Judgment*, the same argument holds for the "moral world."

The theoretical efforts to connect morality and happiness led the Hellenistic schools to endless discussions about the idea of the highest good or *summum bonum* (*Fin.* I, 4).[15] Kant criticized the ancient philosophical attempts to define the highest good by conflating the two terms of the problem, that is, by reducing happiness to virtue or virtue to happiness. In doing so, on the one hand, the Epicureans overlooked the true motives of morality, arguing that acting virtuously amounts to seeking happiness through a rational calculation of means and ends. On the other, the Stoics confused happiness with moral self-satisfaction, claiming that happiness could be reduced to exercising moral duty (*KpV*, *Ak.* V, 111). In contrast, Kant neither believes that being happy is acting morally and taking pleasure in virtue (Recki 2016, p. 75) nor that virtue can be reduced to calculating the best means to obtain maximum pleasure and minimum pain. These two ends are inherently different but equally essential for human beings. Thus, the "highest good" should be understood as the state of maximum morality joined with the greatest possible happiness: "In the highest good which is practical for us, that is, to be made through our will, virtue and happiness are thought as necessarily combined so that the one cannot be assumed by pure practical reason without the other also belonging to it" (*KpV*, *Ak.* V, 113). As a result, one of life's most pressing problems is how to make both ends compatible without reducing one to the other, or, in other words, how to attain the "highest good."

Kant seemed to have entertained momentarily the idea that the "moral world" involves the realization of the "highest good":

> Now in an intelligible world, i.e., in the moral world, in the concept of which we have abstracted from all hindrances to morality (of the inclinations), such a system of happiness proportionately combined with morality can also be thought as necessary, since freedom, partly moved and partly restricted by moral laws, would itself be the cause of general happiness, and rational beings, under the guidance of such principles, would themselves be the author of their own enduring welfare and at the same time that of others" (A809/B837).

For Kant, a world in which we can count on the total cooperation of others to achieve our own ends, or, at least, on others not unnecessarily restricting our actions and desires, would be a world of happier individuals. As Kleingeld claims:

> In the moral world, I promote the happiness of others, and others promote mine. The happiness of each is good in the eyes of the other and *vice versa*. This means that the virtuous agents in this world collectively aim at the happiness of all (Kleingeld 2016, p. 40).[16]

[15] At the beginning of *De Finibus*, Cicero sets the problem in this way: "What is the End, the final and ultimate aim, which gives the standard for all principles of well-being and of right conduct? What does Nature pursue as the thing supremely desirable, what does she avoid as the ultimate evil? It is a subject on which the most learned philosophers disagree profoundly; who then can think it derogatory to such esteem as each may assign to me, to investigate what is the highest good and the truest rule in every relationship of life?" (*Fin.*, I, 4).

[16] Kleingeld identifies the moral world with the highest good on these grounds: "Kant argues that the resulting notion of *a moral world, as the highest good*, has two elements. First, it includes virtue,

Nonetheless, such a world would not lead to the complete realization of happiness, since morality itself restricts the "satisfaction of all our inclinations," which, according to Kant, is included in the very concept of happiness (*KpV*, *Ak.* V. 112).[17] In consequence, it seems that, although in the moral world individuals would be "happier", they still would not be absolutely happy. Thus, happiness and morality cannot be reconciled, even in the "moral world."

Since the moral world does not provide by itself a solution to the problem, Kant's definitive answer has to rely on a third term that connects synthetically the two concepts of morality and happiness—namely, the ideas of God and the soul's immortality (Recki 2016, p. 84). The argument goes like this: Reason can connect morality and happiness without reducing one to the other only by believing that, through our moral behavior, one can become *worthy* of happiness:

> Happiness alone is far from the complete good for our reason. Reason does not approve of it (however much inclination may wish for it) where it is not united with the worthiness to be happy, i.e., with morally good conduct. Yet morality alone, and with it, the mere worthiness to be happy is also far from being the complete good. In order to complete the latter, he who has not conducted himself so as to be unworthy of happiness must be able to hope to partake of it (A813/B841).[18]

But, even if this link between happiness and morality is theoretically sound and, as Kant argues, practically necessary, it remains a mere idea in the sense that, at least *in this life*, we cannot obtain the happiness we are morally worthy of. Moreover, since moral progress is endless for Kant, an infinite human lifetime would be necessary for our character to determine itself to act only on the grounds of duty. And, given that we do not have infinite lives, both our moral efforts and our being worthy of happiness are interrupted by death. Thus, the only rational possibility for the connection between morality and happiness is that we have an immortal soul,

---

as the "supreme" good, that is, as that which is unconditionally good. In order to be the "complete" good, however, Kant adds, the highest good must also include happiness. The two elements are to be connected in such a way that the first is the cause of the second." (Kleingeld 2016, p. 38, my emphasis). I think Kant plays with the idea that the highest good may be understood in terms of the moral world, but finally abandons the analogy, for the reasons I have mentioned, namely that in the moral world we could not obtain the satisfaction of all our inclinations.

[17]Kant defines happiness as a maximum of satisfaction: "Happiness is the satisfaction of all our inclinations (*extensive*, with regard to their manifoldness, as well as *intensive*, with regard to degree, and also *protensive*, with regard to duration)" (A806/B834). Perhaps a more limited understanding of happiness would not require human beings to postulate the existence of God and the immortal soul in order to attain it. However, Kant's definition is actually appropriate to explain the excessive expectations of pleasure that human beings tend to have. Otherwise such phenomena as the disproportionate zeal for wealth, the search for all kinds of distraction and enjoyment, and the desire for immortality, which are so prominent, particularly in our contemporary societies, could not be accounted for.

[18]Also in *KpV*: "[...] happiness is something that, though always pleasant to the possessor of it, is not of itself absolutely and in all respects good but always presupposes morally lawful conduct as its condition" (*Ak.*V, 111).

capable of such infinite moral perfection and of attaining in another life the happiness it became worthy of in this one:

> This endless progress is, however, possible only on the presupposition of the *existence* and personality of the same rational being continuing *endlessly* (which is called the immortality of the soul). Hence the highest good is practically possible only on the presupposition of the immortality of the soul, so that this, as inseparably connected with the moral law, is a postulate of pure practical reason (*KpV*, *Ak.* V, 122).

Our hope for the highest good is only rationally justifiable if it is to be attained in a future world in which we have become worthy of happiness by our moral merits in this life. Yet again, since happiness is a maximum state of satisfaction, to obtain it requires not only an immortal soul but also the existence of a wise and just God, the guarantor of this reward:

> The same law must also lead to the possibility of the second element of the highest good, namely happiness proportioned to that morality, and must do so as disinterestedly as before, solely from impartial reason; in other words, it must lead to the supposition of the existence of a cause adequate to this effect, that is, it must postulate the *existence of God* (*KpV*, *Ak.* V, 124).[19]

Since we cannot reconcile morality and happiness in *this* lifetime, all we can do is prove ourselves worthy of happiness. And, to be both moral and happy, we need multiple lifetimes – an infinite number – to develop our character. But, for that to be possible, an immortal soul must exist. However, to be able to harvest the benefits of being moral, it is insufficient to have an immortal soul if nothing exists that, eventually, ensures that we become happy. Thus, to reconcile morality and happiness, we need not only an immortal soul but also a God, who would grant us this benefit as well. In this way, Kant has offered a whole chain of arguments to practically justify the beliefs in the existence of both the immortal soul and God.[20]

Allow me to summarize the previous discussion. Kant thought that the beliefs in the immortality of the soul and the existence and providence of God ought to be practically *postulated* to justify rationally the natural human hope for the highest good. He defined a *postulate* of practical reason as a problematic theoretical claim that is logically related to *a priori* practical laws (*KpV*, *Ak.* V, 122). This means that the validity of the beliefs in the existence of God and the soul's immortality derive from their necessary connection to moral imperatives:

---

[19] The argument in the Canon is this: "This, however, is only possible in the intelligible world, under a wise author and regent. Reason sees itself as compelled either to assume such a thing, together with life in such a world, which we must regard as a future one" (A811/B839).

[20] About the existential insufficiency of this answer, see: Recki 2016, pp. 85–86. She asks: "[. . .] is the delegation of this most urgent existential problem to the supersensible, to the mere world of thought, not a depressing result compared to the spirit of optimism engendered by the concept of highest good as a *mixtum compositum* (of morality and happiness)? [. . .] Under the assumption of God and the immortality of the soul, the claim to happiness that was initially taken seriously becomes a promise that threatens to render the nature of its fulfilment unrecognizable."

> These postulates are not theoretical dogmas but *presuppositions* having a necessarily practical reference and thus, although they do not indeed extend speculative cognition, they give objective reality to the ideas of speculative reason in *general* (by means of their reference to what is practical) and justify its holding concepts even the possibility of which it could not otherwise presume to affirm (*KpV*, *Ak.* V, 132).

The beliefs in God and the immortal soul are, thus, theoretical beliefs that can only be justified on practical moral grounds. They rationally support the hope that our moral progress be compensated with a proportional share of happiness in the next life.

In this way, we can see how Kant also offers a practical solution to the fourth antinomy. The postulates supply a foundation for religious belief that is merely practical. We must be careful, once again, not to hypostatize the ideas of God and the immortal God. That is, once we have accepted the postulates for practical purposes, we cannot return to a theoretical point of view and derive morality from God's existence and nature. Kant explicitly forbids this return from the practical to the theoretical realm:

> But now when practical reason has attained this high point, namely the concept of a single original being as the highest good, it must not undertake to start out from this concept and derive the moral laws themselves from it, as if it had elevated itself above all empirical conditions of its application and soared up to an immediate acquaintance with new objects (A819/B847).

To interpret the postulates as a theological foundation to morality would entail giving to the latter a heteronomous grounding—placing its principles outside of morality itself, on a God's will. Kant, very much like Hume, strongly opposes this way of proceeding, and seeks for a way to safeguard a secular or civil foundation for morality. However, he accepts that morality can benefit from religion in significant ways. He thinks that morality must be connected with these beliefs in God and the soul's immortality, or the link between happiness and virtue would be impossible, and the biding force of moral laws could not be explained, as reason would be demanding from us something impossible to achieve (Tomasi 2016, pp. 111–129; González 2011, pp. 235–246). Hence, Kant's famous conclusion:

> Thus without God and a world that is now not visible to us but is hoped for, the majestic ideas of morality are, to be sure, objects of approbation and admiration but not incentives for resolve and realization, because they would not fulfill the whole end that is natural for every rational being and determined *a priori* and necessarily through the same pure reason (A813/B841).

Undoubtedly, Kant's postulates of practical reason require a more critical assessment than I can give here. Many questions about the real necessity of these beliefs for practical purposes remain unanswered (Dörflinger 2008, pp. 207–223; Recki 2016 pp. 71–87; González 2011, pp. 235–246.), since no matter how much Kant insisted that these postulates are necessary from a practical viewpoint, he also maintained that they do not constitute conditions for the possibility of moral action. It seems that they are *necessary* in a different way. Perhaps in the sense that reason necessarily thinks of these entities when it attempts to make sense of the inescapable, and otherwise irresolvable, issue of the compatibility between morality and happiness.

Now, we can accept that reason inevitably leads us to think of the possibility of God's existence and the soul's immortality when we face these practical problems, but this does not mean that we should also *believe* in these postulates. However, Kant's view was precisely this. For him, we have good practical reasons to believe in them, given that they are rational *incentives* to moral action. Thus, in arguing that we must postulate the existence and providence of God and the immortality of the soul from a practical point of view, Kant answered the *apraxia* objection in respect to religious matters. To the questions, what should we believe regarding the existence of God after the antinomic conflict has been shown illusory? Should we believe that there is a god and act in accordance with this belief, or should we deny God's existence, and, then, refrain from any religious practice? Kant responded that insofar as we rationally hope that happiness in proportion to our moral efforts is attainable, we should believe that our soul is immortal and that there is a wise and almighty God who will reward our virtue in this life with future happiness. In other words, he held that we should sustain *practical* or *moral theism*, as this religious creed makes reasonable the promise of the highest good.[21] But we should also always keep in mind that we are entitled to hold these beliefs only as *practical hypotheses* or *postulates* and never dogmatically. To conclude the chapter, I will explain now how this believing in moral theism non-dogmatically can be seen as Kant's Academic response to the *apraxia* objection.

7.6 Academic Skepticism in the Postulates of Practical Reason

In the previous chapter, I discussed the metaphor of the judge Kant frequently uses (A xii; B xiii; A 752/B779) to illustrate one of the most important aspects of his critical adaptation of the skeptical method. After examining the rational opposition between thesis and antithesis, and revealing the illusion at the root of the natural dialectic, the critical philosopher offers a verdict or practical solution to the matter. The regulative use of the idea of the world, practical freedom, and the practical postulates of God and the immortality of the soul constitute this practical verdict.

Also, as we saw in Chap. 3, Cicero employed the metaphor of the judge to introduce the debate among the Hellenistic schools on the nature of the gods. He argued that, after having examined and skeptically refuted all the positions involved, it was necessary to reach a provisional agreement to orient the religious practices of Roman citizens. He situated the skeptical analysis of these metaphysical issues in the symbolic context of a court of justice, with philosophers holding

[21] Kantian moral theism, in my view, does not need to be understood in terms of Christian revelation. However, it does share with the latter the beliefs in a wise, all powerful, and benevolent God, and in the immortality of the soul, as preconditions to the hope of attaining the highest good. For an account of this moral faith as involving Christian revelation, see: Theis 2020, pp. 26–66.

arguments pro and con, and said that he would give a final verdict concerning the religious beliefs that could be held for practical purposes (*DND* I, 13–14). At the end of the dialogue, following the New Academy's method, he decided that the Stoic doctrine of the existence and providence of the gods should be approved on account of its highest degree of persuasiveness and its utility to preserve and unite the Roman Republic.

Given the differences in terminology and approach, it is difficult to observe at first sight any similarity with Kant's solution. Cicero recommended holding the doctrine that seems most persuasive or probable after skeptical examination, while Kant would never have used such terms to speak about metaphysical issues (A293/B349). Moreover, for Kant, no metaphysical doctrine wins the combat theoretically, since neither the metaphysical dogmatism of the theses nor the skeptical empiricism of the antitheses proved more persuasive, verisimilar, or probable in the final examination. Perhaps we could find an analogy between Cicero's concept of persuasiveness and Kant's claim that the dogmatic position in the antinomical debates is more "popular" than the empiricist one. But, since this popularity should not incline the balance of assent to any dogmatic position, properly speaking, we cannot say that one side of the antinomies is more persuasive or probable than the other on account of it.

Despite this, Kant's practical resolution to the antinomies, can be interpreted as an answer to the *apraxia* objection about these metaphysical issues. Specifically, we can act concerning these matters by adopting the idea of the world as a regulative principle for scientific research, restricting the belief in freedom to a practical capacity that enable humans to act morally, and accepting the existence and providence of God, and the soul's immortality as incentives to lead moral lives in the expectation of a future state of happiness. In other words, it seems that Kant saw that it was necessary to give a practical solution to these matters while avoiding metaphysical dogmatism.

Let me summarize what I have identified as the connections between Kant's practical solution to the antinomies and the New Academy's way of proceeding. First, using the metaphor of the judge, Kant introduced in his skeptical method a final, practical verdict that seems to be in accordance with Cicero's and Hume's Academic practice. The Academics neither recommend a total suspension of judgment nor left action unaided in metaphysical matters. Kant proceeded in the same way. Hume's Academic concern that research be limited to human, practical affairs, indeed, finds an echo in Kant's view that, having unveiled the source of the dialectical illusion theoretically, it is necessary to turn the attention to the practical matters involved. Second, Kant's practical solution intends to preserve a separation of the theoretical and practical domains, that allows for the theoretical examination of metaphysical claims to continue while some of these claims are non-dogmatically accepted for practical purposes. As we saw in Chaps. 2 and 3, an analogous procedure is at the heart of Academic skepticism. Third, this division of the theoretical and practical realms, along with the acceptance of a belief for practical reasons, cannot be identified with Pyrrhonism's answer to the *apraxia* objection for reasons that I will develop now.

The Pyrrhonian answer to the *apraxia* objection, namely, the four observances Sextus recommends to the skeptic to follow in practical matters, differ from Kant's practical resolution of the antinomies in various respects. First, the beliefs Kant accepted for practical reasons are far from being mere "appearances." They involve very complex rational inferences that depart entirely from Pyrrhonian sensible impressions, even if they are only valid at the practical level. Second, and perhaps more importantly, following the traditions and customs of one's society, which is Pyrrhonism's preferred answer in the case of moral judgment and religious practices, would be, for Kant, a heteronomous response. The Pyrrhonists act in a way that I have called voluntarily heteronomous, while Kant's answer involves adopting beliefs autonomously, even if only for practical reasons. This commitment to autonomy points to the Academics' "intellectual integrity" (Maia Neto 1997, p. 202) or "*libertas philosophandi*" (de Olaso 2018, p. 65), and precludes the possibility that Kant had given a Pyrrhonian answer to the *apraxia* objection. Just as the Academic skeptic is accountable for her actions, the Kantian agent would justify her behavior by apppealing to the regulative principle of the idea of the world, practical freedom, and the religious postulates. Regarding freedom, in particular, she would probably hold that while not denying natural causes, she accepts that human beings are free in the sense that they can act rationally or independently of their natural drives. Furthermore, if someone asked about her religious beliefs, she would answer that she engages in religious practices because she believes that there is a God that will reward her immortal soul for her moral efforts in this life with an equal amount of happiness in another life. But, she would also add that she sustains these beliefs undogmatically and only at the practical level, given that she neither has proof of the existence of God or the immortality of the soul, nor believe that such proof is possible on the theoretical level.

While this answer may appear artificial and only possible for a very convinced Kantian (and, let us accept it, there are not many today!), I believe it offers a good portrayal of what Cicero called "approving" an impression after skeptical examination, instead of "assenting" to it. In my view, Kant's adoption of Academic skepticism can be best appreciated in his characterization of the type of epistemic attitude that accompanies these beliefs. Let us recall that, for Carneades, to approve an impression and its derived judgment was to hold it only as a ground for action, while the theoretical examination continued. Holding beliefs for practical purposes was permitted so long as the Academic skeptic recognized that they lacked a satisfactory theoretical justification and kept searching for this justification to approach truth as much as possible (*Acad.* II, 7).

Kant's proposal is an excellent example of this epistemic attitude. Although in a different context of discussion, that is, speaking about Kant's moral proof of the existence of God in the third *Critique*, G. Tomasi has claimed that he gives this proof a particular "doxastic status," which is less robust than a theoretical "taking as true." She affirms:

> Kant distinguishes between the mental state of holding that 'God exists' is true and the mental state of taking 'God exists' to be true in one's practical reasoning when the question whether God exists becomes salient [...]. A proof can have an effect on theoretical conviction only if its basis is an objectively valid ground or what Kant calls "a logical ground of cognition" (*KU. Ak.* V, 461). As a matter of fact, the moral proof, resting on our commitment to the final end, lacks such basis, and therefore it cannot have an effect on theoretical conviction. However, this weakness does not particularly affect it, since it aims not at knowledge or theoretical conviction but at practical conviction (Tomasi 2016, p. 122).

I suggest that this "taking 'God exists' to be true" in one's practical reasonings also involves submitting this very belief to permanent theoretical doubt. This attitude corresponds to the Academic "approval" of impressions and judgments, which makes it possible to keep one's beliefs at the practical level while skeptically maintaining the theoretical examination. Affirming that one acts according to a belief in a God that would reward one's moral efforts with happiness in a future life, while one continues to theoretically doubt God's existence, providence, and the possibility of an after-life, is not, as it would appear, a sign of inconsistency or sheer religious irrationality. It is instead a very sophisticated critical attitude. To better explain how Kant understood this attitude, I will now briefly turn to the final section of the Canon of Pure Reason, where he refers to the difference between "knowledge," "belief," and "opinion."

7.6.1 Knowledge, Belief, and Opinion

When we make a judgment, Kant argues, we may take it to be true (*Fürwahlhalten*), depending on the justification we have for it. If this justification rests on "objective grounds," namely determinate judgments of experience that are valid for everyone, the epistemic attitude accompanying it is "conviction" (*Überzeugung*). However, if the justification rests on "subjective grounds," that is, reasons that are only valid for an individual or group of people, it is called "persuasion" (*Überredung*) (A820/B849). Now, conviction has three degrees, according to Kant—knowledge, belief, and opinion. When we assert something as *knowledge*, the grounds of the judgment are both objectively and subjectively sufficient. In other words, the claim not only has objective validity but is also sustained with subjective certainty, namely, with the highest degree of assurance (A822/B850). In the case of *belief*, although the grounds of the judgment are objectively insufficient, they are subjectively sufficient—the objective reasons for holding the belief are doubtful, but the confidence with which one sustains it is still very firm. Finally, when we have a mere *opinion*, our judgment's justification is both objectively and subjectively insufficient, that is, we are aware that we do not have enough objective grounds for holding it and neither are entirely sure about it (A822/B85).

Kant argues that since metaphysical claims possess no objective basis whatsoever – either sufficient or insufficient – *from a theoretical point of view*, they cannot be held as knowledge, belief, or opinion (A823/B851).[22] However, on a *practical* level, they can be sustained as beliefs: "Only in a practical relation, however, can taking something that is theoretically insufficient to be true be called believing" (A823/B851). In this way, the claims about the existence of God and the immortality of the soul count as *beliefs* from a practical standpoint.

A practical belief or a belief from a practical point of view[23] is the peculiar epistemic attitude that we have been discussing up to this point. Perhaps it would be easier to understand it if we focused on one particular kind of belief adopted for practical reasons, which Kant discusses in some detail in the Canon—the "pragmatic belief" (A824/B852). This belief involves practically calculating means and ends, guided by a theoretical hypothesis on a possible state of affairs. The example Kant provides is that of a physician who, despite having insufficient knowledge of the patient's illness, has to give a diagnosis and propose a specific course of treatment. The physician's practical end is, of course, the patient's recovery, but the means to achieve it largely depend on the hidden causes of the illness. And, since she does not have sufficient objective grounds to assert them, she is forced to advance a hypothesis *as if* true; otherwise, she would not be able to decide a particular treatment for the patient. According to this characterization, then, to believe something pragmatically is to accept that a specific state of affairs is plausible, in order to set oneself particular goals and choose among possible means to attain them (A824/B852). Despite involving theoretical elements, these beliefs are *practical* because they are adopted solely for the purpose of acting in a particular way.

[22] It is difficult to reconcile this strict claim, according to which on a theoretical level there can be no metaphysical beliefs or opinions, with the notion of "doctrinal belief," which Kant also developed in the Canon. Kant asserts that the "thesis of the existence of God belongs to doctrinal belief. For although with regard to theoretical knowledge of the world I have nothing at my command that necessarily presupposes this thought as a condition of my explanations of the appearances of the world [...] purposive unity is still so important a condition of the application of reason to nature that I cannot pass it by [...]. But I know no other condition for this unity that could serve me as a clue for the investigation of nature except insofar as I presuppose that a highest intelligence has arranged everything in accordance with the wisest ends" (A826/B854). Here, Kant appears to consider that one can believe in God, despite the theoretical insufficiency of the claim, on the grounds of the unity of natural purposes we can observe in the universe, that is, based on the argument from design of the universe or, in Kantian terms, the "physico-theological proof" (A621/B649 – A630/B658). Perhaps it could be said that Kant refers here to a regulative use of the belief in God, as a presupposition for advancing scientific enquiry. However, the term "doctrinal belief" does not indicate this meaning. Other than this, the concept of a "doctrinal belief" may be an indication that, in the first Critique, Kant still entertained the possibility that the belief in God had insufficient objective grounds yet was valid from a theoretical point of view. However, since in the *Critique of Judgment* he addressed the matter in the context of reflective judgment (*KU*, *Ak.* V., 436–422), it is likely that at that point he had abandoned this rather ambiguous notion.

[23] There is a scholarly discussion about the special status Kant seems to give to "belief" in this section of the *Critique*, when he argues that "Only in a practical relation, however, can taking something that is theoretically insufficient to be true be called believing" (A823/B851). Commentators seem to find incompatible this practical foundation of belief, with the previous characterization of it as a sort of middle ground between knowledge and opinion—a characterization made in theoretical terms (Chignell 2007, pp. 323–360; Höwing 2016, pp. 201–222; Pasternack 2011, pp. 290–315).

Admittedly, the Postulates of Practical Reason are similar to these pragmatic beliefs. However, unlike the physician's hypothesis, which is, of course, empirical and contingent, Kant claims that the postulates involve a *necessary*, albeit metaphysical, hypothesis. Again, Kant's use of the term "necessary" here may bring to mind all kinds of metaphysical connotations that we should resist. In this context, "necessary" should be interpreted as meaning that, given both our moral vocation – expressed in the categorical imperative – and our natural desire to attain happiness, the only hypotheses that can make both ends compatible are that God exists and the soul is immortal (A828/B856). Put differently, we must presuppose both hypotheses as true if we are to act morally in the hope of attaining the highest good (Beck 1960, pp. 27–28). The contingent character of the physician's pragmatic belief is based on the empirical fact that she could accept different hypotheses about the patient's illness, while the necessity of the Postulates rests on the inference that, according to Kant, we can only conceive *one* hypothesis to link our desire for happiness with our moral objectives:

> The end here is inescapably fixed, and according to all my insight there is possible only a single condition under which this end is consistent with all ends together and thereby has practical validity, namely, that there be a God and a future world; I also know with complete certainty that no one else knows of any other conditions that lead to this same unity of ends under the moral law. But since the moral precept is thus at the same time my maxim (as reason commands that it ought to be), I will inexorably believe in the existence of God and a future life (A828/B856).

Kant's vocabulary in the above passage is, undoubtedly, a little extreme ("I will inexorably believe. . ."), but this should not make us forget that he had recommended holding these metaphysical beliefs cautiously, that is, only for practical purposes and never as theoretically valid claims. This concept of a metaphysical belief that is only justified from a practical standpoint – the concept of a practical belief – is, in my opinion, Kant's development of the Carneadean idea of approving impressions for practical reasons without assenting to them. The critical philosopher – and more broadly, perhaps, the Enlightened autonomous person – recognizes the limits of human understanding and abandons any theoretical ambition to acquire metaphysical knowledge. Nevertheless, she can still hold some of these beliefs as a basis for action, recognizing that they provide the necessary link between happiness and morality, yet entertaining at the same time theoretical doubts on the matter.

If this is the case, Kant's idea of what constitutes a religious belief rests entirely on practical rationality, not on irrational assent to theoretical dogma. At first sight, this may appear to be at odds with his affirmation in the Prologue to the second edition of the *Critique*:

> [. . .] I had to deny knowledge in order to make room for faith; and the dogmatism of metaphysics, i.e., the prejudice that without criticism reason can make progress in metaphysics, is the true source of all unbelief conflicting with morality, which unbelief is always very dogmatic (Bxxx).

In my opinion, even though the assertion has fideistic undertones, Kant does not affirm here that religion should be founded on irrational motives, like some modern fideists did, such as Kant's friend, J. G. Hamman. On the contrary, he says that theoretical reason has to be limited to make room for *practical faith* or, in his words, for "moral theism." As Beck argued, Kant's religious beliefs are:

> [. . .] postulates of a practical but rational faith, necessary because, without them, moral experience could not be made fully intelligible. There is nothing in them of a subjective will to believe or an emotional need. They are rational but not cognitive (1960, p. 27).

Following this line of interpretation, other commentators, such as Tomasi, have seen Kant's moral theism as a commitment to practical rationality:

> The way reason promises the attainability of the highest good is by permitting assent to the only conditions under which it can conceive of it, namely the existence of God and the immortality of the soul. [. . .] The point is that, for a subject committed to morality, belief in God may have non-epistemic merit, since such a belief plays a role in her efforts at making sense (or a certain sense) of her engagement with the demands of practical reason (Tomasi 2016, p. 128).

Although Tomasi speaks of "assent" in this paragraph, her characterization of the doxastic attitude recommended by Kant's moral theism is very similar to my characterization of Academic "approval." As we have seen, this approval is the product of an autonomous reason that recognizes that some religious beliefs have a valid rational justification, even if solely on the practical level. It is a rational, practical faith that can be interpreted as Kant's Academic response to the *apraxia* objection in metaphysical questions, and constitutes a middle ground between the dogmatism of Christian orthodoxy and the fideism of modern Pyrrhonism. Accordingly, the dogmatist would argue that it is possible to theoretically know these metaphysical entities and, consequently, moral imperatives should be derived from this knowledge, while the fideist would claim that, given the limits of human reason, religious belief should be founded merely on irrational motives, and, once again, morality should be the result of abiding by revealed precepts. Both extremes are, for Kant, "the true source of all unbelief conflicting with morality" (Bxxx), the former, because it is liable to the antinomical conflicts, which leave morality without a solid foundation, and the latter, because it is, once again, the euthanasia of reason, which promotes a falsified, heteronomous, morality. Thus, limiting reason to "make room for faith" really meant for Kant avoiding both theological dogmatism and extreme skepticism and giving religious beliefs their proper place and function in human rationality.

* * *

In conclusion, I have argued that Kant's theoretical solution to the antinomies, while showing how transcendental idealism is the adequate philosophical position to explain the illusions of the natural dialectic, actually leaves reason without practical guidance concerning the metaphysical issues involving the ideas of the word, freedom, and God. However, I believe that Kant gave a practical resolution to the antinomies, which can be interpreted as a response to the *apraxia* objection, and consists of accepting (1) the idea of the world as a regulative principle for empirical research; (2) "practical freedom" as the human capacity to act on the grounds of reason and independently of natural causes; and (3) the beliefs in God and the immortality of the soul as practical postulates that make possible the agreement between our moral vocation and our desire for happiness. After his application of the skeptical method to examining the antinomies, this solution was a sort of final verdict, specifying the metaphysical beliefs that could be held for practical purposes.

Kant's debt to Academic skepticism may be seen in some of the following features of his argument: First, the separation of the realms of theoretical knowledge and practical belief, along with the idea that some judgments that cannot be held from the theoretical point of view, have, nonetheless, practical validity. Second, the epistemic attitude he accorded to these beliefs, namely their adoption only for practical reasons, which is very close to the Carneadean non-dogmatic approval of persuasive impressions. And, finally, the goal of preserving rational, practical autonomy in these matters, despite the realization that human reason is limited and can never attain metaphysical knowledge. Kant's "faith," then, can be interpreted as a set of beliefs held autonomously, but non-dogmatically, and entirely based on practical rationality.

Bibliography

Allais, Lucy. 2015. *Manifest Reality. Kant's Idealism and His Realism*. Oxford: Oxford University Press.

Allison, Henry E. 1990. *Kant's Theory of Freedom*. Cambridge: Cambridge University Press.

———. 2004. *Kant's Transcendental Idealism*. New Haven/London: Yale University Press.

Aramayo, Roberto, and Faviola Rivera, eds. 2017. *La Filosofía Práctica de Kant*. México: Universidad Nacional Autónoma de México.

Baldacchino, Lewis. 1984. Strawson on the Antinomy. *Mind* 93 (369): 91–97.

Beade, Ileana. 2010. Acerca del carácter cosmológico-práctico de la tercera antinomia de la razón pura. *Anales del seminario de historia de la filosofía* 27: 189–216.

Beck, Lewis White. 1960. *A Commentary on Kant's Critique of Practical Reason*. Chicago: The University of Chicago Press.

Carnois, Bernard. 1987. *The Coherence of Kant's Doctrine of Freedom*. Trans. David Booth. Chicago: The University of Chicago Press.

Chignell, Andrew. 2007. Belief in Kant. *Philosophical Review* 116: 323–360.

Cicero, Marcus Tullius. 2000. *De Natura Deorum* (*On the Nature of the Gods*). Trans. H. Rackham. Cambridge/London: Harvard University Press.

———. 2001. *De Finibus* (*On Moral Ends*). Trans. Julia Annas. Cambridge: Cambridge University Press.

———. 2006. *Academica* (*On Academic Scepticism*). Trans. Charles Brittain. Indianapolis/Cambridge: Hackett Publishing Company.

de Olaso, Ezequiel. 2018. *Escepticismo e Ilustración. La crisis pirrónica de Hume y Rousseau*. Buenos Aires: Centro de Investigaciones Filosóficas.

Dörflinger, Bernd. 2008. Führt Moral unausbleiblich zur Religion? Überlegungen zu einer These Kants. In *Kants Metaphysik und Religionsphilosophie*, ed. N. Fischer, 207–223. Hamburg: Felix Meiner Verlag.

Esteves, Julio. 2014. The Alleged Incompatibility Between the Concepts of Practical Freedom in the Dialectic and in the Canon of the Critique of Pure Reason. *Kant Studien* 105 (3): 336–371.

Fried, Milton. 1940. Kant's First Antinomy: A Logical Analysis. *Mind* 49 (194): 204–218.

González, Catalina. 2011. Creencia subjetiva y creencia práctica: las creencias en Dios y en la inmortalidad del alma en el Canon de la razón pura de Kant. In *Perspectivas de la modernidad. Siglos XVI, XVII–XVIII*, ed. Jean Paul Margot and Mauricio Zuluaga, 235–246. Cali: Universidad del Valle.

Grier, M. 2006. The Logic of Illusion and the Antinomies. In *A Companion to Kant*, ed. G. Bird, 192–206. Oxford: Blackwell Publishing Ltd.

Guéroult, Martial. 1954. Canon de la raison pure et Critique de la raison pratique. *Revue Internationale de Philosophie* 30: 331–357.

Guyer, Paul. 2006. *Kant*. London/New York: Routledge.

———. 2014. *Kant*. London: Routledge.

Höwing, Thomas. 2016. Kant on Opinion, Belief, and Knowledge. In *The Highest Good in Kant's Philosophy*, ed. Thomas Höwing, 201–221. Berlin: Walter de Gruyter.

Hume, David. 1902. *Enquiries Concerning the Human Understanding and Concerning the Principles of Morals*, ed. L.A. Selby-Bigge. Oxford: Clarendon Press.

———. 2007. *Dialogues concerning Natural Religion and Other Writings*, ed. Dorothy Coleman. Cambridge: Cambridge University Press.

Josifović, Saša. 2015. Das Kanon-Problem in Kants Kritik der reinen Vernunft. *Kant Studien* 106 (3): 487–506.

Kant, Immanuel. 1996. *Critique of Practical Reason* (*KpV*). In *Practical Philosophy*. Trans. Mary J. Gregor, 133–271. Cambridge/New York: Cambridge University Press.

———. 1998. *Critique of Pure Reason* (*KrV*). Trans. Paul Guyer and Allen Wood. Cambridge/New York: Cambridge University Press.

———. 2000. *Critique of the Power of Judgment* (*KU*). Trans. Paul Guyer. Cambridge: Cambridge University Press.

Kleingeld, Pauline. 1998. Kant on the Unity of Theoretical and Practical Reason. *The Review of Metaphysics* 52 (2): 311–339.

———. 2016. Kant on Good, the Good and the Duty to Promote the Highest Good. In *The Highest Good in Kant's Philosophy*, ed. Thomas Höwing, 33–49. Berlin: Walter de Gruyter.

Kohl, Markus. 2014. Transcendental and Practical Freedom in the *Critique of Pure Reason*. *Kant Studien* 105 (3): 313–335.

Korsgaard, Christine. 1996. *The Sources of Normativity*. Cambridge: Cambridge University Press.

Langton, Rae. 1998. *Kantian Humility: Our Ignorance of Things in Themselves*. Oxford: Clarendon Press.

Maia Neto, Jose R. 1997. Academic Skepticism in Early Modern Philosophy. *Journal of the History of Ideas* 58 (2): 199–220.

Nuzzo, Angelica. 2005. *Kant and the Unity of Reason*. West Lafayette: Purdue University Press.

Parra, Lisímaco. 2007. *Estética y modernidad: un estudio sobre la teoría de la belleza de Immanuel Kant*. Bogotá: Universidad Nacional de Colombia.

———. 2017. Filosofía versus Barroco en la Fundamentación de la Metafísica de las Costumbres. In *La filosofía práctica de Kant*, ed. R. Aramayo and F. Rivera, 19–54. México: Universidad Nacional Autónoma de México.

Pasternack, Lawrence. 2011. The Development and Scope of Kantian Belief: The Highest Good, the Practical Postulates, and the Fact of Reason. *Kant Studien* 102 (3): 290–315.

Recki, Birgit. 2016. "Mixtum Compositum": On the Persistence of Kant's Dualism in the Doctrine of the Highest Good. In *The Highest Good in Kant's Philosophy*, ed. Thomas Höwing, 71–87. Berlin: Walter de Gruyter.

Schönecker, Dieter. 2005. *Kants Begriff transzendentaler und praktischer Freiheit*. Berlin: De Gruyter.

Smith, Norman Kemp. 1950. *A commentary to kant's critique of pure reason*. New York: Humanities Press.

Smith, Plínio J. 2008. La Critique de la raison pure face aux scepticismes cartésien, baylien et humien. *Dialogue* 47: 463–500.

Theis, Robert. 2020. La religion est-elle purement une affaire de raison? Réflexions sur quelques aspects du traitement de la religion chez Kant. *Kant Studien* 111 (1): 29–69.

Tomasi, Gabriela. 2016. God, the Highest Good, and the Rationality of Faith: Reflections on Kant's Moral Proof of the Existence of God. In *The Highest Good in Kant's Philosophy*, ed. Thomas Höwing, 111–129. Berlin: Walter de Gruyter.

Zimmermann, Stephan. 2016. Kant on "Moral Arguments": What does the Objectivity of a Postulate of Pure Practical Reason Consist In? In *The Highest Good in Kant's Philosophy*, ed. Thomas Höwing, 131–155. Berlin: Walter de Gruyter.

Chapter 8
Conclusion

Abstract In this conclusion, I focus on two of the main arguments of this book and say a few words about their potential philosophical and contemporary significance. First, I explore some contemporary consequences of my argument that Academic skepticism offered a method to critically examine beliefs and autonomously determine which ones are sufficiently justified to guide action. Second, I share some final observations about my claim that Hume and Kant adopted the Academic method to wage a critique of metaphysics that sought to promote a gradual secularization of society. Their rejection of Christian dogmatism, supplemented by the promotion of a non-dogmatic and merely practical adoption of the religious beliefs that were considered reasonable after skeptical examination is, in my view, a decisive step in this process of secularization.

Writing the last words of a book requires just as much energy as writing the first ones. One feels that everything important has been said and that merely summarizing the arguments or illuminating the paths that remain open would be of little use to the reader. But this is what we do when we do philosophy—we go back to revising arguments, finding previously unseen connections, and digging into the suddenly visible cracks of what we thought had been solidly established. This being the case, I want to focus on two of the main arguments of this book and say a few words about their potential philosophical and contemporary significance. First, I want to explore some contemporary consequences of my argument that Academic skepticism offered a method to critically examine beliefs and autonomously determine which ones are sufficiently justified to guide action. Second, I will share some final observations about my claim that Hume and Kant adopted the Academic method to wage a critique of metaphysics that sought to promote a gradual secularization of society. Their rejection of Christian dogmatism, supplemented by the promotion of a non-dogmatic and merely practical adoption of the religious beliefs that were considered reasonable after skeptical examination, is, in my view, a decisive step in this process of secularization. Let us begin with the first set of conclusions.

C. González Quintero, *Academic Skepticism in Hume and Kant*, Synthese Library 449, https://doi.org/10.1007/978-3-030-89750-5_8

As we saw in the previous chapters, the Academic method of examining beliefs based on their persuasiveness or probability led Hume and Kant to accept very different beliefs in metaphysical and religious matters. Hume decided that the only theological stance that withstood skeptical examination was a non-dogmatic *anti-religious deism*, while Kant chose his own form of *moral theism* based on what he saw as the practical advantages of religion to promote morality. Given such divergent results, someone could object that the method of the New Academy is scarcely advisable for philosophy. After all, should reason not clearly and decisively establish what constitutes a *justified belief?* And, should we not question a method that brings about such divergent results?

As reasonable as these questions may appear, they are the questions of a dogmatist, not of a skeptic. Put differently, while the standards of objectivity and consistency are what the dogmatist demands from searching for truth, the skeptic requires different criteria because she aspires to distinct goals. The advantage of the New Academy's method is precisely that it enables the philosopher to be rational and active, while remaining epistemically humble. She can still use her reason to draw conclusions that orient her actions, but maintains an open-minded and critical attitude, because she understands that these conclusions are never definitive, clear-cut, or unquestionable. While the dogmatist seeks for security and a pause in her investigations, the skeptic values the constant search and revision of her epistemic attitudes concerning judgment—namely, knowledge, belief, and opinion. She understands the distinction between the theoretical and practical validity of claims, and acknowledges that expecting theoretical certainty on metaphysical questions is beyond her capacity as a human being.

Of course, these nuances are largely absent from our current reflection, both in philosophy and the public sphere. We seem to have lost the ability to distinguish knowledge from belief, belief from opinion, and opinion from self-deception. We are immersed in a sea of mass- and social-media assertions that conflate scientific knowledge with pseudo-scientific predictions to advance both commercial and political purposes. We live in societies where religious beliefs are identified with esoteric views, which mistake out-of-context scientific findings for spiritual guidance. We confuse factually-based information with fake-news that exploits massive fears and delusions. And, we live among many who confound political ideologies with exacerbated feelings of hatred, xenophobia, and economic greed that lead to increasing social alienation and inequality. To summarize it bluntly—we do not live in a world where revising one's opinions and non-dogmatically approving those that resist the test is encouraged in the least.

Given this bleak diagnosis, it seems that we should take the method of the New Academy seriously. The epistemic attitude that the Academics called "approval" entails a delicate balance between doubt and assent that we sorely lack today. It implies never ceasing to doubt, but also recognizing which beliefs are reasonable enough to act on their basis. In my view, Academic skepticism demands that we give each of our claims their just measure of truth, and never adopt them dogmatically or obstinately. It requires us to be able to say: "this seems probable to me, but I don't hold it as knowledge," "this I do not know with certainty," or "I am acting in this way because I believe this, but I am willing to reconsider it and perhaps change my behavior." Academic skepticism advises us, in other words, to cultivate the attitude that Cicero, Hume, and Kant defended against metaphysical and other sorts of dogmatism that we continue to witness nowadays.

Another advantage of the New Academy's method is that it defends rational, practical autonomy, against both irrational adherence to authorities and quietism. Following the Academics, Hume and Kant imagined rational autonomy as critical thinking, and critical thinking as constant self-examination—the will to permanently revise the grounds of validity of one's claims in order to reflectively avoid dogmatism. In other words, being an autonomous rational agent meant, for them, incorporating the skeptical viewpoint—cultivating that annoying internal voice that keeps asking whether we are indeed justified in believing what we believe and in acting as we act. Moreover, this practical autonomy also involves critically examining the rules or criteria one prescribes for oneself when holding something as true. The result of such meta-examination may be a never-ending endeavor of abandoning some beliefs and rules of judgment and acquiring new ones throughout one's life. But, far from being a sign of inconsistency, this process of constant evaluation is a path for overcoming dogmatism and prejudice, and possibly, as Cicero hoped, for gradually appraching truth. And, given the strong connection between belief and action, it enriches particularly our processes of moral deliberation and allows us to take moral responsibility more seriously.

I also want to highlight some core elements of my second central claim—namely that Hume's and Kant's Enlightened critiques of metaphysics sought to promote a gradual secularization of society. I see their critiques as aiming to offer a middle-ground solution between the dogmatism of Christian theological orthodoxy and the Pyrrhonism of some modern fideists, who advocated an irrational "leap of faith" to revealed religion. For Hume and Kant, both extremes were equally conducive to religious fanaticism—the former because it exerted ecclesiastical censure and promoted a blind obedience to authority, the latter because it encouraged irrational enthusiasm and fanaticism. Faced with these extremes, Hume's and Kant's Enlightened views endeavored to severely critique metaphysics and give a civil foundation to morality, but, at the same time, advocated for the free profession of religious beliefs, insofar as they were non-dogmatically adopted and the rational agent recognized her own epistemic limitations. In sum, they thought that to advance in the direction of a secular and tolerant society, their critique should curtail all sorts of religious dogmatism without completely undermining religion itself. More importantly, they thought that this could only occur without enthroning reason, as orthodox theologians had done, or abandoning it altogether, as the fideists advocated. They were convinced that reason should be used to evaluate religious beliefs, while always having its own limits in mind—namely the impossibility of attaining knowledge of metaphysical entities.

Again, far from being an eighteenth-century accomplishment, this goal remains an unfinished effort, as we can observe in everyday global events. On the one hand, most Western societies enjoy a freedom from religious censure that would have been unimaginable for both Hume and Kant. On the other, the reappearance of religious dogmatism and superstition is flagrant, and our societies do not seem to have available and adequate means to combat it. Had they faced a situation like ours, Kant would have been astonished by the easy dismissal of religious beliefs on the ground that they are irrational or self-deceiving fantasies, while Hume would have been surprised with the prominence of some religious sects in the handling of national and transnational political issues. Given this challenging religious context, I believe Academic skepticism should rightly be brought to the fore in current philosophical discussions. Needless to insist on it, the Academics' attitude of critically evaluating metaphysical or religious beliefs while recognizing their practical significance – both positive and negative – may help avoid falling into the rampant dogmatism and extremism of contemporary religious attitudes.

Unfortunately, by losing sight of this ancient school during the history of Western philosophy, we have forgotten that skepticism not only provides an abstract point of reference for epistemological debates, but actually affords guidelines to think through urgent moral and political problems. Thus, I want to invite philosophers engaged with the history of skepticism to continue exploring the Academy and its modern adaptations, and apply its critical spirit to current problems. I believe that acknowledging the philosophical value of this skeptical perspective is an important task, as crucial for the history of skeptical ideas as for contemporary philosophical reflections. I hope, with this book, to have contributed to this objective.